AF689499

CLIMATOLOGY: AN ATMOSPHERIC SCIENCE

Edited by
Daniel Dela Torre

ArclerPress

Climatology: An Atmospheric Science

Edited by Daniel Dela Torre

© 2017

Edition 2018

ArclerPress

Published by
Arcler Press LLC
708 3rd Avenue, 6th Floor
New York
NY 10017
United States of America
www.arclerpress.com

ISBN: 978-1-68094-617-8 (Hardback)

Printed in United States of America

Library of Congress Control Number: 2017933211

This book contains information obtained from highly regarded resources. Some content has been published with permission under Creative Commons License or similar. A Wide variety of references are listed. Reasonable efforts have been made to publish reliable data and views articulated in the chapters are those of the individual contributors, and not necessarily those of the editors or publishers. Editors or publishers are not responsible for the accuracy of the information in the published chapters or consequences of their use. The publisher believes no responsibility for any damage or grievance to the persons or property arising out of the use of any materials, instructions, methods or thoughts in the book. The editors and the publisher have attempted to trace the copyright holders of all material reproduced in this publication and apologize to copyright holders if permission has not been obtained. If any copyright holder has not been acknowledged, please write to us so we may rectify.

Notice: Registered trademark of products or corporate names are used only for explanation and identification without intent of infringement. Although care has been taken to check accuracy of formulas and procedures, the detailed methods should be tested further on a small scale before being adopted commercially.

Arcler Press publishes wide variety of books and eBooks. For more information about Arcler Press and its products, visit our website at www.arclerpress.com.

About the Editor

Daniel Dela Torre

Daniel completed his MSc in Environmental Science from the University of the Philippines Diliman in 2016. His primary interests are ecological monitoring using remote sensing and the effects of climate variability on vegetation. He is engaged in the University of the Philippines as a Senior Science Research Specialist working on LiDAR mapping of agricultural resources.

Table of Contents

2 CLIMATE AND CLIMATE CHANGE

3 GLOBAL WARMING AND CLIMATE CHANGE

4 CLIMATE CHANGE AND MITIGATION

5 CLIMATE MODELS

6 REGIONAL CLIMATOLOGY

7 WEATHER FORECASTING

8 PALEOCLIMATOLOGY

List of Abbreviations

BAMS	BAM shallow
CRC	Carboniferous Rainforest Collapse
CLIPER	Climatology and Persistence
COV	Coefficient of variation
EDP	Environmentally displaced person
EPICA	European Project for Ice Coring in Antarctica
Florida	State Super Ensemble
GFDL	Geophysical Fluid Dynamics Laboratory
GEM	Global Environmental Multiscale Model
GFS	Global Forecasting System
GFS	Global Forecast System
GHG	Greenhouse Gas
HURRAN	Hurricane Analog Technique
HWRF	Hurricane Weather Research and Forecasting
ICPMS	Inductively coupled mass spectrometry
IOM	International Organization for Migration
IPICS	International Partnerships in Ice Core Sciences,
JMA	Japan Meteorological Agency
LCA	Life cycle Assessment
LDV q	Light duty Vehicles
LBAR	Limited Area Sine Transform Barotropic
LGEM	Logistic Growth Equation Model
MAAT	Mean annual air Temperatures

MFM	Movable fine-mesh
NCMWRF	National Centre for Medium Range Weather Forecasting
NOGAPS	Navy Operational Global Prediction System
ORSM	Operational Regional Spectral Model
QLM	Quasi-Lagrangian Limited Area
SLOSH	Sea Lake and Overland Surges from Hurricanes
SSC	Spatial Synoptic Classification system
SHIPS	Statistical Hurricane Intensity Prediction Scheme
STIFOR	Statistical Typhoon Intensity Forecast
SEB	Surface energy balance
SAFA	Systematic Approach Forecast Aid
TEPS	Typhoon Ensemble Prediction System
TYM	Typhoon Model
UNFCCC	UN Framework Convention on Climate Change
UNFCCC	United Nations Framework Convention on Climate Change
UTC	Universal Coordinated Time
VICBAR	Vic Ooyama Barotropic
WAIS	West Antarctic ice sheet
WMO	World Meteorological Organization,
XRF	X-ray fluorescence

Preface

Weather is the condition of the atmosphere over a brief period of time. Climate represents the composite of day-to-day weather over a longer period of time. Climate is the statistics of weather, usually over a 30-year interval. It is measured by assessing the patterns of variation in temperature, humidity, atmospheric pressure, wind, precipitation, atmospheric particle count and other meteorological variables in a given region over long periods of time. Climate differs from weather, in that weather only describes the short-term conditions of these variables in a given region. A region's climate is generated by the climate system, which has five components: atmosphere, hydrosphere, cryosphere, lithosphere, and biosphere. The earth-ocean-atmosphere system may be divided into a number of zones with each traditionally studied by a separate scientific discipline. The atmosphere is the component of the system studied by climatologists and meteorologists. Climatology is the scientific study of climates, which is defined as the mean weather conditions over a period of time. A branch of study within atmospheric sciences, it also takes into account the variables and averages of short-term and long-term weather conditions. Climatology is different than meteorology and can be divided into different areas of study. Various approaches to this field can be taken, including paleoclimatology, which focuses on studying the climate over the course of the Earth's existence by examining records of tree rings, rocks and sediment, and ice cores. Historical climatology focuses primarily on climate changes throughout history and the effects of the climate on people and events over time. Though both climatology and meteorology are areas of study that are considered branches of similar areas of study, climatology differs from meteorology because its focus is on averages of weather and climatic conditions over a long period of time. Meteorology focuses more on current weather conditions such as humidity, air pressure, and temperatures and forecasting the short-term weather conditions to come.

Content Coverage

Chapters One and *Two* present the introduction of atmosphere and climate, correspondingly. Climate differs from weather, in that weather only describes the short-term conditions of these variables in a given region. Climate differs from weather, in that weather only describes the short-term conditions of these variables in a given region.

Chapters Three and *Four* present a coverage of global warming and climate change, consistently. Global warming and climate change are terms for the observed century-scale rise in the average temperature of the Earth's climate system and its related effects.

Chapters Five and *Six* are intended to focus on climate models and regional climatology, distinctly. Climate models use quantitative methods to simulate the interactions of the important drivers of climate, including atmosphere, oceans, land surface and ice.

Chapters Seven and *Eight* elucidate the details of weather forecasting and paleoclimatology, separately. Weather forecasting is the application of science and technology to predict the state of the atmosphere for a given location, but paleoclimatology focuses on studying the climate over the course of the earth's existence by examining records of tree rings, rocks and sediment, and ice cores.

CHAPTER 1

Atmosphere: An Introduction

Chapter Contents

- **Introduction**
- **Pressure**
- **Atmospheric Escape**
- **Terrain**
- **Composition**
- **Structure**
- **Circulation**
- **Importance**
- **Cryosphere**
- **Structure**
- **Geosphere**
- **Pedosphere**
- **Biosphere**
- **Hydrosphere**

The atmosphere is generally considered to be the region of the thermosphere, which is the thin, outer layer of Earth's atmosphere that starts out around 56 miles (90 kilometers) up and stretches all the way up to about 375 miles (604 kilometers). Spacecraft such as the International Space Station (ISS) or US Space Shuttle typically orbit in the upper atmosphere at a range of about 140 miles (225 kilometers). By contrast, commercial aircraft travel in the much lower level stratosphere that extends to a maximum height of 31 miles (50 kilometers) where the Earth's ozone layer exists.

INTRODUCTION

An **atmosphere** is a layer of gases surrounding a planet or other material body, that is held in place by the gravity of that body. An atmosphere is more likely to be retained if the gravity it is subject to is high and the temperature of the atmosphere is low.

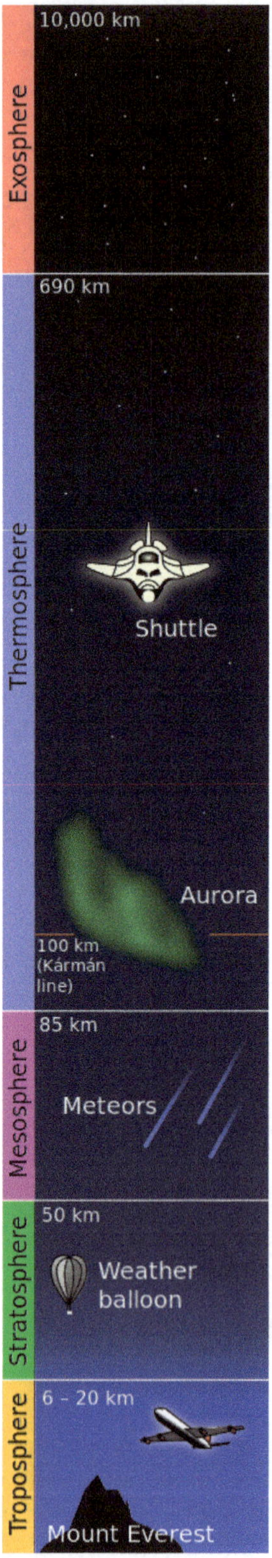

The layers of Earth's atmosphere.

The atmosphere of Earth is mostly composed of nitrogen (about 78%), oxygen (about 21%), argon (about 0.9%) with carbon dioxide and other gases in trace amounts. Oxygen is used by most organisms for respiration, nitrogen is fixed by bacteria and lightning to produce ammonia used in the construction of nucleotides and amino acids and carbon dioxide is used by plants, algae and cyanobacteria for photosynthesis. The atmosphere helps protect living organisms from genetic damage by solar ultraviolet radiation, solar wind and cosmic rays. Its current composition is the product of billions of years of biochemical modification of the paleoatmosphere by living organisms.

Mars's thin atmosphere.

The term stellar atmosphere describes the outer region of a star, and typically includes the portion starting from the opaquephotosphere outwards. Stars with sufficiently low temperatures may form compound molecules in their outer atmosphere.

PRESSURE

Atmospheric pressure is the force per unit area that is applied perpendicularly to a surface by the surrounding gas. It is determined by a planet's gravitational force in combination with the total mass of a column of gas above a location. On Earth, units of air pressure are based on the internationally recognized standard atmosphere (atm), which is defined as 101.325 kPa (760 Torr or 14.696 psi). It is measured with a barometer.

The pressure of an atmospheric gas decreases with altitude due to the diminishing mass of gas above. The height at which the pressure from an atmosphere declines by a factor of e (an irrational number with a value of 2.71828..) is called the scale height and is denoted by H. For an atmosphere with a uniform temperature, the scale height is proportional to the temperature and inversely proportional to the product of the mean molecular mass of dry air and the local acceleration of gravity at that location. For such a model atmosphere, the pressure declines exponentially with increasing altitude. However, atmospheres are not uniform in temperature, so the exact determination of the atmospheric pressure at any particular altitude is more complex.

ATMOSPHERIC ESCAPE

Atmospheric escape is the loss of planetary atmospheric gases to outer space. A number of different mechanisms can be responsible for atmospheric escape, operating at different time scales; the most prominent is **Jeans Escape,** named after British astronomer Sir James Jeans, who described the process of atmospheric loss to the molecular kinetic energy..

Thermal Escape Mechanisms

One classical thermal escape mechanism is **Jeans escape**. In a quantity of gas, the average velocity of a molecule is determined by temperature, but the velocities of individual molecules change as they collide with one another, gaining and losing kinetic energy. The variation in kinetic energy among the molecules is described by the Maxwell distribution. The kinetic energy and mass of a molecule determine its velocity by $E_{kin} = \frac{1}{2}mv^2$.

Individual molecules in the high tail of the distribution may reach escape velocity, at a level in the atmosphere where the mean free path is comparable to the scale height, and leave the atmosphere.

The more massive the molecule of a gas is, the lower the average velocity of molecules of that gas at a given temperature, and the less likely it is that any of them reach escape velocity.

This is why hydrogen escapes from an atmosphere more easily than carbon dioxide. Also, if the planet has a higher mass, the escape velocity is greater, and fewer particles will escape. This is why the gas giant planets still retain significant amounts of hydrogen and helium, which have largely escaped from Earth's atmosphere. The distance a planet orbits from a star also plays a part; a close planet has a hotter atmosphere, with a range of velocities shifted into the higher end of the distribution, hence, a greater likelihood of escape. A distant body has a cooler atmosphere, with a range of lower velocities, and less chance of escape. This helps Titan, which is small compared to Earth but further from the Sun, retain its atmosphere.

An atmosphere with a high enough pressure and temperature can undergo a different escape mechanism - hydrodynamic escape. In this situation the atmosphere simply flows off like a wind into space, due to pressure gradients initiated by thermal energy deposition. Here it is possible to lose heavier molecules that would not normally be lost. Hydrodynamic escape has been observed for exoplanets close-to their host star, including several hot Jupiters (HD 209458b, HD 189733b) and a hot Neptune (GJ 436b).

Significance of Solar Winds

The relative importance of each loss process is a function of planet mass, its atmosphere composition, and its distance from its sun. A common erroneous belief is that the primary non-thermal escape mechanism is atmospheric stripping by a solar wind in the absence of a magnetosphere. Excess kinetic energy from solar winds can impart sufficient energy to the atmospheric particles to allow them to reach escape velocity, causing atmospheric escape. The solar wind, composed of ions, is deflected by magnetic fields because the charged particles within the wind flow along magnetic field lines. The presence of a magnetic field thus deflects solar winds, preventing the loss of atmosphere. On Earth, for instance, the interaction between the solar wind and earth's magnetic field deflects the solar wind about the planet, with near total deflection at a distance of 10 Earth radii. This region of deflection is called a bow shock.

Depending on planet size and atmospheric composition, however, a lack of magnetic field does not determine the fate of a planet's atmosphere. Venus, for instance, has no powerful magnetic field. Its close proximity to the Sun also increases the speed and number of particles, and would presumably cause the atmosphere to be stripped almost entirely, much like that of Mars. Despite this, the atmosphere of Venus is two orders of magnitudes denser than Earth's. Recent models indicate that stripping by solar wind accounts for less than 1/3 of total non-thermal loss processes.

While Venus and Mars have no magnetosphere to protect the atmosphere from solar winds, photoionizing radiation (sunlight) and the interaction of the solar wind with the atmosphere of the planets causes ionization of the uppermost part of the atmosphere. This ionized region in turn induces magnetic moments that deflect solar winds much like a magnetic field. This limits solar-wind effects to the uppermost part of the atmosphere, roughly 1.2–1.5 planetary radii away from the planet, or an order of magnitude closer to the surface than Earth's magnetic field creates. Beyond this region, called a bow shock, the solar wind is slowed to subsonic velocities. Nearer to the surface, solar-wind dynamic pressure reaches a balance with the pressure from the ionosphere, in a region called the ionopause. This interaction typically prevents solar-wind stripping from being the dominant loss process of the atmosphere.

Comparison of non-Thermal loss Processes Based on Planet and Particle Mass

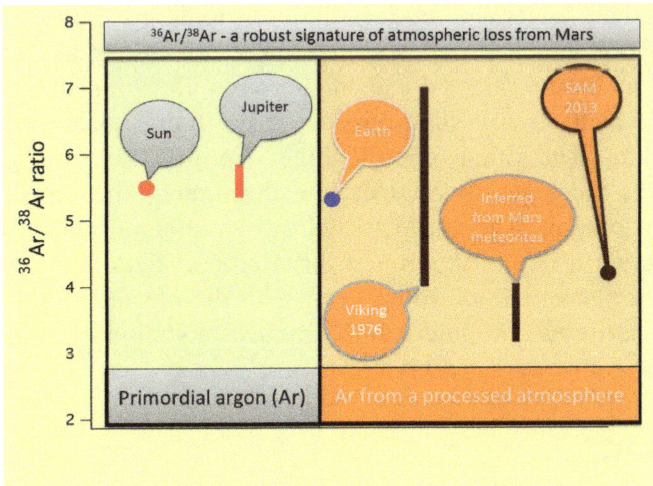

Argon Isotope Ratios are a signature of atmospheric escape on Mars (Curiosity rover) (April, 2013)

The dominant non-thermal loss processes depends on the planetary body. The relative significance of each process depends on planetary mass, atmospheric composition, and distance from the sun. The dominant non-thermal loss processes for Venus and Mars, two terrestrial planets neither with magnetic fields, are dissimilar. The dominant non-thermal loss process on Mars is from solar winds, as the atmosphere is not dense enough to shield itself from the winds during peak solar activity. Venus is somewhat shielded from solar winds because of its denser atmosphere and as a result, solar pick-up is not its dominant non-thermal loss process. Smaller bodies without magnetic fields are more likely to suffer from solar winds, as the planet is too small to have sufficient gravity to produce a dense enough atmosphere and stop solar wind pick-up.

The dominant loss process for Venus' atmosphere is through electric force field acceleration. As electrons are less massive than other particles, they are more likely to escape from the top of Venus's ionosphere. As a result, a minor net positive charge develops. That net positive charge, in turn, creates an electric field that can accelerate other positive charges out of the atmosphere. As a result, H^+ ions are accelerated beyond escape velocity. Other important loss processes on Venus are photochemical reactions driven by Venus's proximity to the Sun. Photochemical reactions rely on the splitting of molecules into constituent atoms, often with a significant portion of the kinetic energy carried off in the less massive particle with sufficiently high kinetic energy to escape. Oxygen, relative to hydrogen, is not of sufficiently low mass to escape through this mechanism.

Phenomena of non-thermal Loss Processes on Moons with Atmospheres

Several natural satellites in the Solar System have atmospheres and are subject to atmospheric loss processes. They typically have no magnetic fields of their own, but orbit planets with powerful magnetic fields. Many of these moons lie within the magnetic fields generated by the planets and are less likely to undergo sputtering and pick-up. The shape of the bow shock, however, allows for some moons, such as Titan, to pass through the bow shock when their orbits take them between the Sun and their primary. Titan spends roughly half of its transit time outside of the bow-shock and being subjected to unimpeded solar winds. The kinetic energy gained from pick-up and sputtering associated with the solar winds increases thermal escape throughout the transit of Titan, causing neutral hydrogen to escape. The escaped hydrogen maintains an orbit following in the wake of Titan, creating a neutral hydrogen torus around Saturn. Io, in its transit around Jupiter, encounters a plasma cloud. Interaction with the plasma cloud induces sputtering, kicking off sodium particles. The interaction produces a stationary banana-shaped charged sodium cloud along a part of the orbit of Io.

Impact Erosion

The impact of a large meteoroid can lead to the loss of atmosphere. If a collision is energetic enough, it is possible for ejecta, including atmospheric molecules, to reach escape velocity. Just one impact such as the Chicxulub event does not lead to a significant loss, but the terrestrial planets went through enough impacts when they were forming for this to matter.

Sequestration

Sequestration is not a form of escape from the planet, but a loss of molecules from the atmosphere and into the planet. It occurs on Earth when water vapor condenses to form rain or glacial ice. It also occurs on Earth when carbon dioxide is sequestered in sediments, or cycled through the oceans. The dry ice caps on Mars are also an example of sequestration.

One mechanism for sequestration is chemical; for example, most of the carbon dioxide of the Earth's original atmosphere has been chemically sequestered into carbonate rock. Very likely a similar process has occurred on Mars. Oxygen can be sequestered by oxidation of rocks; for example, by increasing the oxidation states of ferric rocks from Fe^{2+} to Fe^{3+}. Gases can also be sequestered by adsorption, where fine particles in the regolith capture gas which adheres to the surface particles.

Dominant Atmospheric Escape and Loss Processes on Earth

Earth is too large to lose a significant proportion of its atmosphere through Jeans escape. The current rate of loss is about three kilograms (3 kg) of hydrogen and 50 grams (50 g) of helium per second. The exosphere is the high-altitude region where atmospheric density is sparse and Jeans escape occurs. Jeans escape calculations assuming an exosphere temperature of 1,800 K show that to deplete O^+ ions by a factor of *e* (2.718...) would take nearly a billion years. 1,800 K is higher than the actual observed exosphere temperature; at the actual average exosphere temperature, depletion of O^+ ions would not occur even over a trillion years. Furthermore, most oxygen on Earth is bound as O_2, which is too massive to escape Earth by Jeans escape.

Earth's magnetic field protects it from solar winds and prevents escape of ions, except near the magnetic poles where charged particles stream towards the earth along magnetic field lines. The gravitational attraction of Earth's mass prevents other non-thermal loss processes from appreciably depleting the atmosphere. Yet Earth's atmosphere is two orders of magnitude less dense than that of Venus at the surface. Because of the temperature regime of Earth, CO_2 and H_2O are sequestered in the hydrosphere and lithosphere. H_2O vapor is sequestered as liquid H_2O in oceans, greatly decreasing the atmospheric density. With liquid water running over the surface of Earth, CO_2 can be drawn down from the atmosphere and sequestered in sedimentary rocks. Some estimates indicate that nearly all carbon on Earth is contained in sedimentary rocks, with the atmospheric portion being approximately 1/250,000 of Earth's CO_2 reservoir. If both of the reservoirs were released to the atmosphere, Earth's atmosphere would be even denser than Venus's atmosphere. Therefore, the dominant "loss" mechanism of Earth's atmosphere is not escape to space, but sequestration.

TERRAIN

Atmospheres have dramatic effects on the surfaces of rocky bodies. Objects that have no atmosphere, or that have only an exosphere, have terrain that is covered in craters. Without an atmosphere, the planet has no protection from meteors, and all of them collide with the surface and create craters.

A rocky body with a thick atmosphere does not have significant craters on its surface. The friction generated when a meteor enters an atmosphere causes the vast majority of it to burn up before hitting the surface. When meteors do impact, the effects are often erased by the action of wind. As a result, craters are rare on objects with atmospheres.

All objects with atmospheres have wind and weather. Wind erosion is a significant factor in shaping the terrain of rocky planets with atmospheres, and over time can erase the effects of both craters and volcanoes. In addition, since liquids can not exist without pressure, an atmosphere allows liquid to be present at the surface, resulting in lakes, rivers and oceans. Earth and Titan are known to have liquids at their surface and terrain on the planet suggests that Mars had liquid on its surface in the past.

COMPOSITION

Earth's atmospheric gases scatter blue light more than other wavelengths, giving Earth a blue halo when seen from space.

Initial atmospheric composition is generally related to the chemistry and temperature of the local solar nebula during planetary formation and the subsequent escape of interior gases. The original atmospheres started with the radially local rotating gases that collapsed to the spaced rings that formed the planets. They were then modified over time by various complex factors, resulting in quite different outcomes.

The atmospheres of the planets Venus and Mars are primarily composed of carbon dioxide, with small quantities of nitrogen, argon, oxygen and traces of other gases.

The atmospheric composition on Earth is largely governed by the by-products of the life that it sustains. Dry air from Earth's atmosphere contains 78.08% nitrogen, 20.95% oxygen, 0.93% argon, 0.038% carbon dioxide, and traces of hydrogen, helium, and other "noble" gases (by volume), but generally a variable amount of water vapour is also present, on average about 1% at sea level.

The low temperatures and higher gravity of the Solar System's giant planets—Jupiter, Saturn, Uranus and Neptune—allow them more readily to retain gases with low molecular masses. These planets have hydrogen–helium atmospheres, with trace amounts of more complex compounds.

Two satellites of the outer planets possess significant atmospheres. Titan, a moon of Saturn, and Triton, a moon of Neptune, have atmospheres mainly of nitrogen. When in the part of its orbit closest to the Sun, Pluto has an atmosphere of nitrogen and methane similar to Triton's, but these gases are frozen when it is farther from the Sun.

Other bodies within the Solar System have extremely thin atmospheres not in equilibrium. These include the Moon (sodium gas), Mercury (sodium gas), Europa (oxygen), Io (sulfur), and Enceladus (water vapor).

The first exoplanet whose atmospheric composition was determined is HD 209458b, a gas giant with a close orbit around a star in the constellation Pegasus. Its atmosphere is heated to temperatures over 1,000 K, and is steadily escaping into space. Hydrogen, oxygen, carbon and sulfur have been detected in the planet's inflated atmosphere.

STRUCTURE

Earth

Earth's atmosphere consists of a number of layers, summarised in the diagram above which explains what the layers are , that differ in properties such as composition, temperature and pressure. The lowest layer is the troposphere, which extends from the surface to the bottom of the stratosphere. Three quarters of the atmosphere's mass resides within the troposphere, and is the layer within which the Earth's weather develops. The depth of this layer varies between 17 km at the equator to 7 km at the poles. The stratosphere, extending from the top of the troposphere to the bottom of the mesosphere, contains the ozone layer. The ozone layer ranges in altitude between 15 and 35 km, and is where most of the ultraviolet radiation from the Sun is absorbed. The top of the mesosphere, ranges from 50 to 85 km, and is the layer wherein most meteors burn up. The thermosphere extends from 85 km to the base of the exosphere at 690 km and contains the ionosphere, a region where the atmosphere is ionised by incoming solar radiation. The ionosphere increases in thickness and moves closer to the Earth during daylight and rises at night allowing certain frequencies of radio communication a greater range. The Kármán line, located within the thermosphere at an altitude of 100 km, is commonly used to define the boundary between Earth's atmosphere and outer space. The exosphere begins variously from about 690 to 1,000 km above the surface, where it interacts with the planet's magnetosphere, to space. Each of the layers has a different lapse rate, defining the rate of change in temperature with height.

Others

Other astronomical bodies such as these listed have known atmospheres.

In the Solar System

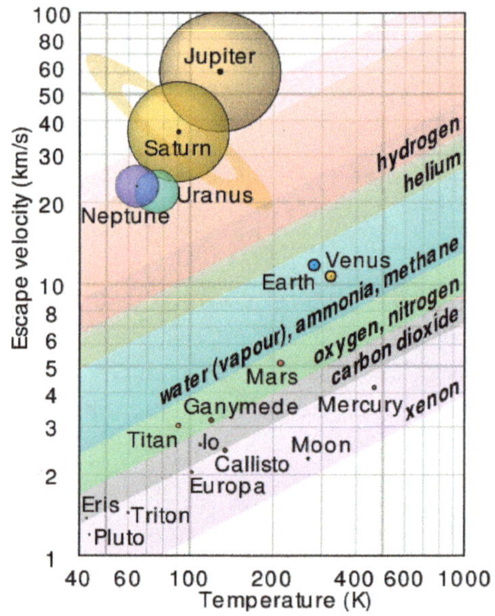

Graphs of escape velocity against surface temperature of some Solar System objects showing which gases are retained. The objects are drawn to scale, and their data points are at the black dots in the middle.

- Atmosphere of the Sun
- Atmosphere of Mercury
- Atmosphere of Venus
- Atmosphere of Earth
- Atmosphere of the Moon
- Atmosphere of Mars
- Atmosphere of Ceres

- Atmosphere of Jupiter
- Atmosphere of Io
- Atmosphere of Callisto
- Atmosphere of Europa
- Atmosphere of Ganymede
- Atmosphere of Saturn
- Atmosphere of Titan
- Atmosphere of Enceladus
- Atmosphere of Uranus
- Atmosphere of Titania
- Atmosphere of Neptune
- Atmosphere of Triton
- Atmosphere of Pluto

Outside the Solar System

- Atmosphere of HD 209458 b

CIRCULATION

Atmospheric circulation is the large-scale movement of air, and together with ocean circulation is the means by which thermal energy is redistributed on the surface of the Earth.

The Earth's atmospheric circulation varies from year to year, but the large scale structure of its circulation remains fairly constant. The smaller scale weather systems – mid-latitude depressions, or tropical convective cells – occur randomly, and long range weather predictions of those cannot be made beyond ten days in practice, or a month in theory (see Chaos theory and Butterfly effect). The large scale atmospheric circulation of the Earth, however, is an average of its systems and patterns, and is considered stable over longer periods of time.

The Earth's weather is a consequence of its illumination by the Sun, and the laws of thermodynamics. The atmospheric circulation can be viewed, from that standpoint, as a heat engine driven by the Sun's energy, and whose energy sink, ultimately, is the blackness of space. The work produced by that engine causes the motion of the masses of air and in that process it redistributes the energy absorbed by the Earth's surface near the tropics to space and incidentally to the latitudes nearer the poles.

The large scale atmospheric circulation cells shift polewards in warmer periods (e.g. interglacials compared to glacials), but remain largely constant as they are, fundamentally, a property of the Earth's size, rotation rate, heating and atmospheric depth, all of which change little. Over very long time periods (hundreds of millions of years), a tectonic uplift can significantly alter their major elements, such as the jet stream, and plate tectonics may shift ocean currents. During the extremely hot climates of the Mesozoic, a third desert belt may have existed at the Equator. But, the overall latitudinal pattern of Earth's climate has not changed.

Latitudinal Circulation Features

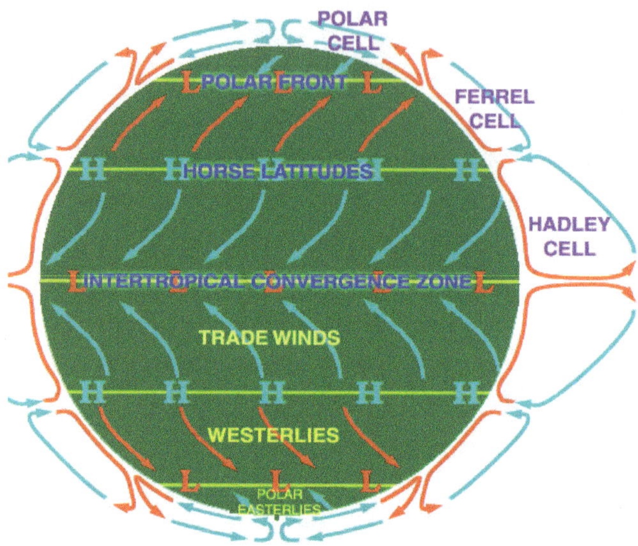

An idealised view of three large circulation cells showing surface winds.

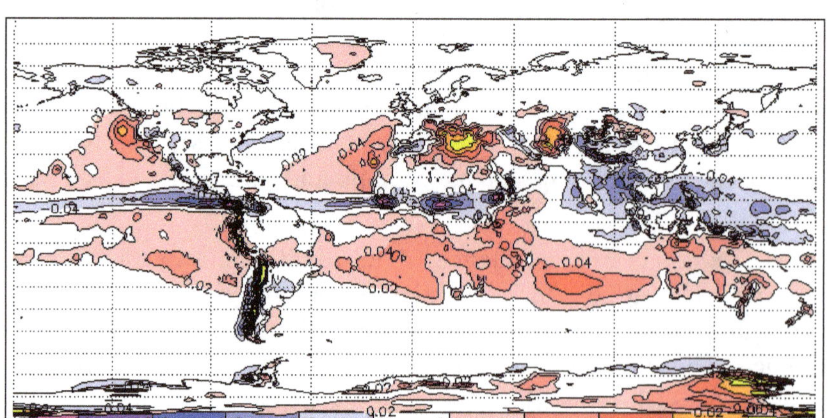

Vertical velocity at 500 hPa, July average. Ascent (negative values) is concentrated close to the solar equator; descent (positive values) is more diffuse but also occurs mainly in the Hadley cell.

The wind belts girdling the planet are organised into three cells; they are from the equator toward the poles, the Hadley cell, the Ferrel cell, and the Polar cell. Those cells exist in both the northern and southern hemispheres. The vast bulk of the atmospheric motion occurs in the Hadley cell. Note that there is one discrete Hadley cell that may split, shift and merge in a complicated process over time. The high pressure systems acting on the Earth's surface is balanced by the low pressure systems elsewhere. As a result, there a balance of forces acting on the Earth's surface.

Hadley Cell

The ITCZ's band of clouds over the Eastern Pacific and the Americas as seen from space.

The atmospheric circulation pattern that George Hadley described was an attempt to explain the trade winds. The Hadley cell is a closed circulation loop, which begins at the equator. There, moist air is warmed by the Earth's surface, decreases in density and rises. A similar air mass rising on the other side of the equator forces those rising air masses to move poleward. The rising air creates a low pressure zone near the equator. As the air moves poleward, it cools, becomes more dense, and descends at about 30th parallel, creating a high-pressure area. The descended air then travels toward the equator along the surface, replacing the air that rose from the equatorial zone, closing the loop of the Hadley cell. The poleward movement of the air in the upper part of the troposphere deviates toward the east due to the coriolis acceleration (a manifestation of conservation of energy). At the ground level however, the movement of the air toward the equator in the lower troposphere deviates toward the west producing a wind from the east (again due to conservation of energy). The winds that flow to the west (from the east, easterly wind) at the ground level in the Hadley cell, are called the Trade Winds.

Though the Hadley cell is described as located at the equator, in the northern hemisphere, it shifts to higher latitudes June and July and toward lower latitudes December and January as it is caused by the Sun's heating of the surface. The zone where the greatest heating takes place is called the thermal equator. As the southern hemisphere summer is December to March, the movement of the thermal equator to higher southern latitudes takes place then.

The Hadley system provides an example of a thermally direct circulation. The thermodynamic efficiency and power of the Hadley system, considered as a heat engine, is estimated as 2.6% and 200×10^{12}(TW).

Polar Cell

The **Polar cell**, likewise, is a simple system. Though cool and dry relative to equatorial air, the air masses at the 60th parallel are still sufficiently warm and moist to undergo convection and drive a thermal loop. At the 60th parallel, the air rises to the tropopause (about 8 km at this latitude) and moves poleward. As it does so, the upper level air mass deviates toward the east. When the air reaches the polar areas, it has cooled and is considerably denser than the underlying air. It descends, creating a cold, dry high-pressure area. At the polar surface level, the mass of air is driven toward the 60th parallel, replacing the air that rose there, and the Polar circulation cell is complete. As the air at the surface moves equatorward, it deviates toward the west. Again, the deviations of the air masses is due to conservation of energy, which is also referred to as the Coriolis effect. The air flows at the surface are called the Polar easterlies (easterly from the east).

The outflow of air mass from the cell creates harmonic waves in the atmosphere known as Rossby waves. These ultra-long waves determine the path of the jet stream, which travels within the transitional zone between the tropopause and the Ferrel cell. By acting as a heat sink, the Polar cell moves the abundant heat from the equator toward the polar regions.

The Hadley cell and the Polar cell are similar in that they are thermally direct; in other words, they exist as a direct consequence of surface temperatures. Their thermal characteristics drive the weather in their domain. The sheer volume of energy that the Hadley cell transports, and the depth of the heat sink that is the Polar cell, ensures that the effects of transient weather phenomena are not only not felt by the system as a whole, but — except under unusual circumstances — do not form. The endless chain of passing highs and lows which is part of everyday life for mid-latitude dwellers, at latitudes between 30 and 60° latitude, is unknown above the 60th and below the 30th parallels. There are some notable exceptions to this rule. In Europe, unstable weather extends to at least 70° north.

These atmospheric features are stable. Even though they may strengthen or weaken regionally over time, they do not vanish entirely.

The Polar cell, orography and Katabatic winds in Antarctica, can create very cold conditions at the surface, for instance the coldest temperature recorded on Earth: −89.2 °C at Vostok Station in Antarctica, measured 1983.

Ferrel Cell

Part of the air rising at 60° latitude diverges at high altitude toward the poles and creates the polar cell. The rest moves toward the equator where it collides at 30° latitude with the high level air of the Hadley cell. There it subsides and strengthens the high pressure ridges beneath. A large part of the energy that drives the Ferrel cell is provided by the Polar and Hadley cells circulating on either side and that drag the Ferrel cell with it. The Ferrel cell, theorized by William Ferrel (1817–1891), is therefore a secondary circulation feature, whose existence depends upon the Hadley and Polar cells on either side of it; it behaves much as an atmospheric ball bearing between the two. It might be thought of as an eddy created by the Hadley and Polar cells. The Ferrel cell is weak and the air flow and temperatures within are variable. For this reason, the mid latitudes are sometimes known as the **"zone of mixing."** At high altitudes, the Ferrel cell overrides the Hadley and Polar cells. The air of the Ferrel cell that descends at 30° latitude returns poleward at the ground level and as it does so it deviates toward the east. In the upper atmosphere of the Ferrel cell the air moving equatorward, deviates toward the west. Both of those deviations, as in the case of the Hadley and Polar cells, are driven by conservation of energy. As a result, just as the Easterly Trade Winds are found below the Hadley cell, the Westerlies are found beneath the Ferrel cell. The forces driving the flow in the Ferrel cell are weak and so the weather in that zone is variable. Thus, strong high-pressure areas which divert the prevailing westerlies, such as a Siberian high, can override the Ferrel cell, making it discontinuous.

While the Hadley and Polar cells are truly closed loops, the Ferrel cell is not, and the telling point is in the Westerlies, which are more formally known as the Prevailing Westerlies. The Easterly Trade Winds and the Polar Easterlies have nothing over which to prevail, as their parent circulation cells are strong enough, and face few obstacles either in the form of massive terrain features or high pressure zones. The weaker Westerlies of the Ferrel cell, however, can be disrupted. The local passage of a cold front may change that in a matter of minutes, and frequently does. As a result, at the surface, winds can vary abruptly in direction. But the winds above the surface, where they are less disrupted by terrain, are essentially westerly. A low pressure zone at 60° latitude that moves equator wards, or a high pressure zone at 30° latitude that moves poleward, will accelerate the Westerlies of the Ferrel cell. A strong high, moving polewards may bring Westerly winds for days. As the high and low pressure zones are established by the Hadley and Polar cells, it can be said that they drive the weather of the middle latitude Ferrel cell.

The Ferrel cell is driven by the Hadley and Polar cells. It has neither a strong source of heat nor a strong cold sink to drive convection. As a result, the weather within the Ferrel cell is highly variable and is influenced by changes to the Hadley and Polar cells. The base of the Ferrel cell is characterized by the movement of air masses, and the location of those air masses is influenced in part by the location of the jet stream, even though it flows near the tropopause. Overall, the movement of surface air is from the 30th latitude to the 60th. However, the upper flow of the Ferrel cell is weak and not well defined.

In contrast to the Hadley and Polar systems, the Ferrel system provides an example of a thermally indirect circulation. The Ferrel system acts as a heat pump with a coefficient of performance of 12.1, consuming kinetic energy at an approximate rate of 275 TW.

Longitudinal Circulation Features

While the Hadley, Ferrel, and Polar cells (whose axes are oriented along parallels i.e.

latitudes) are the major features of global heat transport, they do not act alone. Temperature differences also drive a set of circulation cells, whose axis of circulation are longitudinally oriented. This atmospheric motion is known as **zonal overturning circulation**.

Latitudinal circulation is a result of the highest solar radiation per unit area (solar intensity) falling on the tropics. The solar intensity decreases as the latitude increases, reaching essentially zero at the poles. Longitudinal circulation, however, is a result of the heat capacity of water, it's absorptivity, and it's mixing. Water absorbs more heat than does the land, but its temperature does not rise as greatly as does the land. As a result, temperature variations on land are greater than on water. The Hadley, Ferrel, and Polar cells operate at the largest scale of thousands of kilometers (synoptic scale). But, even at mesoscales (a horizontal range of 5 to several hundred kilometres), this effect is noticeable. During the day, air warmed by the relatively hotter land rises, and as it does so it draws a cool breeze from the sea that replaces the risen air. At night, the relatively warmer water and cooler land reverses the process, and a breeze from the land, of air cooled by the land, is carried offshore by night. This described effect is daily (diurnal).

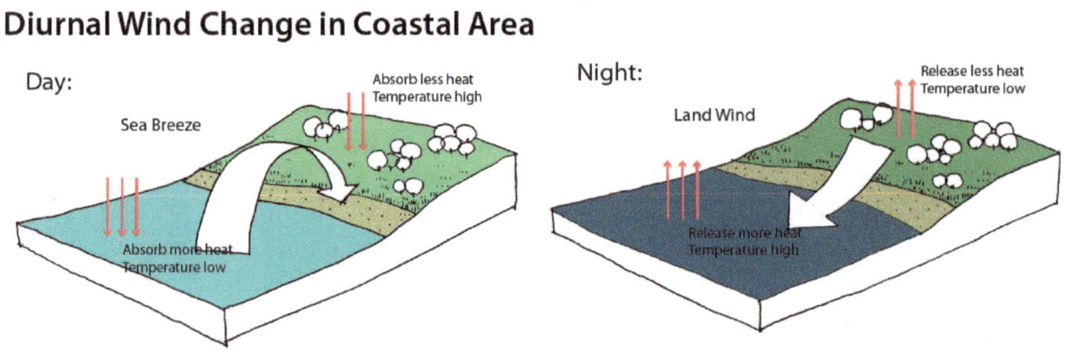

Diurnal wind change in local coastal area, also applies on the continental scale.

At the larger, synoptic, scale of oceans and continents, this effect is seasonal or even decadal. Warm air rising over the equatorial, continental, and western Pacific Ocean regions. When it reaches the tropopause, it cools and subsides in a region of relatively cooler water mass.

The Pacific Ocean cell plays a particularly important role in Earth's weather. This entirely ocean-based cell comes about as the result of a marked difference in the surface temperatures of the western and eastern Pacific. Under ordinary circumstances, the western Pacific waters are warm and the eastern waters are cool. The process begins when strong convective activity over equatorial East Asia and subsiding cool air off South America's west coast creates a wind pattern which pushes Pacific water westward and piles it up in the western Pacific. (Water levels in the western Pacific are about 60 cm higher than in the eastern Pacific, a difference due entirely to the force of moving air.)

Walker Circulation

The Pacific cell is of such importance that it has been named the **Walker circulation** after Sir Gilbert Walker, an early-20th-century director of British observatories in India, who sought a means of predicting when the monsoon winds of India would fail. While he was never successful in doing so, his work led him to the discovery of a link between the periodic pressure variations in the Indian Ocean, and those between the eastern and western Pacific, which he termed the "Southern Oscillation.

The movement of air in the Walker circulation affects the loops on either side. Under normal circumstances, the weather behaves as expected. But every few years, the winters become unusually warm or unusually cold, or the frequency of hurricanes increases or decreases, and the pattern sets in for an indeterminate period. The Walker Cell plays a key role in this and in the El Niño phenomenon. If convective activity slows in the Western Pacific for some reason (this reason is not currently known), the climate dominoes next to it begin to topple. First, the upper-level westerly winds fail. This cuts off the source of returning, cool air that would normally subside at about 30° north latitude, and therefore the air returning as surface Easterlies ceases.

The consequence of this is twofold. Warm water ceases to surge into the eastern Pacific from the west (it was piled by past easterly winds) since there is no longer a surface wind to push it into the area of the west pacific. This and the corresponding effects of the Southern Oscillation result in long-term unseasonable temperatures and precipitation patterns in North and South America, Australia, and Southeast Africa, and the disruption of ocean currents.

Meanwhile, in the Atlantic, fast-blowing upper level Westerlies of the Hadley cell, which would ordinarily be blocked by the Walker circulation and unable to reach such intensities, form. These winds tear apart the tops of nascent hurricanes and greatly diminish the number which are able to reach full strength.

El Niño – Southern Oscillation

El Niño and *La Niña* are opposite surface temperature anomalies of the Southern Pacific, which heavily influence the weather on a large scale. In the case of El Niño, warm surface water approaches the coasts of South America which results in blocking the upwelling of nutrient-rich deep water. This has serious impacts on the fish populations.

In the La Niña case, the convective cell over the western Pacific strengthens inordinately, resulting in colder than normal winters in North America, and a more robust cyclone season in South-East Asia and Eastern Australia. There is also an increased upwelling of deep cold ocean waters and more intense uprising of surface air near South America, resulting in increasing numbers of drought occurrences, although it is often argued that fishermen reap benefits from the more nutrient-filled eastern Pacific waters.

The neutral part of the cycle – the normal component – has been referred to humorously by some as La Nada, which means the nothing in Spanish.

IMPORTANCE

From the perspective of a planetary geologist, the atmosphere is an evolutionary agent essential to the shaping of a planet. The wind picks up dust and other particles which when they collide with the terrain erodes the relief and leaves deposits (eolian processes). Frost and precipitations, which depend on the atmospheric composition, also influence the relief. Climate changes can influence a planet's geological history. Conversely, studying surface of Earth leads to an understanding of the atmosphere and climate of a planet — both its present state and its past.

For a meteorologist, the composition of the atmosphere determines the climate and its variations.

For a biologist or paleontologist, the atmospheric composition is closely dependent on the appearance of the life and its evolution.

CRYOSPHERE

This high resolution image, designed for the Fifth Assessment Report of the IPCC, shows the extent of the regions affected by components of the cryosphere around the world. Over land, continuous permafrost is shown in a dark pink while discontinuous permafrost is shown in a lighter shade of pink. Over much of the northern hemisphere's land area, a semi-transparent white veil depicts the regions that were affected by snowfall at least one day during the period 2000-2012. The bright green line along the southern border of this region shows the maximum snow extent while a black line across the North America, Europe and Asia shows the 50% snow extent line. Glaciers are shown as small golden dots in mountainous areas and in the far northern and southern latitudes. Over the water, ice shelves are shown around Antarctica along with sea ice surrounding the ice shelves. Sea ice is also shown at the North Pole. For both poles. the 30 year average sea ice extent is shown by a yellow outline. In addition, the ice sheets of Greenland and Antarctica are clearly visible.

The **cryosphere** is those portions of Earth's surface where water is in solid form, including sea ice, lake ice, river ice, snow cover, glaciers, ice caps, ice sheets, and frozen ground (which includes permafrost). Thus, there is a wide overlap with the hydrosphere. The cryosphere is an integral part of the global climate system with important linkages and feedbacks generated through its influence on surface energy and moisture fluxes, clouds, precipitation, hydrology, atmospheric and oceanic circulation. Through these feedback processes, the cryosphere plays a significant role in the global climate and in climate model response to global changes. The term deglaciation describes the retreat of cryospheric features. Cryology is the study of cryospheres.

STRUCTURE

Frozen water is found on the Earth's surface primarily as snow cover, freshwater ice in lakes and rivers, sea ice, glaciers, ice sheets, and frozen ground and permafrost (permanently frozen ground). The residence time of water in each of these cryospheric sub-systems varies widely. Snow cover and freshwater ice are essentially seasonal, and most sea ice, except for ice in the central Arctic, lasts only a few years if it is not seasonal. A given water particle in glaciers, ice sheets, or ground ice, however, may remain frozen for 10-100,000 years or longer, and deep ice in parts of East Antarctica may have an age approaching 1 million years.

Most of the world's ice volume is in Antarctica, principally in the East Antarctic Ice Sheet. In terms of areal extent, however, Northern Hemisphere winter snow and ice extent comprise the largest area, amounting to an average 23% of hemispheric surface area in January. The large areal extent and the important climatic roles of snow and ice, related to their unique physical properties, indicate that the ability to observe and model snow and ice-cover extent, thickness, and physical properties (radiative and thermal properties) is of particular significance for climate research.

There are several fundamental physical properties of snow and ice that modulate energy exchanges between the surface and the atmosphere. The most important properties are the surface reflectance (albedo), the ability to transfer heat (thermal diffusivity), and the ability to change state (latent heat). These physical properties, together with surface roughness, emissivity, and dielectric characteristics, have important implications for observing snow and ice from space. For example, surface roughness is often the dominant factor determining the strength of radar backscatter . Physical properties such as crystal structure, density, length, and liquid water content are important factors affecting the transfers of heat and water and the scattering of microwave energy.

The surface reflectance of incoming solar radiation is important for the surface energy balance (SEB). It is the ratio of reflected to incident solar radiation, commonly referred to as albedo. Climatologists are primarily interested in albedo integrated over the shortwave portion of the electromagnetic spectrum (~300 to 3500 nm), which coincides with the main solar energy input. Typically, albedo values for non-melting snow-covered surfaces are high (~80-90%) except in the case of forests. The higher albedos for snow and ice cause rapid shifts in surface reflectivity in autumn and spring in high latitudes, but the overall climatic significance of this increase is spatially and temporally modulated by cloud cover. (Planetary albedo is determined principally by cloud cover, and by the small amount of total solar radiation received in high latitudes during winter months.) Summer and autumn are times of high-average cloudiness over the Arctic Ocean so the albedo feedback associated with the large seasonal changes in sea-ice extent is greatly reduced. Groisman *et al.* (1994a) observed that snow cover exhibited the greatest influence on the Earth radiative balance in the spring (April to May) period when incoming solar radiation was greatest over snow-covered areas.

The thermal properties of cryospheric elements also have important climatic consequences. Snow and ice have much lower thermal diffusivities than air. Thermal diffusivity is a measure of the speed at which temperature waves can penetrate a substance. Snow and ice are many orders of magnitude less efficient at diffusing heat than air. Snow cover insulates the ground surface, and sea ice insulates the underlying ocean, decoupling the surface-atmosphere interface with respect to both heat and moisture fluxes. The flux of moisture from a water surface is eliminated by even a thin skin of ice, whereas the flux of heat through thin ice continues to be substantial until it attains a thickness in excess of 30 to 40 cm. However, even a small amount of snow on top of the ice will dramatically reduce the heat flux and slow down the rate of ice growth. The insulating effect of snow also has major implications for the hydrological cycle. In non-permafrost regions, the insulating effect of snow is such that only near-surface ground freezes and deep-water drainage is uninterrupted.

While snow and ice act to insulate the surface from large energy losses in winter, they also act to retard warming in the spring and summer because of the large amount of energy required to melt ice (the latent heat of fusion, 3.34×10^5 J/kg at 0 °C). However, the strong static stability of the atmosphere over areas of extensive snow or ice tends to confine the immediate cooling effect to a relatively shallow layer, so that associated atmospheric anomalies are usually short-lived and local to regional in scale. In some areas of the world such as Eurasia, however, the cooling associated with a heavy snowpack and moist spring soils is known to play a role in modulating the summer monsoon circulation. Gutzler and Preston (1997) recently presented evidence for a similar snow-summer circulation feedback over the southwestern United States.

The role of snow cover in modulating the monsoon is just one example of a short-term cryosphere-climate feedback involving the land surface and the atmosphere. From Figure 1 it can be seen that there are numerous cryosphere-climate feedbacks in the global climate system. These operate over a wide range of spatial and temporal scales from local seasonal cooling of air temperatures to hemispheric-scale variations in ice sheets over time-scales of thousands of years. The feedback mechanisms involved are often complex and incompletely understood. For example, Curry *et al.* (1995) showed that the so-called "simple" sea ice-albedo feedback involved complex interactions with lead fraction, melt ponds, ice thickness, snow cover, and sea-ice extent.

Snow

Snow cover has the second-largest areal extent of any component of the cryosphere, with a mean maximum areal extent of approximately 47 million km^2. Most of the Earth's snow-covered area (SCA) is located in the Northern Hemisphere, and temporal variability is dominated by the seasonal cycle; Northern Hemisphere snow-cover extent ranges from 46.5 million km^2 in January to 3.8 million km^2 in August. North American winter SCA has exhibited an increasing trend over much of this century (Brown and Goodison 1996; Hughes *et al.* 1996) largely in response to an increase in precipitation.However, the available satellite data show that the hemispheric winter snow cover has exhibited little interannual variability over the 1972-1996 period, with a coefficient of variation (COV=s.d./mean) for January Northern Hemisphere snow cover of < 0.04. According to Groisman *et al.* (1994a) Northern Hemisphere spring snow cover should exhibit a decreasing trend to explain an observed increase in Northern Hemisphere spring air temperatures this century. Preliminary estimates of SCA from historical and reconstructed in situ snow-cover data suggest this is the case for Eurasia, but not for North America, where spring snow cover has remained close to current levels over most of this century. Because of the close relationship observed between hemispheric air temperature and snow-cover extent over the period of satellite data (IPCC 1996), there is considerable interest in monitoring Northern Hemisphere snow-cover extent for detecting and monitoring climate change.

Snow cover is an extremely important storage component in the water balance, especially seasonal snowpacks in mountainous areas of the world. Though limited in extent, seasonal snowpacks in the Earth's mountain ranges account for the major source of the runoff for stream flow and groundwater recharge over wide areas of the midlatitudes. For example, over 85% of the annual runoff from the Colorado River basin originates as snowmelt. Snowmelt runoff from the Earth's mountains fills the rivers and recharges the aquifers that over a billion people depend on for their water resources. Further, over 40% of the world's protected areas are in mountains, attesting to their value both as unique ecosystems needing protection and as recreation areas for humans. Climate warming is expected to result in major changes to the partitioning of snow and rainfall, and to the timing of snowmelt, which will have important implications for water use and management. These changes also involve potentially important decadal and longer time-scale feedbacks to the climate system through temporal and spatial changes in soil moisture and runoff to the oceans.(Walsh 1995). Freshwater fluxes from the snow cover into the marine environment may be important, as the total flux is probably of the same magnitude as desalinated ridging and rubble areas of sea ice. In addition, there is an associated pulse of precipitated pollutants which accumulate over the Arctic winter in snowfall and are released into the ocean upon ablation of the sea-ice .

Sea Ice

Sea ice covers much of the polar oceans and forms by freezing of sea water. Satellite data since the early 1970s reveal considerable seasonal, regional, and interannual variability in the sea-ice covers of both hemispheres. Seasonally, sea-ice extent in the Southern Hemisphere varies by a factor of 5, from a minimum of 3-4 million km^2 in February to a maximum of 17-20 million km^2 in September. The seasonal variation is much less in the Northern Hemisphere where the confined nature and high latitudes of the Arctic Ocean result in a much larger perennial ice cover, and the surrounding land limits the equatorward extent of wintertime ice. Thus, the seasonal variability in Northern Hemisphere ice extent varies by only a factor of 2, from a minimum of 7-9 million km^2 in September to a maximum of 14-16 million km^2 in March.

The ice cover exhibits much greater regional-scale interannual variability than it does hemispherical. For instance, in the region of the Sea of Okhotsk and Japan, maximum ice extent decreased from 1.3 million km^2 in 1983 to 0.85 million km^2 in 1984, a decrease of 35%, before rebounding the following year to 1.2 million km^2. The regional fluctuations in both hemispheres are such that for any several-year period of the satellite record some regions exhibit decreasing ice coverage while others exhibit increasing ice cover. The overall trend indicated in the passive microwave record from 1978 through mid-1995 shows that the extent of Arctic sea ice is decreasing 2.7% per decade. Subsequent work with the satellite passive-microwave data indicates that from late October 1978 through the end of 1996 the extent of Arctic sea ice decreased by 2.9% per decade while the extent of Antarctic sea ice increased by 1.3% per decade. The Intergovernmental Panel on Climate Change publication *Climate change 2013: The Physical Science Basis* stated that sea ice extent for the Northern Hemisphere showed a decrease of 3.8% ± 0.3% per decade from November 1978 to December 2012.

Lake Ice and River Ice

Ice forms on rivers and lakes in response to seasonal cooling. The sizes of the ice bodies involved are too small to exert other than localized climatic effects. However, the freeze-up/break-up processes respond to large-scale and local weather factors, such that considerable interannual variability exists in the dates of appearance and disappearance of the ice. Long series of lake-ice observations can serve as a proxy climate record, and the monitoring of freeze-up and break-up trends may provide a convenient integrated and seasonally specific index of climatic perturbations. Information on river-ice conditions is less useful as a climatic proxy because ice formation is strongly dependent on river-flow regime, which is affected by precipitation, snow melt, and watershed runoff as well as being subject to human interference that directly modifies channel flow, or that indirectly affects the runoff via land-use practices.

Lake freeze-up depends on the heat storage in the lake and therefore on its depth, the rate and temperature of any inflow, and water-air energy fluxes. Information on lake depth is often unavailable, although some indication of the depth of shallow lakes in the Arctic can be obtained from airborne radar imagery during late winter (Sellman *et al.* 1975) and spaceborne optical imagery during summer (Duguay and Lafleur 1997). The timing of breakup is modified by snow depth on the ice as well as by ice thickness and freshwater inflow.

Frozen Ground and Permafrost

Frozen ground (permafrost and seasonally frozen ground) occupies approximately 54 million km^2 of the exposed land areas of the Northern Hemisphere (Zhang et al., 2003) and therefore has the largest areal extent of any component of the cryosphere. Permafrost (perennially frozen ground) may occur where mean annual air temperatures (MAAT) are less than -1 or -2 °C and is generally continuous where MAAT are less than -7 °C. In addition, its extent and thickness are affected by ground moisture content, vegetation cover, winter snow depth, and aspect. The global extent of permafrost is still not completely known, but it underlies approximately 20% of Northern Hemisphere land areas. Thicknesses exceed 600 m along the Arctic coast of northeastern Siberia and Alaska, but, toward the margins, permafrost becomes thinner and horizontally discontinuous. The marginal zones will be more immediately subject to any melting caused by a warming trend. Most of the presently existing permafrost formed during previous colder conditions and is therefore relic. However, permafrost may form under present-day polar climates where glaciers retreat or land emergence exposes unfrozen ground. Washburn (1973) concluded that most continuous permafrost is in balance with the present climate at its upper surface, but changes at the base depend on the present climate and geothermal heat flow; in contrast, most discontinuous permafrost is probably unstable or “in such delicate equilibrium that the slightest climatic or surface change will have drastic disequilibrium effects”.

Under warming conditions, the increasing depth of the summer active layer has significant impacts on the hydrologic and geomorphic regimes. Thawing and retreat of permafrost have been reported in the upper Mackenzie Valley and along the southern margin of its occurrence in Manitoba, but such observations are not readily quantified and generalized. Based on average latitudinal gradients of air temperature, an average northward displacement of the southern permafrost boundary by 50-to-150 km could be expected, under equilibrium conditions, for a 1 °C warming.

Only a fraction of the permafrost zone consists of actual ground ice. The remainder (dry permafrost) is simply soil or rock at subfreezing temperatures. The ice volume is generally greatest in the uppermost permafrost layers and mainly comprises pore and segregated ice in Earth material. Measurements of bore-hole temperatures in permafrost can be used as indicators of net changes in temperature regime. Gold and Lachenbruch (1973) infer a 2-4 °C warming over 75 to 100 years at Cape Thompson, Alaska, where the upper 25% of the 400-m thick permafrost is unstable with respect to an equilibrium profile of temperature with depth (for the present mean annual surface temperature of -5 °C). Maritime influences may have biased this estimate, however. At Prudhoe Bay similar data imply a 1.8 °C warming over the last 100 years (Lachenbruch *et al.* 1982). Further complications may be introduced by changes in snow-cover depths and the natural or artificial disturbance of the surface vegetation.

The potential rates of permafrost thawing have been established by Osterkamp (1984) to be two centuries or less for 25-meter-thick permafrost in the discontinuous zone of interior Alaska, assuming warming from -0.4 to 0 °C in 3–4 years, followed by a further 2.6 °C rise. Although the response of permafrost (depth) to temperature change is typically a very slow process (Osterkamp 1984; Koster 1993), there is ample evidence for the fact that the active layer thickness quickly responds to a temperature change (Kane *et al.* 1991). Whether, under a warming or cooling scenario, global climate change will have a significant effect on the duration of frost-free periods in both regions with seasonally and perennially frozen ground.

Glaciers and Ice Sheets

Ice sheets and glaciers are flowing ice masses that rest on solid land. They are controlled by snow accumulation, surface and basal melt, calving into surrounding oceans or lakes and internal dynamics. The latter results from gravity-driven creep flow ("glacial flow) within the ice body and sliding on the underlying land, which leads to thinning and horizontal spreading. Any imbalance of this dynamic equilibrium between mass gain, loss and transport due to flow results in either growing or shrinking ice bodies.

Ice sheets are the greatest potential source of global freshwater, holding approximately 77% of the global total. This corresponds to 80 m of world sea-level equivalent, with Antarctica accounting for 90% of this. Greenland accounts for most of the remaining 10%, with other ice bodies and glaciers accounting for less than 0.5%. Because of their size in relation to annual rates of snow accumulation and melt, the residence time of water in ice sheets can extend to 100,000 or 1 million years. Consequently, any climatic perturbations produce slow responses, occurring over glacial and interglacial periods. Valley glaciers respond rapidly to climatic fluctuations with typical response times of 10–50 years. However, the response of individual glaciers may be asynchronous to the same climatic forcing because of differences in glacier length, elevation, slope, and speed of motion. Oerlemans (1994) provided evidence of coherent global glacier retreat which could be explained by a linear warming trend of 0.66 °C per 100 years.

While glacier variations are likely to have minimal effects upon global climate, their recession may have contributed one third to one half of the observed 20th Century rise in sea level (Meier 1984; IPCC 1996). Furthermore, it is extremely likely that such extensive glacier recession as is currently observed in the Western Cordillera of North America, where runoff from glacierized basins is used for irrigation and hydropower, involves significant hydrological and ecosystem impacts. Effective water-resource planning and impact mitigation in such areas depends upon developing a sophisticated knowledge of the status of glacier ice and the mechanisms that cause it to change. Furthermore, a clear understanding of the mechanisms at work is crucial to interpreting the global-change signals that are contained in the time series of glacier mass balance records.

Combined glacier mass balance estimates of the large ice sheets carry an uncertainty of about 20%. Studies based on estimated snowfall and mass output tend to indicate that the ice sheets are near balance or taking some water out of the oceans. Marinebased studies suggest sea-level rise from the Antarctic or rapid ice-shelf basal melting. Some authors (Paterson 1993; Alley 1997) have suggested that the difference between the observed rate of sea-level rise (roughly 2 mm/y) and the explained rate of sea-level rise from melting of mountain glaciers, thermal expansion of the ocean, etc. (roughly 1 mm/y or less) is similar to the modeled imbalance in the Antarctic (roughly 1 mm/y of sea-level rise; Huybrechts 1990), suggesting a contribution of sea-level rise from the Antarctic.

Relationships between global climate and changes in ice extent are complex. The mass balance of land-based glaciers and ice sheets is determined by the accumulation of snow, mostly in winter, and warm-season ablation due primarily to net radiation and turbulent heat fluxes to melting ice and snow from warm-air advection,(Munro 1990). However, most of Antarctica never experiences surface melting. Where ice masses terminate in the ocean, iceberg calving is the major contributor to mass loss. In this situation, the ice margin may extend out into deep water as a floating ice shelf, such as that in the Ross Sea. Despite the possibility that global warming could result in losses to the Greenland ice sheet being offset by gains to the Antarctic ice sheet, there is major concern about the possibility of a West Antarctic Ice Sheet collapse. The West Antarctic Ice Sheet is grounded on bedrock below sea level, and its collapse has the potential of raising the world sea level 6–7 m over a few hundred years.

Most of the discharge of the West Antarctic Ice Sheet is via the five major ice streams (faster flowing ice) entering the Ross Ice Shelf, the Rutford Ice Stream entering Ronne-Filchner shelf of the Weddell Sea, and the Thwaites Glacier and Pine Island Glacier entering the Amundsen Ice Shelf. Opinions differ as to the present mass balance of these systems (Bentley 1983, 1985), principally because of the limited data. The West Antarctic Ice Sheet is stable so long as the Ross Ice Shelf is constrained by drag along its lateral boundaries and pinned by local grounding.

GEOSPHERE

There are several conflicting definitions for geosphere.

The **geosphere** may be taken as the collective name for the lithosphere, the hydrosphere, the cryosphere, and the atmosphere.

In Aristotelian physics, the term was applied to four spherical *natural places*, concentrically nested around the center of the Earth, as described in the lectures *Physica* and *Meteorologica*. They were believed to explain the motions of the four *terrestrial elements: Earth*, *Water*, *Air* and *Fire*.

In modern texts and in Earth system science, geosphere refers to the solid parts of the Earth; it is used along with atmosphere, hydrosphere, and biosphere to describe the systems of the Earth (the interaction of these systems with the magnetosphere is sometimes listed). In that context, sometimes the term lithosphere is used instead of geosphere or solid Earth. The lithosphere, however, only refers to the uppermost layers of the solid Earth (oceanic and continental crustal rocks and uppermost mantle).

Since space exploration began, it has been observed that the extent of the ionosphere or plasmasphere is highly variable, and often much larger than previously appreciated, at times extending to the boundaries of the Earth's magnetosphere or geomagnetosphere. This highly variable outer boundary of *geogenic* matter has been referred to as the "geopause, to suggest the relative scarcity of such matter beyond it, where the solar wind dominates.

PEDOSPHERE

The **pedosphere** (from Greek πέδον pedon soil or earth and σφαίρα sfaíra sphere) is the outermost layer of the Earth that is composed of soil and subject to soil formation processes. It exists at the interface of the lithosphere, atmosphere, hydrosphere and biosphere. The sum total of all the organisms, soils, water and air is termed as the pedosphere. The pedosphere is the skin of the Earth and only develops when there is a dynamic interaction between the atmosphere (air in and above the soil), biosphere (living organisms), lithosphere (unconsolidated regolith and consolidated bedrock) and the hydrosphere (water in, on and below the soil). The pedosphere is the foundation of terrestrial life on this planet. There is a realization that the pedosphere needs to be distinctly recognized as a dynamic interface of all terrestrial ecosystems and be integrated into the Earth system science knowledge base.

The pedosphere acts as the mediator of chemical and biogeochemical flux into and out of these respective systems and is made up of gaseous, mineralic, fluid and biologic components. The pedosphere lies within the Critical Zone, a broader interface that includes vegetation, pedosphere, groundwater aquifer systems, regolith and finally ends at some depth in the bedrock where the biosphere and hydrosphere cease to make significant changes to the chemistry at depth. As part of the larger global system, any particular environment in which soil forms is influenced solely by its geographic position on the globe as climatic, geologic, biologic and anthropogenic changes occur with changes in longitude and latitude.

The pedosphere lies below the vegetative cover of the biosphere and above the hydrosphere and lithosphere. The soil forming process (pedogenesis) can begin without the aid of biology but is significantly quickened in the presence of biologic reactions. Soil formation begins with the chemical and/or physical breakdown of minerals to form the initial material that overlies the bedrock substrate. Biology quickens this by secreting acidic compounds (dominantly fulvic acids) that help break rock apart. Particular biologic pioneers are lichen, mosses and seed bearing plants but many other inorganic reactions take place that diversify the chemical makeup of the early soil layer. Once weathering and decomposition products accumulate, a coherent soil body allows the migration of fluids both vertically and laterally through the soil profile causing ion exchange between solid, fluid and gaseous phases. As time progresses, the bulk geochemistry of the soil layer will deviate away from the initial composition of the bedrock and will evolve to a chemistry that reflects the type of reactions that take place in the soil.

Lithosphere

The primary conditions for soil development are controlled by the chemical composition of the rock that the soil will eventually be forming on. Rock types that form the base of the soil profile are often either sedimentary (carbonate or siliceous), igneous or metaigneous (metamorphosed igneous rocks) or volcanic and metavolcanic rocks. The rock type and the processes that lead to its exposure at the surface are controlled by the regional geologic setting of the specific area under study, which revolve around the underlying theory of plate tectonics, subsequent deformation, uplift, subsidence and deposition.

Metaigneous and metavolcanic rocks form the largest component of cratons and are high in silica. Igneous and volcanic rocks are also high in silica but with non-metamorphosed rock, weathering becomes faster and the mobilization of ions is more widespread. Rocks high in silica produce silicic acid as a weathering product. There are few rock types that lead to localized enrichment of some of the biologically limiting elements like phosphorus (P) and nitrogen (N). Phosphatic shale (<15% P_2O_5) and phosphorite (>15% P_2O_5) form in anoxic deep water basins that preserve organic material.Greenstone (metabasalt), phyllite and schist release up to 30-50% of the nitrogen pool. Thick successions of carbonate rocks are often deposited on craton margins during sea level rise. The widespread dissolution of carbonate and evaporate minerals leads to elevated levels of Mg^{2+}, HCO_3^-, Sr^{2+}, Na^+, Cl^- and SO_4^{2-} ions in aqueous solution.

Weathering and Dissolution of Minerals

The process of soil formation is dominated by chemical weathering of silicate minerals, aided by acidic products of pioneering plants and organisms as well as carbonic acid inputs from the atmosphere. Carbonic acid is produced in the atmosphere and soil layers through the carbonation reaction.

$$H_2O + CO_2 \longrightarrow H^+ + HCO_3^- \longrightarrow H_2CO_3$$

This is the dominant form of chemical weathering and aides in the breakdown of carbonate minerals like calcite and dolomite and silicate minerals like feldspar. The breakdown of the Na-feldspar, albite, by carbonic acid to form kaolinite clay is as follows:

$$2\,NaAlSi_3O_8 + 2\,H_2CO_3 + 9\,H_2O \longrightarrow 2\,Na^+ + 2\,HCO_3^- + 4\,H_4SiO_4 + Al_2Si_2O_5(OH)_4$$

Evidence of this reaction in the field would be elevated levels of bicarbonate (HCO_3^-), sodium and silica ions in the water runoff. The breakdown of carbonate minerals:

$$CaCO_3 + H_2CO_3 \longrightarrow Ca^{2+} + 2\,HCO_3^- \quad \text{or } CaCO_3 \longrightarrow Ca^{2+} + CO_3^{2-}$$

The further dissolution of carbonic acid (H_2CO_3) and bicarbonate (HCO_3) produces CO_2 gas. Oxidization is also a major contributor to the breakdown of many silicate minerals and formation of secondary minerals (diagenesis) in the early soil profile. Oxidation of Olivine ($FeMgSiO_2$) releases Fe, Mg and Si ions. The Mg is soluble in water and is carried in the runoff but the Fe often reacts with oxygen to precipitate Fe_2O_3 (hematite), the oxidized state of iron oxide. Sulfur, a byproduct of decaying organic material will also react to Fe to form pyrite (FeS_2) but often in reducing environments. Pyrite dissolution leads to high pH levels due to elevated H+ ions and further precipitation of Fe_2O_3 ultimately changing the redox conditions of the environment.

Biosphere

Inputs from the biosphere may begin with lichen and other microorganisms that secrete oxalic acid. These microorganisms, associated with the lichen community or independently inhabiting rocks, include a number of blue-green algae, green algae, various fungi, and numerous bacteria. Lichen has long been viewed as the pioneers of soil development as the following statement suggests:

"The initial conversion of rock into soil is carried on by the pioneer lichens and their successors, the mosses, in which the hair-like rhizoids assume the role of roots in breaking down the surface into fine dust"

However, lichens are not necessarily the only pioneering organisms nor the earliest form of soil formation as it has been documented that seed-bearing plants may occupy an area and colonize quicker than lichen. Also, eolian sedimentation can produce high rates of sediment accumulation. Nonetheless, lichen can certainly withstand harsher conditions than most vascular plants and although they have slower colonization rates, do form the dominant group in alpine regions.

Acids released from plant roots include acetic and citric acids. During the decay of organic matter Phenolic acids are released from plant matter and humic and fulvic acids are released by soil microbes. These organic acids speed up chemical weathering by combining with some of the weathering products in a process known as chelation. In the soil profile, the organic acids are often concentrated at the top while carbonic acid plays a larger role towards the bottom or below in the aquifer.

As the soil column develops further into thicker accumulations, larger animals come to inhabit the soil and continue to alter the chemical evolution of their respective niche. Earthworms aerate the soil and convert large amounts of organic matter into rich humus, improving soil fertility. Small burrowing mammals store food, grow young and may hibernate in the pedosphere altering the course of soil evolution. Large mammalian herbivores above ground transport nutrients in form of nitrogen-rich waste and phosphorus-rich antlers while predators leave phosphorus-rich piles of bones on the soil surface, leading the localized enrichment of the soil below.

Redox Conditions in Wetland Soils

Nutrient cycling in lakes and freshwater wetlands depends heavily on redox conditions. Under a few millimeters of water heterotrophic bacteria metabolize and consume oxygen. They therefore deplete the soil of oxygen and create the need for anaerobic respiration. Some anaerobic microbial processes include denitrification, sulfate reduction and methanogenesis and are responsible for the release of N_2(nitrogen), H_2S (hydrogen sulfide) and CH_4 (methane). Other anaerobic microbial processes are linked to changes in the oxidation state of iron and manganese. As a result of anaerobic decomposition, the soil stores large amounts of organic carbon because decomposition is incomplete.

The redox potential describes which way chemical reactions will proceed in oxygen deficient soils and controls the nutrient cycling in flooded systems. Redox potential, or reduction potential, is used to express the likelihood of an environment to receive electrons and therefore become reduced. For example, if a system already has plenty of electrons (anoxic, organic-rich shale) it is reduced and will likely donate electrons to a part of the system that has a low concentration of electrons, or an oxidized environment, to equilibrate to the chemical gradient. The oxidized environment has high redox potential, whereas the reduced environment has a low redox potential.

The redox potential is controlled by the oxidation state of the chemical species, pH and the amount of oxygen (O_2) there is in the system. The oxidizing environment accepts electrons because of the presence of O_2, which acts as electron acceptors:

$$O_2 + 4\,e^- + 4\,H^+ \longrightarrow H_2O$$

This equation will tend to move to the right in acidic conditions which causes higher redox potentials to be found at lower pH levels. Bacteria, heterotrophic organisms, consume oxygen while decomposing organic material which depletes the soils of oxygen, thus increasing the redox potential. In low redox conditions the deposition of ferrous iron (Fe^{2+}) will increase with decreasing decomposition rates, thus preserving organic remains and depositing humus. At high redox potential, the oxidized form of iron, ferric iron (Fe^{3+}), will be deposited commonly as hematite. By using analytical geochemical tools such as x-ray fluorescence (XRF) or inductively coupled mass spectrometry (ICP-MS) the two forms of Fe ($Fe^{2+}$and Fe^{3+}) can be measured in ancient rocks therefore determining the redox potential for ancient soils.

Such a study was done on Permian through Triassic rocks (300-200 million years old) in Japan and British Columbia. The geologists found hematite throughout the early and middle Permian but began to find the reduced form of iron in pyrite within the ancient soils near the end of the Permian and into the Triassic. This suggests that conditions became less oxygen rich, even anoxic, during the late Permian, which eventually led to the greatest extinction in earth's history, the P-T extinction.

Decomposition in anoxic or reduced soils is also carried out by sulfur-reducing bacteria which, instead of O_2 use SO_4^{2-} as an electron acceptor and produce hydrogen sulfide (H_2S) and carbon dioxide in the process:

$$2\,H^+ + SO_4^{2-} + 2(CH_2O) \longrightarrow 2\,CO_2 + H_2S + 2\,H_2O$$

The H_2S gas percolates upwards and reacts with Fe^{2+} and precipitates pyrite, acting as a trap for the toxic H_2S gas. However, H_2S is still a large fraction of emissions from wetland soils. In most freshwater wetlands there is little sulfate (SO_4^{2-}) so methanogenesis becomes the dominant form of decomposition by methanogenic bacteria only when sulfate is depleted. Acetate, a compound that is a byproduct of fermenting cellulose is split by methanogenic bacteria to produce methane (CH_4) and carbon dioxide (CO_2), which are released to the atmosphere. Methane is also released during the reduction of CO_2 by the same bacteria.

Atmosphere

In the pedosphere it is safe to assume that gases are in equilibrium with the atmosphere. Because plant roots and soil microbes release CO_2 to the soil, the concentration of bicarbonate(HCO_3) in soil waters is much greater than that in equilibrium with the atmosphere, the high concentration of $CO_2$and the occurrence of metals in soil solutions results in lower pH levels in the soil. Gases that escape from the pedosphere to the atmosphere include the gaseous byproducts of carbonate dissolution, decomposition, redox reactions and microbial photosynthesis. The main inputs from the atmosphere are aeolian sedimentation, rainfall and gas diffusion. Eolian sedimentation includes anything that can be entrained by wind or that stays suspended, seemingly indefinitely, in air and includes a wide variety of aerosol particles, biological particles like pollen and dust to pure quartz sand. Nitrogen is the most abundant constituent in rain (after water), as water vapor utilizes aerosol particles to nucleate rain droplets.

Soil in forests

Soil is well developed in the forest as suggested by the thick humus layers, rich diversity of large trees and animals that live there. In forests, precipitation exceeds evapotranspiration which results in an excess of water that percolates downward through the soil layers. Slow rates of decomposition leads to large amounts of fulvic acid, greatly enhancing chemical weathering. The downward percolation, in conjunction with chemical weathering leaches magnesium (Mg), iron (Fe), and aluminum (Al) from the soil and transports them downward, a process known as podzolization. This process leads to marked contrasts in the appearance and chemistry of the soil layers.

Soil in the Tropics

Tropical forests (rainforests) receive more insolation and rainfall over longer growing seasons than any other environment on earth. With these elevated temperatures, insolation and rainfall, biomass is extremely productive leading to the production of as much as 800 grams of carbon per square meter per year. Higher temperatures and larger amounts of water contribute to higher rates of chemical weathering. Increased rates of decomposition cause smaller amounts of fulvic acid to percolate and leach metals from the zone of active weathering. Thus, in stark contrast to soil in forests, tropical forests have little to no podzolization and therefore do not have marked visual and chemical contrasts with the soil layers. Instead, the mobile metals Mg, Fe and Al are precipitated as oxide minerals giving the soil a rusty red color.

Soil in Grasslands and Deserts

Precipitation in grasslands is equal to or less than evapotranspiration and causes soil development to operate in relative drought. Leaching and migration of weathering products is therefore decreased. Large amounts of evaporation causes buildup of calcium (Ca) and other large cations flocculate clay minerals and fulvic acids in the upper soil profile. Impermeable clay limits downward percolation of water and fulvic acids, reducing chemical weathering and podzolization. The depth to the maximum concentration of clay increases in areas of increased precipitation and leaching. When leaching is decreased, the Ca precipitates as calcite ($CaCO_3$) in the lower soil levels, a layer known as caliche.

Deserts behave similarly to grasslands but operate in constant drought as precipitation is less than evapotranspiration. Chemical weathering proceeds more slowly than in grasslands and beneath the caliche layer may be a layer of gypsum and halite. To study soils in deserts, pedologists have used the concept of chronosequences to relate timing and development of the soil layers. It has been shown that P is leached very quickly from the system and therefore decreases with increasing age. Furthermore, carbon buildup in the soils is decreased due to slower decomposition rates. As a result, the rates of carbon circulation in the biogeochemical cycle is decreased.

BIOSPHERE

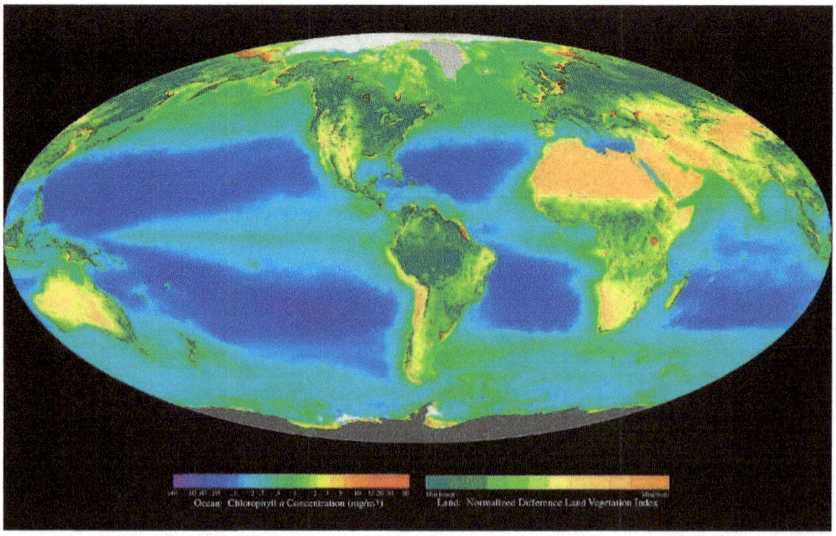

A false-color composite of global oceanic and terrestrial photoautotroph abundance, from September 1997 to August 2016. Provided by the SeaWiFS Project, NASA/Goddard Space Flight Center and ORBIMAGE.

The **biosphere** is the global sum of all ecosystems. The two joined words are bio and sphere. It can also be termed as the zone of life on Earth, a closed system (apart from solar and cosmic radiation and heat from the interior of the Earth), and largely self-regulating. By the most general biophysiological definition, the biosphere is the global ecological system integrating all living beings and their relationships, including their interaction with the elements of the lithosphere, geosphere, hydrosphere, and atmosphere. The biosphere is postulated to have evolved, beginning with a process of biopoesis (life created naturally from non-living matter, such as simple organic compounds) or biogenesis (life created from living matter), at least some 3.5 billion years ago. The earliest evidence for life on Earth includes biogenic graphite found in 3.7 billion-year-old metasedimentary rocks from Western Greenland and microbial mat fossils found in 3.48 billion-year-old sandstone from Western Australia. More recently, in 2015, remains of biotic life were found in 4.1 billion-year-old rocks in Western Australia. According to one of the researchers, If life arose relatively quickly on Earth ... then it could be common in the universe.

In a general sense, biospheres are any closed, self-regulating systems containing ecosystems. This includes artificial biospheres such as Biosphere 2 and BIOS-3, and potentially ones on other planets or moons.

Origin and use of the Term

A beach scene on Earth, simultaneously showing the lithosphere (ground), hydrosphere (ocean) and atmosphere (air)

The term "biosphere" was coined by geologist Eduard Suess in 1875, which he defined as the place on Earth's surface where life dwells.

While the concept has a geological origin, it is an indication of the effect of both Charles Darwin and Matthew F. Maury on the Earth sciences. The biosphere's ecological context comes from the 1920s (see Vladimir I. Vernadsky), preceding the 1935 introduction of the term ecosystem by Sir Arthur Tansley (see ecology history). Vernadsky defined ecology as the science of the biosphere. It is an interdisciplinary concept for integrating astronomy, geophysics, meteorology, biogeography, evolution, geology, geochemistry, hydrology and, generally speaking, all life and Earth sciences.

Narrow Definition

Geochemists define the biosphere as being the total sum of living organisms (the "biomass" or "biota" as referred to by biologists and ecologists). In this sense, the biosphere is but one of four separate components of the geochemical model, the other three being *geosphere*, *hydrosphere*, and *atmosphere*. When these four component spheres are combined into one system, it is known as the Ecosphere. This term was coined during the 1960s and encompasses both biological and physical components of the planet.

The Second International Conference on Closed Life Systems defined *biospherics* as the science and technology of analogs and models of Earth's biosphere; i.e., artificial Earth-like biospheres. Others may include the creation of artificial non-Earth biospheres—for example, human-centered biospheres or a native Martian biosphere—as part of the topic of biospherics.

Extent of Earth's Biosphere

Water covers 71% of the Earth's surface. Image is the Blue Marble photographed from Apollo 17.

Every part of the planet, from the polar ice caps to the equator, features life of some kind. Recent advances in microbiology have demonstrated that microbes live deep beneath the Earth's terrestrial surface, and that the total mass of microbial life in so-called "uninhabitable zones" may, in biomass, exceed all animal and plant life on the surface. The actual thickness of the biosphere on earth is difficult to measure. Birds typically fly at altitudes as high as 1,800 m (5,900 ft; 1.1 mi) and fish live as much as 8,372 m (27,467 ft; 5.202 mi) underwater in the Puerto Rico Trench. There are more extreme examples for life on the planet: Rüppell's vulture has been found at altitudes of 11,300 m (37,100 ft; 7.0 mi); bar-headed geese migrate at altitudes of at least 8,300 m (27,200 ft; 5.2 mi); yaks live at elevations as high as 5,400 m (17,700 ft; 3.4 mi) above sea level; mountain goats live up to 3,050 m (10,010 ft; 1.90 mi). Herbivorous animals at these elevations depend on lichens, grasses, and herbs.

Microscopic organisms live in every part of the biosphere, including soil, hot springs, "seven miles deep" in the ocean, "40 miles high" in the atmosphere and inside rocks far down within the Earth's crust (see also endolith). Microorganisms, under certain test conditions, have been observed to thrive in the vacuum of outer space. The total amount of soil and subsurface bacterial carbon is estimated as 5 x 10^{17} g, or the "weight of the United Kingdom". The mass of prokaryote microorganisms — which includes bacteria and archaea, but not the nucleated eukaryote microorganisms — may be as much as 0.8 trillion tons of carbon (of the total biosphere mass, estimated at between 1 and 4 trillion tons).Barophilic marine microbes have been found at more than a depth of 10,000 m (33,000 ft; 6.2 mi) in the Mariana Trench, the deepest spot in the Earth's oceans. In fact, single-celled life forms have been found in the deepest part of the Mariana Trench, by the Challenger Deep, at depths of 11,034 m (36,201 ft; 6.856 mi). Other researchers reported related studies that microorganisms thrive inside rocks up to 580 m (1,900 ft; 0.36 mi) below the sea floor under 2,590 m (8,500 ft; 1.61 mi) of ocean off the coast of the northwestern United States, as well as 2,400 m (7,900 ft; 1.5 mi) beneath the seabed off Japan. Culturable thermophilic microbes have been extracted from cores drilled more than 5,000 m (16,000 ft; 3.1 mi) into the Earth's crust in Sweden, from rocks between 65–75 °C (149–167 °F). Temperature increases with increasing depth into the Earth's crust. The rate at which the temperature increases depends on many factors, including type of crust (continental vs. oceanic), rock type, geographic location, etc. The greatest known temperature at which microbial life can exist is 122 °C (252 °F) (*Methanopyrus kandleri* Strain 116), and it is likely that the limit of life in the "deep biosphere" is defined by temperature rather than absolute depth. On 20 August 2014, scientists confirmed the existence of microorganisms living 800 m (2,600 ft; 0.50 mi) below the ice of Antarctica. According to one researcher, You can find microbes everywhere — they're extremely adaptable to conditions, and survive wherever they are.

Our biosphere is divided into a number of biomes, inhabited by fairly similar flora and fauna. On land, biomes are separated primarily by latitude. Terrestrial biomes lying within the Arctic and Antarctic Circles are relatively barren of plant and animal life, while most of the more populous biomes lie near the equator.

Specific Biospheres

For this list, if a word is followed by a number, it is usually referring to a specific system or number. Thus:

- Biosphere 1, the planet Earth.
- Biosphere 2, laboratory in Arizona, United States, which contains 3.15 acres (13,000 m^2) of closed ecosystem.
- BIOS-3, a closed ecosystem at the Institute of Biophysics in Krasnoyarsk,

Siberia, in what was then the Soviet Union.

- Biosphere J (CEEF, Closed Ecology Experiment Facilities), an experiment in Japan.

Extraterrestrial Biospheres

No biospheres have been detected beyond the Earth; therefore, the existence of extraterrestrial biospheres remains hypothetical. The rare Earth hypothesis suggests they should be very rare, save ones composed of microbial life only. On the other hand, Earth analogs may be quite numerous, at least in the Milky Way galaxy. Given limited understanding of abiogenesis, it is currently unknown what percentage of these planets actually develop biospheres.

It is also possible that artificial biospheres will be created during the future, for example on Mars. The process of creating an uncontained system that mimics the function of Earth's biosphere is called terraforming.

HYDROSPHERE

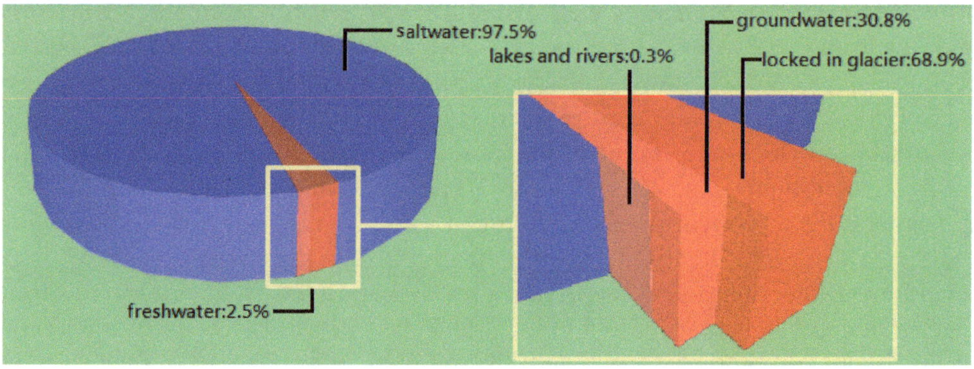

The **hydrosphere** is the combined mass of water found on, under, and above the surface of a planet, minor planet or natural satellite.

It has been estimated that there are 1386 million cubic kilometres of water on Earth. This includes water in liquid and frozen forms in groundwater, oceans, lakes and streams. Saltwater accounts for 97.5% of this amount. Fresh water accounts for only 2.5%. Of this fresh water, 68.9% is in the form of ice and permanent snow cover in the Arctic, the Antarctic, and mountain glaciers. 30.8% is in the form of fresh groundwater. Only 0.3% of the fresh water on Earth is in easily accessible lakes, reservoirs and river systems. The total mass of the Earth's hydrosphere is about 1.4×10^{18} tonnes, which is about 0.023% of Earth's total mass. About 20×10^{12} tonnes of this is in Earth's atmosphere (for practical purposes, 1 cubic metre of water weighs one tonne). Approximately 75% of Earth's surface, an area of some 361 million square kilometers (139.5 million square miles), is covered by ocean. The average salinity of Earth's oceans is about 35 grams of salt per kilogram of sea water (3.5%).

Water Cycle

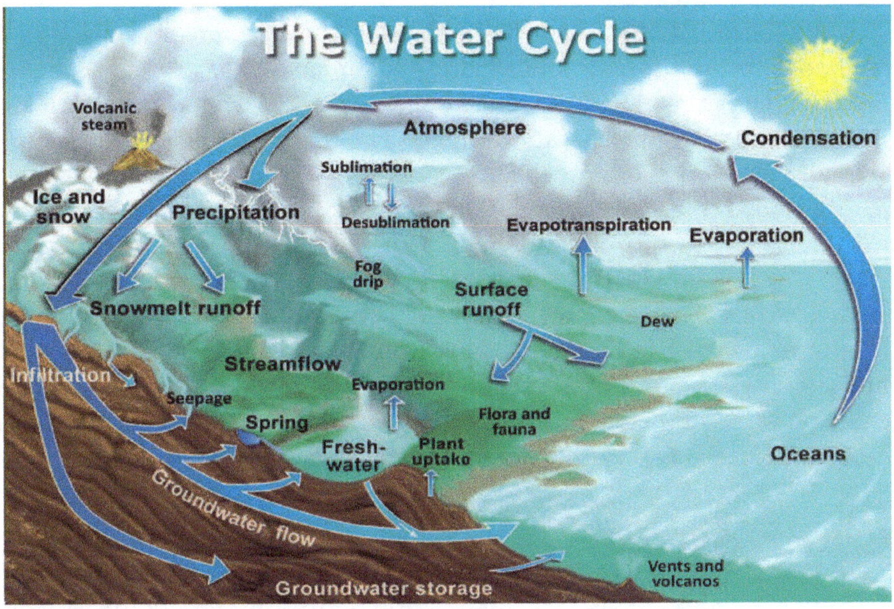

The hydrological cycle transfers water from one state or reservoir to another. Reservoirs include atmospheric moisture (snow, rain and clouds),streams, oceans, rivers, lakes, groundwater, subterranean aquifers, polar icecaps and saturated soil. Solar energy, in the form of heat and light (insolation), and gravity cause the transfer from one state to another over periods from hours to thousands of years. Most evaporation comes from the oceans and is returned to the earth as snow or rain (page 27). Sublimation refers to evaporation from snow and ice. Transpiration refers to the expiration of water through the minute pores or stomata of trees. Evapotranspiration is the term used by hydrologists in reference to the three processes together, transpiration, sublimation and evaporation.

In his book *Water*, Marq de Villiers described the hydrosphere as a closed system in which water exists. The hydrosphere is intricate, complex, interdependent, all-pervading and stable and "seems purpose-built for regulating life (de Villiers 2003:26)." De Villiers claimed that, "On earth, the total amount of water has almost certainly not changed since geological times: what we had then we still have. Water can be polluted, abused, and misused but it is neither created nor destroyed, it only migrates. There is no evidence that water vapor escapes into space (page 26)."

"Every year the turnover of water on Earth involves 577,000 km^3 of water. This is water that evaporates from the oceanic surface (502,800 km3) and from land (74,200 km^3). The same amount of water falls as atmospheric precipitation, 458,000 km^3 on the ocean and 119,000 km^3 on land. The difference between precipitation and evaporation from the land surface (119,000 - 74,200 = 44,800 km^3/year) represents the total runoff of the Earth's rivers (42,700 km^3/year) and direct groundwater runoff to the ocean (2100 km^3/year). These are the principal sources of fresh water to support life necessities and man's economic activities."

Water is a basic necessity of life. Since 2/3 of the Earth is covered by water, the Earth is also called the blue planet and the watery planet. Hydrosphere plays an important role in the existence of the atmosphere in its present form. Oceans are important in this regard. When the Earth was formed it had only a very thin atmosphere rich in hydrogen and helium similar to the present atmosphere of Mercury. Later the gases hydrogen and helium were expelled from the atmosphere. The gases and water vapor released as the Earth cooled became our present atmosphere. Other gases and water vapor released by volcanoes also entered the atmosphere. As the Earth cooled the water vapor in the atmosphere condensed and fell as rain. The atmosphere cooled further as atmospheric carbon dioxide dissolved in to rain water. In turn this further caused the water vapor to condense and fall as rain. This rain water filled the depressions on the Earth's surface and formed the oceans. It is estimated that this occurred about 4000 million years ago. The first life forms began in the oceans. These organisms did not breathe oxygen. Later, when cyanobacteria evolved, the process of conversion of carbon dioxide into food and oxygen began. As a result, our atmosphere has a distinctly different composition from that of the other planets; it is a fundamental requirement for life on Earth.

Recharging Reservoirs

According to Igor A. Shiklomanov, it takes 2500 years for the complete recharge and replenishment of oceanic waters, 10,000 years for permafrost and ice, 1500 years for deep groundwater and mountainous glaciers, 17 years in lakes and 16 days in rivers.

Specific Fresh Water Availability

"Specific water availab*i*lity is the residual (after use) per capita quantity of fresh water." Fresh water resources are unevenly distributed in terms of space and time and can go from floods to water shortages within months in the same area. In 1998 76% of the total population had a specific water availability of less than 5.0 thousand m3 per year per capita. Already by 1998, 35% of the global population suffered "very low or catastrophically low water supplies" and Shiklomanov predicted that the situation would deteriorate in the twenty-first century with "most of the Earth's population will be living under the conditions of low or catastrophically low water supply" by 2025. There is only 2.5% of fresh water in the hydrosphere.

REFERENCES

1. Henry George Liddell, Robert Scott, A Greek-English Lexicon, on Perseus Digital Library
2. Henry George Liddell, Robert Scott, A Greek-English Lexicon, on Perseus Digital Library
3. Seki, K.; Elphic, R. C.; Hirahara, M.; Terasawa, T.; Mukai, T. (2001). "On Atmospheric Loss of Oxygen Ions from Earth Through Magnetospheric Processes". Science. **291** *(5510): 1939–1941. Bibcode*:2001Sci...291.1939S. doi:10.1126/science.1058913. PMID 11239148. Retrieved 2007-03-07.

4. Weaver, D.; Villard, R. (2007-01-31). Hubble Probes Layer-cake Structure of Alien World's Atmosphere. Hubble News Center. Retrieved 2007-03-11.

5. Muriel Gargaud, Encyclopedia of Astrobiology, Volume 3, Springer Science & Business Media, May 26, 2011, p. 879.

6. David C. Catling and Kevin J. Zahnle, The Planetary Air Leak, Scientific American, May 2009, p. 26 (accessed 25 July 2012)

7. Shizgal, B. D.; Arkos, G. G. (1996). Nonthermal escape of the atmospheres of Venus, Earth, and Mars. Reviews of Geophysics. **34** *(4): 483–505. Bibcode*:1996RvGeo..34..483S. doi:10.1029/96RG02213.

8. Lammer, H.; Lichtenegger, H. I. M.; Biernat, H. K.; Erkaev, N. V.; Arshukova, I. L.; Kolb, C.; Gunell, H.; Lukyanov, A.; Holmstrom, M.; Barabash, S.; Zhang, T. L.; Baumjohann, W. (2006). Loss of hydrogen and oxygen from the upper atmosphere of Venus. Planetary and Space Science. **54** *(13–14): 1445–1456. Bibcode*:2006P&SS...54.1445L. doi:10.1016/j.pss.2006.04.022.

9. Webster, Guy (April 8, 2013). Remaining Martian Atmosphere Still Dynamic. NASA. Retrieved April 9, 2013.

10. Wall, Mike (April 8, 2013). Most of Mars' Atmosphere Is Lost in Space. Space.com. Retrieved April 9, 2013.

11. Lammer, H.; Stumptner, W.; Bauer, S. J. (1998). Dynamic escape of H from Titan as consequence of sputtering induced heating. Planetary and Space Science. **46** *(9–10): 1207–1213. Bibcode*:1998P&SS...46.1207L. doi:10.1016/S0032-0633(98)00050-6.

12. Wilson, J. K.; Mendillo, M.; Baumgardner, J.; Schneider, N. M.; Trauger, J. T.; Flynn, B. (2002). The dual sources of Io's sodium clouds. Icarus. **157** *(2): 476–489. Bibcode*:2002Icar..157..476W. doi:10.1006/icar.2002.6821.

13. Melosh, H.J.; Vickery, A.M. (April 1989). Impact erosion of the primordial atmosphere of Mars. Nature. **338**: 487–489. Bibcode:1989Natur.338..487M. doi:10.1038/338487a0.

14. Space Studies Board, Division on Engineering and Physical Sciences (Jan 15, 1961). The Atmospheres of Mars and Venus. National Academies Press.

15. Henry George Liddell, Robert Scott, *A Greek-English Lexicon*, on Perseus

16. Hall, D. K., 1996: Remote sensing applications to hydrology: imaging radar. Hydrological Sciences, 41, 609-624.

17. Groisman, P. Ya, T. R. Karl, and R. W. Knight, 1994a: Observed impact of snow cover on the heat balance and the rise of continental spring temperatures. Science, 363, 198-200.

18. Lynch-Stieglitz, M., 1994: The development and validation of a simple snow model for the GISS GCM. J. Climate, 7, 1842-1855.

19. Cohen, J., and D. Rind, 1991: The effect of snow cover on the climate. J. Climate, 4, 689-706.

20. Vernekar, A. D., J. Zhou, and J. Shukla, 1995: The effect of Eurasian snow cover on the Indian monsoon. J. Climate, 8, 248-266.

21. Gutzler, D. S., and J. W. Preston, 1997: Evidence for a relationship between spring snow cover in North America and summer rainfall in New Mexico. Geophys. Res. Lett., 24, 2207-2210.

22. Robinson, D. A., K. F. Dewey, and R. R. Heim, 1993: Global snow cover monitoring: an update. Bull. Amer. Meteorol. Soc., 74, 1689-1696.

23. Groisman, P. Ya, and D. R. Easterling, 1994: Variability and trends of total precipitation and snowfall over the United States and Canada. J. Climate, 7, 184-205.

24. Brown, R. D., 1997: Historical variability in Northern Hemisphere spring snow covered area. Annals of Glaciology, 25, 340-346.

25. Prinsenberg, S. J. 1988: Ice-cover and ice-ridge contributions to the freshwater contents of Hudson Bay and Foxe Basin. Arctic, 41, 6-11.

26. Zwally, H. J., J. C. Comiso, C. L. Parkinson, W. J. Campbell, F. D. Carsey, and P. Gloersen, 1983: Antarctic Sea Ice, 1973-1976: Satellite Passive-Microwave Observations. NASA SP-459, National Aeronautics and Space Administration, Washington, D.C., 206 pp.

27. Gloersen, P., W. J. Campbell, D. J. Cavalieri, J. C. Comiso, C. L. Parkinson, and H. J. Zwally, 1992: Arctic and Antarctic Sea Ice, 1978-1987: Satellite Passive-Microwave Observations and Analysis. NASA SP-511, National Aeronautics and Space Administration, Washington, D.C., 290 pp.

28. Parkinson, C. L., J. C. Comiso, H. J. Zwally, D. J. Cavalieri, P. Gloersen, and W. J. Campbell, 1987: Arctic Sea Ice, 1973-1976: Satellite Passive-Microwave Observations, NASA SP-489, National Aeronautics and Space Administration, Washington, D.C., 296 pp.

29. Parkinson, C. L., 1995: Recent sea-ice advances in Baffin Bay/Davis Strait and retreats in the Bellinshausen Sea. Annals of Glaciology, 21, 348-352.

30. Johannessen, O. M., M. Miles, and E. Bjørgo, 1995: The Arctic's shrinking sea ice. Nature, 376, 126-127.

31. Cavalieri, D. J., P. Gloersen, C. L. Parkinson, J. C. Comiso, and H. J. Zwally, 1997: Observed hemispheric asymmetry in global sea ice changes. Science, 278, 1104-1106.

32. *Climate Change 2013: The Physical Science Basis (PDF). ipcc. Intergovernmental Panel on Climate Change. p. 324. Retrieved 16 June 2015.*

33. Washburn, A. L., 1973: Periglacial processes and environments. Edward Arnold, London, 320 pp. p.48

34. *Greve, R.; Blatter, H. (2009). Dynamics of Ice Sheets and Glaciers. Springer. doi:10.1007/978-3-642-03415-2. ISBN 978-3-642-03414-5.*

35. Oerlemans, J., 1994: Quantifying global warming from the retreat of glaciers. Science, 264, 243-245.

36. Pelto, M. S., 1996: Annual net balance of North Cascade Glaciers, 1984-94. J. Glaciology, 42, 3-9.

37. Bentley, C. R., and M. B. Giovinetto, 1991: Mass balance of Antarctica and sea level change. In: G. Weller, C. L. Wilson and B. A. B. Severin (eds.), Polar regions and climate change. University of Alaska, Fairbanks, p. 481-488.

38. Jacobs, S. S., H. H. Helmer, C. S. M. Doake, A. Jenkins, and R. M. Frohlich, 1992: Melting of ice shelves and the mass balance of Antarctica. J. Glaciology, 38, 375-387.

39. Paterson, W. S. B., 1993: World sea level and the present mass balance of the Antarctic ice sheet. In: W.R. Peltier (ed.), Ice in the Climate System, NATO ASI Series, I12, Springer-Verlag, Berlin, 131-140.

40. Van den Broeke, M. R., 1996: The atmospheric boundary layer over ice sheets and glaciers. Utrecht, Universitiet Utrecht, 178 pp..

41. Van den Broeke, M. R., and R. Bintanja, 1995: The interaction of katabatic wind and the formation of blue ice areas in East Antarctica. J. Glaciology, 41, 395-407

42. Ohmura, A., M. Wild, and L. Bengtsson, 1996: A possible change in mass balance of the Greenland and Antarctic ice sheets in the coming century. J. Climate, 9, 2124-2135.

CHAPTER 2

Climate and Climate Change

Chapter Contents

Climate is the average weather in a place over many years. While the weather can change in just a few hours, climate takes hundreds, thousands, even millions of years to change.

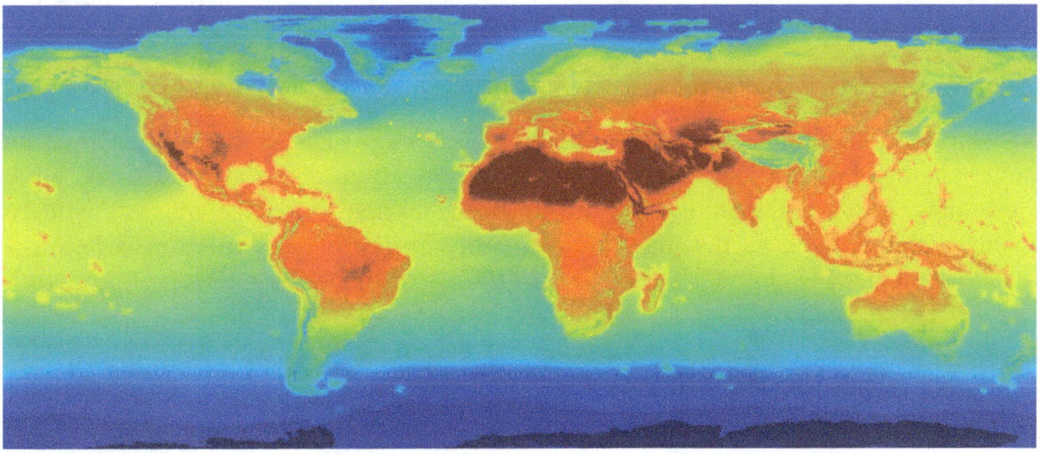

Sometimes the climate of a place is described with graphs like this. This graph shows how temperature usually changes over a year for a particular place on Earth.

INTRODUCTION

Climate is the statistics of weather, usually over a 30-year interval. It is measured by assessing the patterns of variation in temperature, humidity, atmospheric pressure, wind, precipitation, atmospheric particle count and other meteorological variables in a given region over long periods of time. Climate differs from weather, in that weather only describes the short-term conditions of these variables in a given region.

A region›s climate is generated by the **climate system**, which has five components: atmosphere, hydrosphere, cryosphere, lithosphere, and biosphere.

The climate of a location is affected by its latitude, terrain, and altitude, as well as nearby water bodies and their currents. Climates can be classified according to the average and the typical ranges of different variables, most commonly temperature and precipitation. The most commonly used classification scheme was Köppen climate classification originally developed by Wladimir Köppen. The Thornthwaite system, in use since 1948, incorporates evapotranspiration along with temperature and precipitation information and is used in studying biological diversity and the potential effects on it of climate changes. The Bergeron and Spatial Synoptic Classification systems focus on the origin of air masses that define the climate of a region.

Paleoclimatology is the study of ancient climates. Since direct observations of climate are not available before the 19th century, paleoclimates are inferred from *proxy variables* that include non-biotic evidence such as sediments found in lake beds and ice cores, and biotic evidence such as tree rings and coral. Climate models are mathematical models of past, present and future climates. Climate change may occur over long and short timescales from a variety of factors; recent warming is discussed in global warming. Global warming results in redistributions. For example, "a 3°C change in mean annual temperature corresponds to a shift in isotherms of approximately 300–400 km in latitude (in the temperate zone) or 500 m in elevation. Therefore, species are expected to move upwards in elevation or towards the poles in latitude in response to shifting climate zones".

DEFINITION

Climate (from Ancient Greek *klima*, meaning *inclination*) is commonly defined as the weather averaged over a long period. The standard averaging period is 30 years, but other periods may be used depending on the purpose. Climate also includes statistics other than the average, such as the magnitudes of day-to-day or year-to-year variations. The Intergovernmental Panel on Climate Change (IPCC) 2001 glossary definition is as follows:

Climate in a narrow sense is usually defined as the "average weather," or more rigorously, as the statistical description in terms of the mean and variability of relevant quantities over a period ranging from months to thousands or millions of years. The classical period is 30 years, as defined by the World Meteorological Organization (WMO). These quantities are most often surface variables such as temperature, precipitation, and wind. Climate in a wider sense is the state, including a statistical description, of the climate system.

The World Meteorological Organization (WMO) describes climate "normals" as "reference points used by climatologists to compare current climatological trends to that of the past or what is considered ‹normal›. A Normal is defined as the arithmetic average of a climate element (e.g. temperature) over a 30-year period. A 30 year period is used, as it is long enough to filter out any interannual variation or anomalies, but also short enough to be able to show longer climatic trends." The WMO originated from the International Meteorological Organization which set up a technical commission for climatology in 1929. At its 1934 Wiesbaden meeting the technical commission designated the thirty-year period from 1901 to 1930 as the reference time frame for climatological standard normals. In 1982 the WMO agreed to update climate normals, and in these were subsequently completed on the basis of climate data from 1 January 1961 to 31 December 1990.

The difference between climate and weather is usefully summarized by the popular phrase "Climate is what you expect, weather is what you get." Over historical time spans there are a number of nearly constant variables that determine climate, including latitude, altitude, proportion of land to water, and proximity to oceans and mountains. These change only over periods of millions of years due to processes such as plate tectonics. Other climate determinants are more dynamic: the thermohaline circulation of the ocean leads to a 5 °C (9 °F) warming of the northern Atlantic Ocean compared to other ocean basins. Other ocean currents redistribute heat between land and water on a more regional scale. The density and type of vegetation coverage affects solar heat absorption, water retention, and rainfall on a regional level. Alterations in the quantity of atmospheric greenhouse gases determines the amount of solar energy retained by the planet, leading to global warming or global cooling. The variables which determine climate are numerous and the interactions complex, but there is general agreement that the broad outlines are understood, at least insofar as the determinants of historical climate change are concerned.

CLIMATE CLASSIFICATION

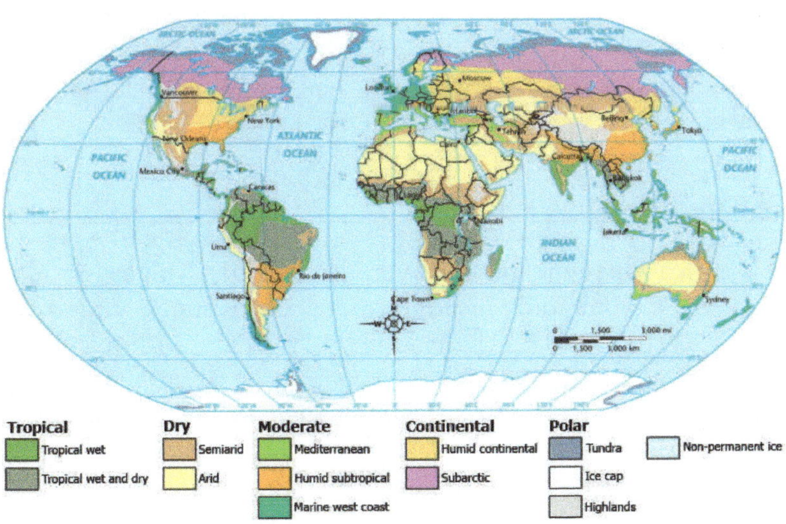

Worldwide climate classifications.

There are several ways to classify climates into similar regimes. Originally, climes were defined in Ancient Greece to describe the weather depending upon a location's latitude. Modern climate classification methods can be broadly divided into *genetic* methods, which focus on the causes of climate, and *empiric* methods, which focus on the effects of climate. Examples of genetic classification include methods based on the relative frequency of different air mass types or locations within synoptic weather disturbances. Examples of empiric classifications include climate zones defined by plant hardiness, evapotranspiration, or more generally the Köppen climate classification which was originally designed to identify the climates associated with certain biomes. A common shortcoming of these classification schemes is that they produce distinct boundaries between the zones they define, rather than the gradual transition of climate properties more common in nature.

Bergeron and Spatial Synoptic

The simplest classification is that involving air masses. The Bergeron classification is the most widely accepted form of air mass classification. Air mass classification involves three letters. The first letter describes its moisture properties, with c used for continental air masses (dry) and m for maritime air masses (moist). The second letter describes the thermal characteristic of its source region: T for tropical, P for polar, A for Arctic or Antarctic, M for monsoon, E for equatorial, and S for superior air (dry air formed by significant downward motion in the atmosphere). The third letter is used to designate the stability of the atmosphere. If the air mass is colder than the ground below it, it is labeled k. If the air mass is warmer than the ground below it, it is labeled w. While air mass identification was originally used in weather forecasting during the 1950s, climatologists began to establish synoptic climatologies based on this idea in 1973.

Based upon the Bergeron classification scheme is the Spatial Synoptic Classification system (SSC). There are six categories within the SSC scheme: Dry Polar (similar to continental polar), Dry Moderate (similar to maritime superior), Dry Tropical (similar to continental tropical), Moist Polar (similar to maritime polar), Moist Moderate (a hybrid between maritime polar and maritime tropical), and Moist Tropical (similar to maritime tropical, maritime monsoon, or maritime equatorial).

Köppen

The Köppen classification depends on average monthly values of temperature and precipitation. The most commonly used form of the Köppen classification has five primary types labeled A through E. These primary types are A) tropical, B) dry, C) mild mid-latitude, D) cold mid-latitude, and E) polar. The five primary classifications can be further divided into secondary classifications such as rain forest, monsoon, tropical savanna, humid subtropical, humid continental, oceanic climate, Mediterranean climate, desert, steppe, subarctic climate, tundra, and polar ice cap.

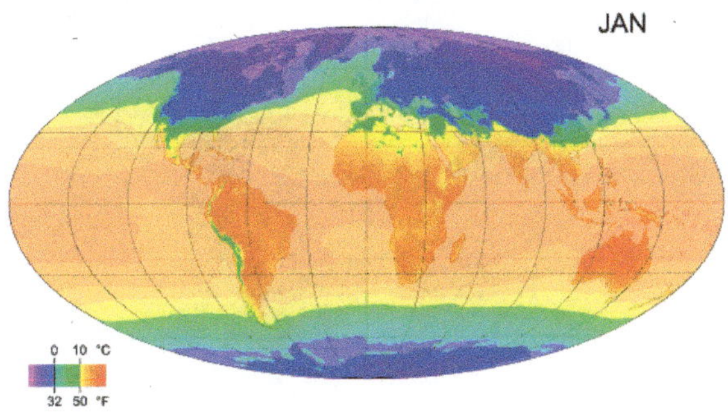

Monthly average surface temperatures from 1961–1990. This is an example of how climate varies with location and season.

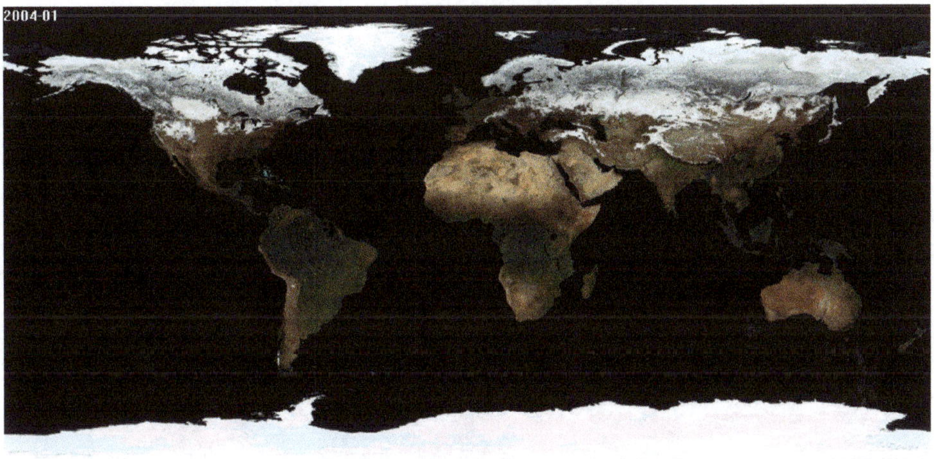

Monthly global images from NASA Earth Observatory (interactive SVG).

Rain forests are characterized by high rainfall, with definitions setting minimum normal annual rainfall between 1,750 millimetres (69 in) and 2,000 millimetres (79 in). Mean monthly temperatures exceed 18 °C (64 °F) during all months of the year.

A **monsoon** is a seasonal prevailing wind which lasts for several months, ushering in a region's rainy season. Regions within North America, South America, Sub-Saharan Africa, Australia and East Asia are monsoon regimes.

The world's cloudy and sunny spots. NASA Earth Observatory map using data collected between July 2002 and April 2015.

A **tropical savanna** is a grassland biome located in semiarid to semi-humid climate regions of subtropical and tropical latitudes, with average temperatures remain at or above 18 °C (64 °F) year round and rainfall between 750 millimetres (30 in) and 1,270 millimetres (50 in) a year. They are widespread on Africa, and are found in India, the northern parts of South America, Malaysia, and Australia.

Cloud cover by month for 2014. NASA Earth Observatory

The **humid subtropical** climate zone where winter rainfall (and sometimes snowfall) is associated with large storms that the westerlies steer from west to east. Most summer rainfall occurs during thunderstorms and from occasional tropical cyclones.Humid subtropical climates lie on the east side continents, roughly between latitudes 20° and 40° degrees away from the equator.

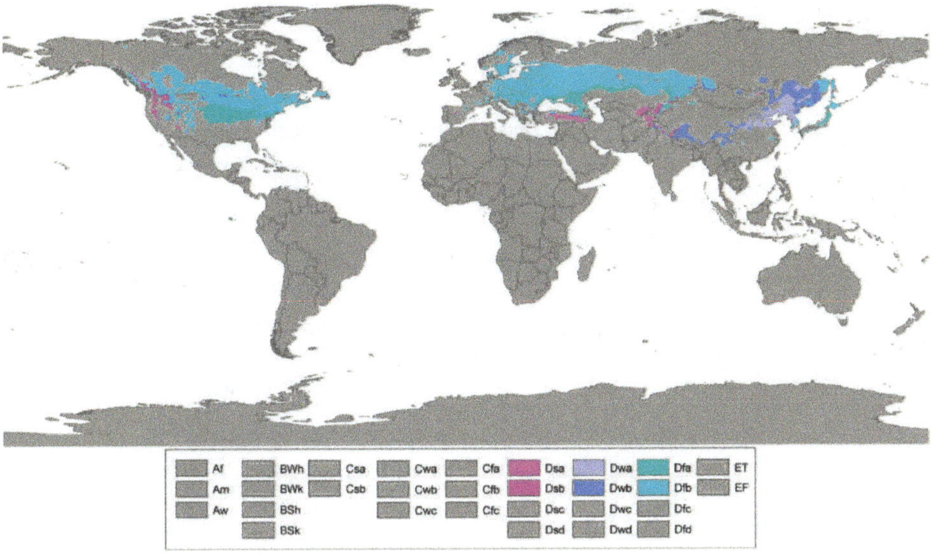

Humid continental climate, worldwide.

A **humid continental** climate is marked by variable weather patterns and a large seasonal temperature variance. Places with more than three months of average daily temperatures above 10 °C (50 °F) and a coldest month temperature below −3 °C (27 °F) and which do not meet the criteria for an arid or semiarid climate, are classified as continental.

An **oceanic climate** is typically found along the west coasts at the middle latitudes of all the world's continents, and in southeastern Australia, and is accompanied by plentiful precipitation year round.

The **Mediterranean climate** regime resembles the climate of the lands in the Mediterranean Basin, parts of western North America, parts of Western and South Australia, in southwestern South Africa and in parts of central Chile. The climate is characterized by hot, dry summers and cool, wet winters.

A **steppe** is a dry grassland with an annual temperature range in the summer of up to 40 °C (104 °F) and during the winter down to −40 °C (−40 °F).

A **subarctic climate** has little precipitation, and monthly temperatures which are above 10 °C (50 °F) for one to three months of the year, with permafrost in large parts of the area due to the cold winters. Winters within subarctic climates usually include up to six months of temperatures averaging below 0 °C (32 °F).

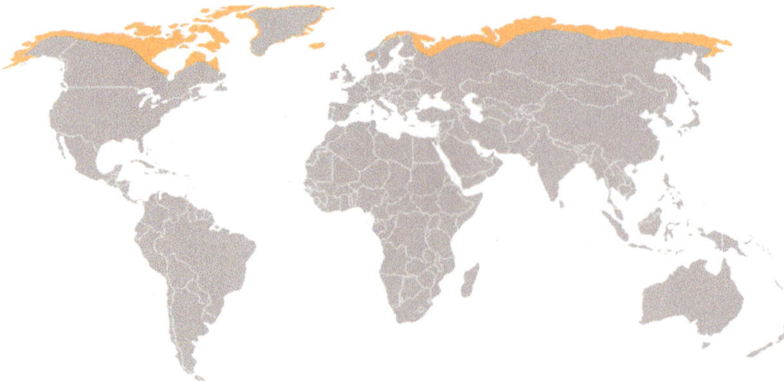

Map of arctic tundra.

Tundra occurs in the far Northern Hemisphere, north of the taiga belt, including vast areas of northern Russia and Canada.

A **polar ice cap**, or polar ice sheet, is a high-latitude region of a planet or moon that is covered in ice. Ice caps form because high-latitude regions receive less energy as solar radiation from the sun than equatorial regions, resulting in lower surface temperatures.

A **desert** is a landscape form or region that receives very little precipitation. Deserts usually have a large diurnal and seasonal temperature range, with high or low, depending on location daytime temperatures (in summer up to 45 °C or 113 °F), and low nighttime temperatures (in winter down to 0 °C or 32 °F) due to extremely low humidity. Many deserts are formed by rain shadows, as mountains block the path of moisture and precipitation to the desert.

Thornthwaite

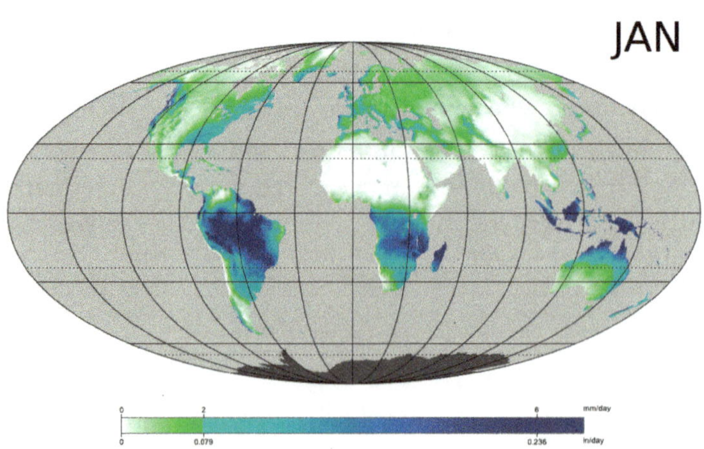

Precipitation by month.

Devised by the American climatologist and geographer C. W. Thornthwaite, this climate classification method monitors the soil water budget using evapotranspiration. It monitors the portion of total precipitation used to nourish vegetation over a certain area. It uses indices such as a humidity index and an aridity index to determine an area's moisture regime based upon its average temperature, average rainfall, and average vegetation type. The lower the value of the index in any given area, the drier the area is.

The moisture classification includes climatic classes with descriptors such as hyperhumid, humid, subhumid, subarid, semi-arid (values of −20 to −40), and arid (values below −40). Humid regions experience more precipitation than evaporation each year, while arid regions experience greater evaporation than precipitation on an annual basis. A total of 33 percent of the Earth's landmass is considered either arid or semi-arid, including southwest North America, southwest South America, most of northern and a small part of southern Africa, southwest and portions of eastern Asia, as well as much of Australia. Studies suggest that precipitation effectiveness (PE) within the Thornthwaite moisture index is overestimated in the summer and underestimated in the winter. This index can be effectively used to determine the number of herbivore and mammal species numbers within a given area. The index is also used in studies of climate change.

Thermal classifications within the Thornthwaite scheme include microthermal, mesothermal, and megathermal regimes. A microthermal climate is one of low annual mean temperatures, generally between 0 °C (32 °F) and 14 °C (57 °F) which experiences short summers and has a potential evaporation between 14 centimetres (5.5 in) and 43 centimetres (17 in). A mesothermal climate lacks persistent heat or persistent cold, with potential evaporation between 57 centimetres (22 in) and 114 centimetres (45 in). A megathermal climate is one with persistent high temperatures and abundant rainfall, with potential annual evaporation in excess of 114 centimetres (45 in).

RECORD

Modern

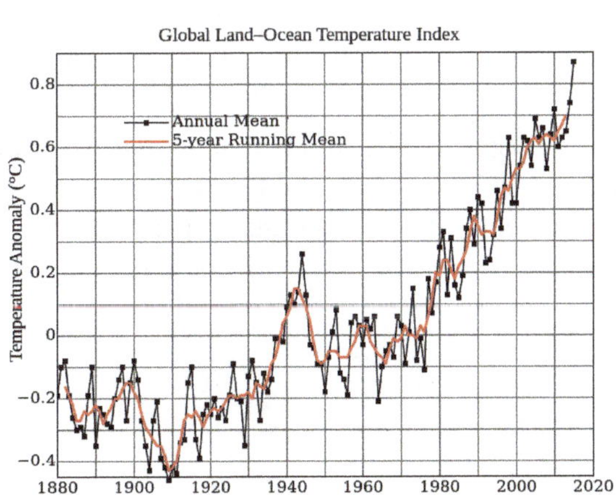

Global mean surface temperature change since 1880. Source: NASA GISS.

Details of the modern climate record are known through the taking of measurements from such weather instruments as thermometers, barometers, and anemometers during the past few centuries. The instruments used to study weather over the modern time scale, their known error, their immediate environment, and their exposure have changed over the years, which must be considered when studying the climate of centuries past.

CLIMATE CHANGE

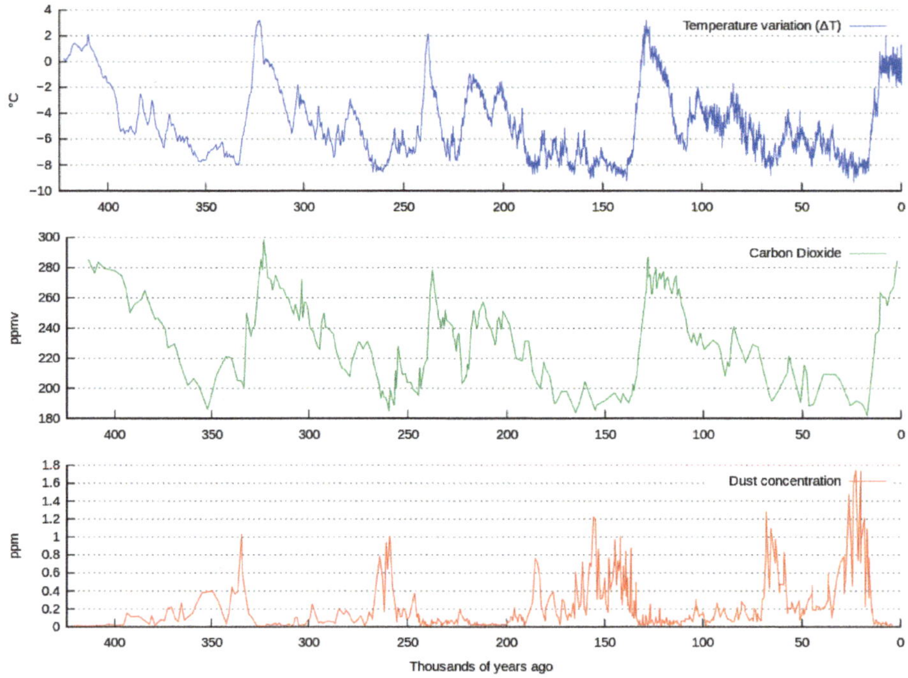

Variations in CO_2, temperature and dust from the Vostok ice core over the past 450,000 years.

See also: Climate change, Global warming, Temperature record, and Attribution of recent climate change

Climate change is the variation in global or regional climates over time. It reflects changes in the variability or average state of the atmosphere over time scales ranging from decades to millions of years. These changes can be caused by processes internal to the Earth, external forces (e.g. variations in sunlight intensity) or, more recently, human activities.

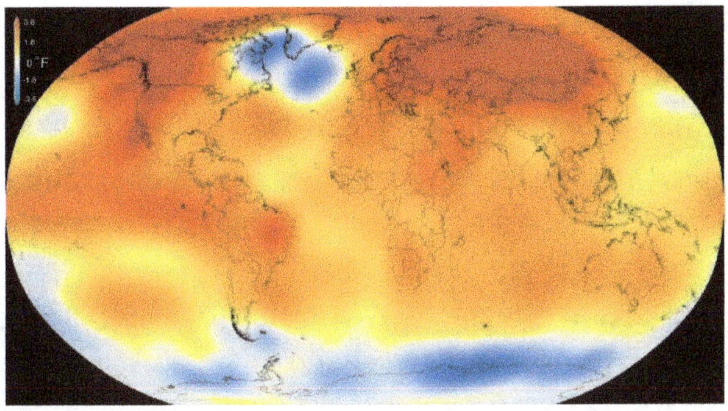

2015 – Warmest Global Year on Record (since 1880) – Colors indicate temperature anomalies (NASA/NOAA; 20 January 2016).

In recent usage, especially in the context of environmental policy, the term "climate change" often refers only to changes in modern climate, including the rise in average surface temperature known as global warming. In some cases, the term is also used with a presumption of human causation, as in the United Nations Framework Convention on Climate Change (UNFCCC). The UNFCCC uses "climate variability" for non-human caused variations.

Earth has undergone periodic climate shifts in the past, including four major ice ages. These consisting of glacial periods where conditions are colder than normal, separated by interglacial periods. The accumulation of snow and ice during a glacial period increases the surface albedo, reflecting more of the Sun›s energy into space and maintaining a lower atmospheric temperature. Increases in greenhouse gases, such as by volcanic activity, can increase the global temperature and produce an interglacial period. Suggested causes of ice age periods include the positions of the continents, variations in the Earth›s orbit, changes in the solar output, and volcanism.

Climate Models

Climate models use quantitative methods to simulate the interactions of the atmosphere, oceans, land surface and ice. They are used for a variety of purposes; from the study of the dynamics of the weather and climate system, to projections of future climate. All climate models balance, or very nearly balance, incoming energy as short wave (including visible) electromagnetic radiation to the earth with outgoing energy as long wave (infrared) electromagnetic radiation from the earth. Any imbalance results in a change in the average temperature of the earth.

The most talked-about applications of these models in recent years have been their use to infer the consequences of increasing greenhouse gases in the atmosphere, primarily carbon dioxide (see greenhouse gas). These models predict an upward trend in the global mean surface temperature, with the most rapid increase in temperature being projected for the higher latitudes of the Northern Hemisphere.

Models can range from relatively simple to quite complex:

- Simple radiant heat transfer model that treats the earth as a single point and averages outgoing energy
- this can be expanded vertically (radiative-convective models), or horizontally
- finally, (coupled) atmosphere–ocean–sea ice global climate models discretise and solve the full equations for mass and energy transfer and radiant exchange.

Climate forecasting is a way by some scientists are using to predict climate change. In 1997 the prediction division of the International Research Institute for Climate and Society at Columbia University began generating seasonal climate forecasts on a real-time basis. To produce these forecasts an extensive suite of forecasting tools was developed, including a multimodel ensemble approach that required thorough validation of each model›s accuracy level in simulating interannual climate variability.

AIR MASS

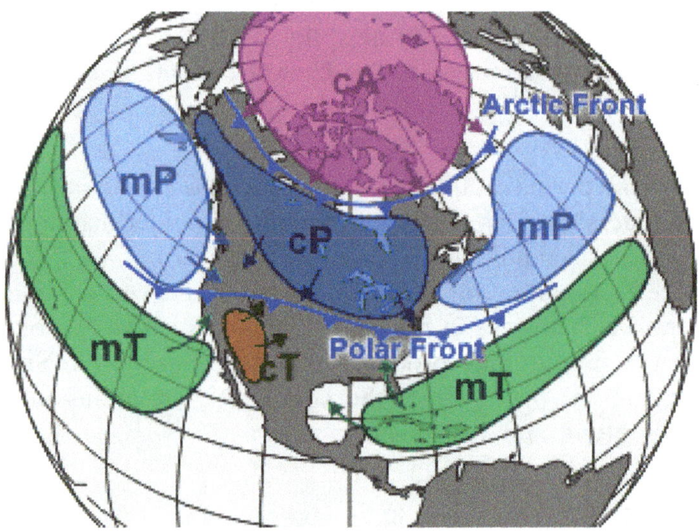

Different air masses which affect North America, as well as other continents, tend to be separated by frontal boundaries

In meteorology, an **air mass** is a volume of air defined by its temperature and water vapor content. Air masses cover many hundreds or thousands of square miles, and adapt to the characteristics of the surface below them. They are classified according to latitude and their continental or maritime source regions. Colder air masses are termed polar or arctic, while warmer air masses are deemed tropical. Continental and superior air masses are dry while maritime and monsoon air masses are moist. Weather fronts separate air masses with different density (temperature and/or moisture) characteristics. Once an air mass moves away from its source region, underlying vegetation and water bodies can quickly modify its character. Classification schemes tackle an air mass' characteristics, and well as modification.

Classification and Notation

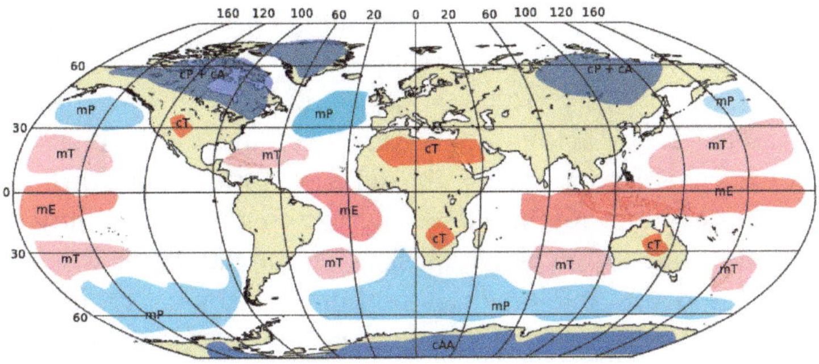

Source regions of global air masses.

The Bergeron classification is the most widely accepted form of air mass classification, though others have produced more refined versions of this scheme over different regions of the globe. Air mass classification involves three letters. The first letter describes its moisture properties, with c used for continental air masses (dry) and m for maritime air masses (moist). The second letter describes the thermal characteristic of its source region: T for Tropical, P for Polar, A for arctic or Antarctic, M for monsoon, E for Equatorial, and S for superior air (an adiabatically drying and warming air formed by significant downward motion in the atmosphere). For instance, an air mass originating over the desert southwest of the United States in summer may be designated "cT". An air mass originating over northern Siberia in winter may be indicated as "cA".

The stability of an air mass may be shown using a third letter, either "k" (air mass colder than the surface below it) or "w" (air mass warmer than the surface below it). An example of this might be a polar air mass blowing over the Gulf Stream, denor it is replacing is colder than the replaced air mass (usually for polar air masses). For example, a series of fronts over the Pacific might show an air mass denoted mPk followed by another denoted mPk›.

Another convention utilizing these symbols is the indication of modification or transformation of one type to another. For instance, an Arctic air mass blowing out over the Gulf of Alaska may be shown as "cA-mPk". Yet another convention indicates the layering of air masses in certain situations. For instance, the overrunning of a polar air mass by an air mass from the Gulf of Mexico over the Central United States might be shown with the notation "mT/cP" (sometimes using a horizontal line as in fraction notation).

Characteristics

Arctic, Antarctic, and polar air masses are cold. The qualities of arctic air are developed over ice and snow-covered ground. Arctic air is deeply cold, colder than polar air masses. Arctic air can be shallow in the summer, and rapidly modify as it moves equatorward. Polar air masses develop over higher latitudes over the land or ocean, are very stable, and generally shallower than arctic air. Polar air over the ocean (maritime) loses its stability as it gains moisture over warmer ocean waters.

Tropical and equatorial air masses are hot as they develop over lower latitudes. Those that develop over land (continental) are drier and hotter than those that develop over oceans, and travel poleward on the western periphery of the subtropical ridge. Maritime tropical air masses are sometimes referred to as trade air masses. Monsoon air masses are moist and unstable. Superior air masses are dry, and rarely reach the ground. They normally reside over maritime tropical air masses, forming a warmer and drier layer over the more moderate moist air mass below, forming what is known as a trade wind inversion over the maritime tropical air mass. Continental Polar air masses (cP) are air masses that are cold and dry due to their continental source region. Continental polar air masses that affect North America form over interior Canada. Continental Tropical air masses (cT) are a type of tropical air produced by the subtropical ridge over large areas of land and typically originate from low-latitude deserts such as the Sahara Desert in northern Africa, which is the major source of these air masses. Other less important sources producing cT air masses are the Arabian Peninsula, the central arid/semi-arid part of Australia and deserts lying in the Southwestern United States. Continental tropical air masses are extremely hot and dry.

Movement and Fronts

A **weather front** is a boundary separating two masses of air of different densities, and is the principal cause of meteorological phenomena. In surface weather analyses, fronts are depicted using various colored lines and symbols, depending on the type of front. The air masses separated by a front usually differ in temperature and humidity. Cold fronts may feature narrow bands of thunderstorms and severe weather, and may on occasion be preceded by squall lines or dry lines. Warm fronts are usually preceded by stratiform precipitation and fog. The weather usually clears quickly after a front›s passage. Some fronts produce no precipitation and little cloudiness, although there is invariably a wind shift.

Cold fronts and occluded fronts generally move from west to east, while warm fronts move poleward. Because of the greater density of air in their wake, cold fronts and cold occlusions move faster than warm fronts and warm occlusions. Mountains and warm bodies of water can slow the movement of fronts. When a front becomes stationary, and the density contrast across the frontal boundary vanishes, the front can degenerate into a line which separates regions of differing wind velocity, known as a shearline. This is most common over the open ocean.

Modification

Air masses can be modified in a variety of ways. Surface flux from underlying vegetation, such as forest, acts to moisten the overlying air mass. Heat from underlying warmer waters can significantly modify an air mass over distances as short as 35 kilometres (22 mi) to 40 kilometres (25 mi). For example, southwest of extratropical cyclones, curved cyclonic flow bringing cold air across the relatively warm water bodies can lead to narrow lake-effect snow bands. Those bands bring strong localized precipitation since large water bodies such as lakes efficiently store heat that results in significant temperature differences (larger than 13 °C or 23 °F) between the water surface and the air above. Because of this temperature difference, warmth and moisture are transported upward, condensing into vertically oriented clouds (see satellite picture) which produce snow showers. The temperature decrease with height and cloud depth are directly affected by both the water temperature and the large-scale environment. The stronger the temperature decrease with height, the deeper the clouds get, and the greater the precipitation rate becomes.

Lake-effect snow bands near the Korean Peninsula.

REFERENCES

1. Planton, Serge (France; editor) (2013). "Annex III. Glossary: IPCC - Intergovernmental Panel on Climate Change" (PDF). IPCC Fifth Assessment Report. p. 1450. Retrieved 25 July 2016.
2. Shepherd, Dr. J. Marshall; Shindell, Drew; O›Carroll, Cynthia M. (1 February 2005). "What›s the Difference Between Weather and Climate?". *NASA. Retrieved 13 November 2015.*
3. AR4 SYR Synthesis Report Annexes. Ipcc.ch. Retrieved on 2011-06-28.
4. C. W. Thornthwaite (1948). "An Approach Toward a Rational Classification of Climate" (PDF). Geographical Review. 38 *(1): 55–94.* doi:10.2307/210739. JSTOR 210739.
5. Hughes, Lesley (2000). Biological consequences of globalwarming: is the signal

already. p. 56.

6. Hughes, Leslie (1 February 2000). "Biological consequences of global warming: is the signal already apparent?". *Trends in Ecology and Evolution.* 15 *(2): 56–61.* doi:10.1016/S0169-5347(99)01764-4. Retrieved November 17, 2016.
7. "Climate". Glossary of Meteorology. American Meteorological Society. Retrieved 2008-05-14.
8. "Climate averages. Met Office. Retrieved 2008-05-17.
9. Intergovernmental Panel on Climate Change. Appendix I: Glossary. Retrieved on 2007-06-01.
10. "Climate Data and Data Related Products". World Meteorological Organization. Retrieved 1 September 2015.
11. "Commission For Climatology: Over Eighty Years of Service" (pdf). World Meteorological Organization. 2011. pp. 6, 8, 10, 21, 26. Retrieved 1 September 2015.
12. National Weather Service Office Tucson, Arizona. Main page. Retrieved on 2007-06-01.
13. Stefan Rahmstorf The Thermohaline Ocean Circulation: A Brief Fact Sheet. Retrieved on 2008-05-02.
14. Gertjan de Werk and Karel Mulder. Heat Absorption Cooling For Sustainable Air Conditioning of Households. Retrieved on 2008-05-02.
15. Ledley, T.S.; Sundquist, E. T.; Schwartz, S. E.; Hall, D. K.; Fellows, J. D.; Killeen, T. L. (1999). "Climate change and greenhouse gases". EOS. 80 *(39): 453. Bibcode*:1999EOSTr..80Q.453L. doi:10.1029/99EO00325. Retrieved 2008-05-17.
16. United States National Arboretum. USDA Plant Hardiness Zone Map. Retrieved on 2008-03-09
17. "Thornthwaite Moisture Index". Glossary of Meteorology. American Meteorological Society. Retrieved 2008-05-21.
18. "Airmass Classification". Glossary of Meteorology. American Meteorological Society. Retrieved 2008-05-22.
19. Schwartz, M.D. (1995). "Detecting Structural Climate Change: An Air Mass-Based Approach in the North Central United States, 1958–1992". Annals of the Association of American Geographers. 85 *(3): 553–568. doi:10.1111/j.1467-8306.1995.tb01812.x.*
20. Robert E. Davis, L. Sitka, D. M. Hondula, S. Gawtry, D. Knight, T. Lee, and J. Stenger. J1.10 A preliminary back-trajectory and air mass climatology for the Shenandoah Valley (Formerly J3.16 for Applied Climatology). Retrieved on 2008-05-21.
21. Susan Woodward. Tropical Broadleaf Evergreen Forest: The Rainforest. Retrieved on 2008-03-14.

22. "Monsoon". Glossary of Meteorology. American Meteorological Society. Retrieved 2008-05-14.
23. International Committee of the Third Workshop on Monsoons. The Global Monsoon System: Research and Forecast. Retrieved on 2008-03-16.
24. Central, Brian. "The Bright Side of 13 Years of Clouds in 1 Map". Retrieved 2015-05-17.
25. Susan Woodward. Tropical Savannas. Retrieved on 2008-03-16.
26. "Cloud Fraction (1 month - Terra/MODIS) - NASA". Cloud Fraction (1 month - Terra/MODIS) - NASA. Retrieved 2015-05-18.
27. Central, Brian. "The Bright Side of 13 Years of Clouds in 1 Map". Retrieved 2015-05-18.
28. "Humid subtropical climate". Encyclopædia Britannica. Encyclopædia Britannica Online. 2008. Retrieved 2008-05-14.
29. Michael Ritter. Humid Subtropical Climate. Archived October 14, 2008, at the Wayback Machine. Retrieved on 2008-03-16.
30. Peel, M. C.; Finlayson B. L. & McMahon, T. A. (2007). "Updated world map of the Köppen-Geiger climate classification". Hydrol. Earth Syst. Sci. 11 *(5): 1633–1644.* doi:10.5194/hess-11-1633-2007. ISSN 1027-5606.
31. Climate. Oceanic Climate. Retrieved on 2008-04-15.
32. Michael Ritter. Mediterranean or Dry Summer Subtropical Climate. Retrieved on 2008-04-15.
33. Blue Planet Biomes. Steppe Climate. Retrieved on 2008-04-15.
34. Michael Ritter. Subarctic Climate. Retrieved on 2008-04-16.
35. Susan Woodward. Taiga or Boreal Forest. Retrieved on 2008-06-06.
36. "The Tundra Biome". The World›s Biomes. Retrieved 2006-03-05.
37. Michael Ritter. Ice Cap Climate. Retrieved on 2008-03-16.
38. San Diego State University. Introduction to Arid Regions: A Self-Paced Tutorial. Retrieved on 2008-04-16. Archived June 12, 2008, at the Wayback Machine.
39. Glossary of Meteorology. Thornthwaite Moisture Index. Retrieved on 2008-05-21.
40. "Moisture Index". Glossary of Meteorology. American Meteorological Society. Retrieved 2008-05-21.
41. Eric Green. Foundations of Expansive Clay Soil. Retrieved on 2008-05-21.
42. Istituto Agronomico per l›Otremare. 3 Land Resources. Retrieved on 2008-05-21.
43. Fredlund, D.G.; Rahardjo, H. (1993). Soil Mechanics for Unsaturated Soils (PDF). Wiley-Interscience. ISBN 978-0-471-85008-3. OCLC 26543184. Retrieved 2008-05-21.
44. ^ Jump up to:[a] [b] Gregory J. McCabe and David M. Wolock. Trends and temperature sensitivity of moisture conditions in the conterminous United States. Retrieved on 2008-05-21.
45. Hawkins, B.A.; Pausas, Juli G. (2004). "Does plant richness influence animal

richness?: the mammals of Catalonia (NE Spain)". Diversity & Distributions. 10 *(4): 247–252. doi:10.1111/j.1366-9516.2004.00085.x. Retrieved 2008-05-21.*

46. "Microthermal Climate". Glossary of Meteorology. American Meteorological Society. Retrieved 2008-05-21.
47. "Mesothermal Climate". Glossary of Meteorology. American Meteorological Society. Retrieved 2008-05-21.
48. "Megathermal Climate". Glossary of Meteorology. American Meteorological Society. Retrieved 2008-05-21.
49. Spencer Weart. The Modern Temperature Trend. Retrieved on 2007-06-01.
50. National Oceanic and Atmospheric Administration. NOAA Paleoclimatology. Retrieved on 2007-06-01.
51. Arctic Climatology and Meteorology. Climate change. Retrieved on 2008-05-19.
52. Gillis, Justin (28 November 2015). "Short Answers to Hard Questions About Climate Change". New York Times. Retrieved 29 November 2015.
53. Brown, Dwayne; Cabbage, Michael; McCarthy, Leslie; Norton, Karen (20 January 2016). "NASA, NOAA Analyses Reveal Record-Shattering Global Warm Temperatures in 2015". NASA. Retrieved 21 January 2016.
54. "Glossary". Climate Change 2001: The Scientific Basis. Contribution of Working Group I to the Third Assessment Report of the Intergovernmental Panel on Climate Change. Intergovernmental Panel on Climate Change. 2001-01-20. Retrieved 2008-05-22.
55. Illinois State Museum (2002). Ice Ages. Retrieved on 2007-05-15.
56. Eric Maisonnave. Climate Variability. Retrieved on 2008-05-02. Archived June 10, 2008, at the Wayback Machine.
57. Climateprediction.net. Modelling the climate. Retrieved on 2008-05-02.
58. http://portal.iri.columbia.edu/portal/server.pt/gateway/PTARGS_0_0_2476_580_1063_43/http%3B/iriportal3.ldeo.columbia.edu%3B7087/publishedcontent/publish/development/climate_activities/prediction/climate_forecasts/climate_forecasts_overview__introduction/cf_intro_learn_more.html

CHAPTER 3

Global Warming and Climate Change

Chapter Contents

Global warming is a phenomenon characterized by a slow warming of the surface temperature of the earth. The basic understanding is usually that the atmosphere immediately surrounding the planet is getting incrementally hotter each year, which in turn leads the surface of both land and sea to warm, as well. The term itself does not imply a cause or speak to a cause; rather, it is a statement of fact that most in the scientific community acknowledge as true and provable. What causes more disagreement is discussions of *why* the temperature seems to be rising, what potential negatives are likely to result, and what human communities can or should do to slow or reverse the trend. Most theories posit that human emissions of carbon dioxide (CO_2) gas, whether through industrial processing or vehicle emissions, are chiefly to blame, while others suggest that the warming is simply part of the earth's natural rhythm and will right itself with time. A lot of research around the world is devoted to the topic, and it's often a hotly debated concept in political arenas and research institutes globally.

INTRODUCTION

Global warming and **climate change** are terms for the observed century-scale rise in the average temperature of the Earth's climate system and its related effects. Multiple lines of scientific evidence show that the climate system is warming. Although the increase of near-surface atmospheric temperature is the measure of global warming often reported in the popular press, most of the additional energy stored in the climate system since

1970 has gone into the oceans. The rest has melted ice and warmed the continents and atmosphere.[a] Many of the observed changes since the 1950s are unprecedented over tens to thousands of years.

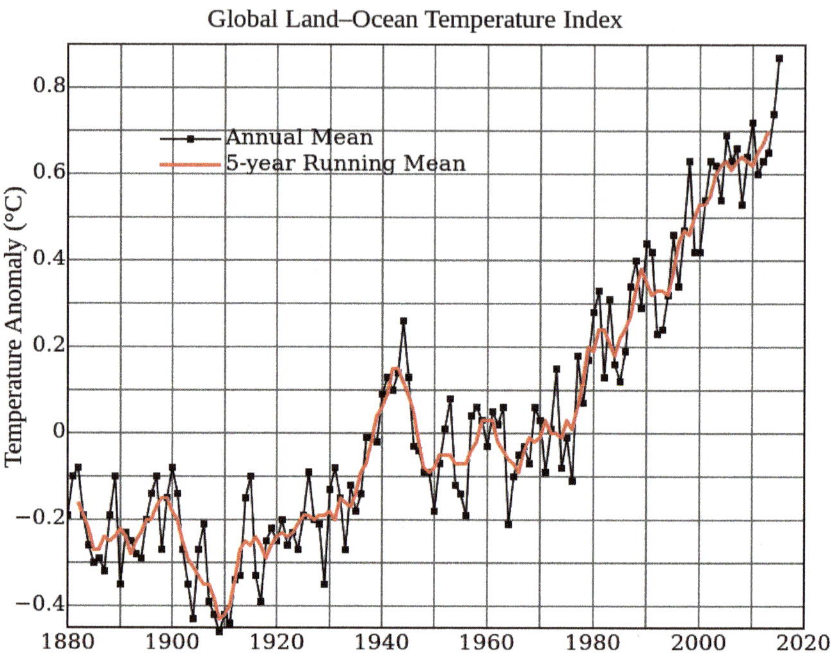

Global mean surface temperature change from 1880 to 2015, relative to the 1951–1980 mean. The black line is the annual mean and the red line is the 5-year running mean. Source: NASA GISS.

World map showing surface temperature trends (°C per decade) between 1950 and 2014. Source: NASA GISS.

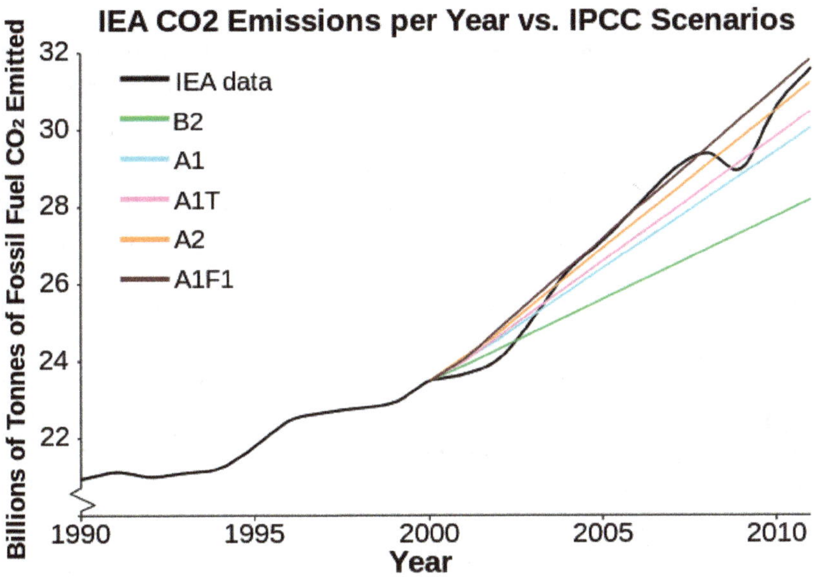

Fossil fuel related carbon dioxide (CO_2) emissions compared to five of the IPCC's "SRES" emissions scenarios, published in 2000. The dips are related to global recessions. Image source: Skeptical Science.

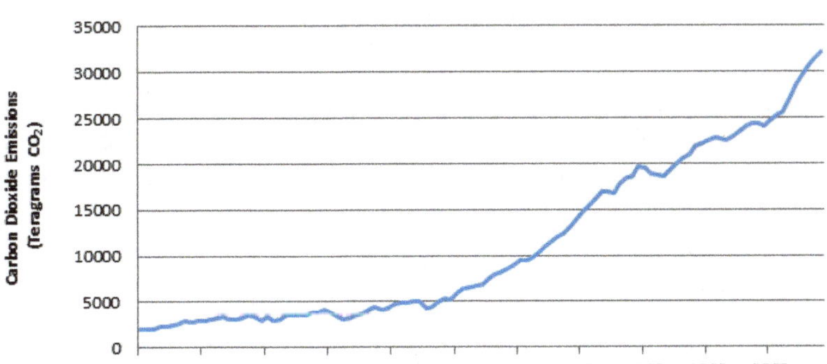

Fossil fuel related carbon dioxide emissions over the 20th century. Image source: EPA.

Scientific understanding of global warming is increasing. The Intergovernmental Panel on Climate Change (IPCC) reported in 2014 that scientists were more than 95% certain that global warming is mostly being caused by human (anthropogenic) activities, mainly increasing concentrations of greenhouse gases such as methane and carbon dioxide (CO_2). Human activities have led to carbon dioxide concentrations above levels not seen in hundreds of thousands of years. Methane and other, often much more potent, greenhouse gases are also rising along with CO_2. Currently, about half of the carbon dioxide released from the burning of fossil fuels remains in the atmosphere. The rest is absorbed by vegetation and the oceans.Climate model projections summarized in the report indicated that during the 21st century the global surface temperature is likely to rise a further 0.3 to 1.7 °C (0.5 to 3.1 °F) for their lowest emissions scenario and 2.6 to 4.8 °C (4.7 to 8.6 °F) for the highest emissions scenario. These findings have been recognized by the national science academies of the major industrialized nations and are not disputed by any scientific body of national or international standing.

Future climate change and associated impacts will differ from region to region around the globe. Anticipated effects include warming global temperature, rising sea levels, changing precipitation, and expansion of deserts in the subtropics. Warming is expected to be greater over land than over the oceans and greatest in the Arctic, with the continuing retreat of glaciers, permafrost and sea ice. Other likely changes include more frequent extreme weather events including heat waves, droughts, heavy rainfall with floods and heavy snowfall; ocean acidification; and species extinctions due to shifting temperature regimes. Effects significant to humans include the threat to food security from decreasing crop yields and the abandonment of populated areas due to rising sea levels. Because the climate system has a large "inertia" and greenhouse gases will stay in the atmosphere for a long time, many of these effects will not only exist for decades or centuries, but will persist for tens of thousands of years.

Possible societal responses to global warming include mitigation by emissions reduction, adaptation to its effects, building systems resilient to its effects, and possible future climate engineering. Most countries are parties to the United Nations Framework Convention on Climate Change (UNFCCC),whose ultimate objective is to prevent dangerous anthropogenic climate change. Parties to the UNFCCC have agreed that deep cuts in emissions are required and that global warming should be limited to well below 2.0 °C (3.6 °F) relative to pre-industrial levels,[c] with efforts made to limit warming to 1.5 °C (2.7 °F).

Public reactions to global warming and concern about its effects are also increasing. A global 2015 Pew Research Center report showed a median of 54% consider it "a very serious problem". There are significant regional differences, with Americans and Chinese (whose economies are responsible for the greatest annual CO_2 emissions) among the least concerned.

OBSERVED TEMPERATURE CHANGES

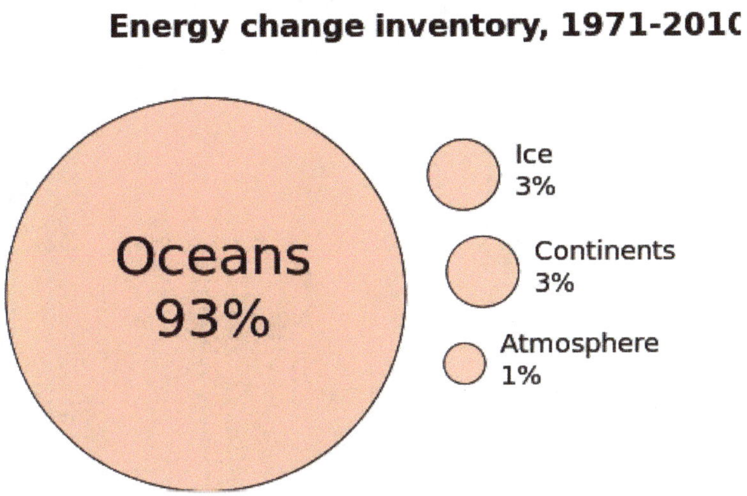

Earth has been in *radiative imbalance* since at least the 1970s, where less energy leaves the atmosphere than enters it. Most of this extra energy has been absorbed by the oceans. It is very likely that human activities substantially contributed to this increase in ocean heat content.

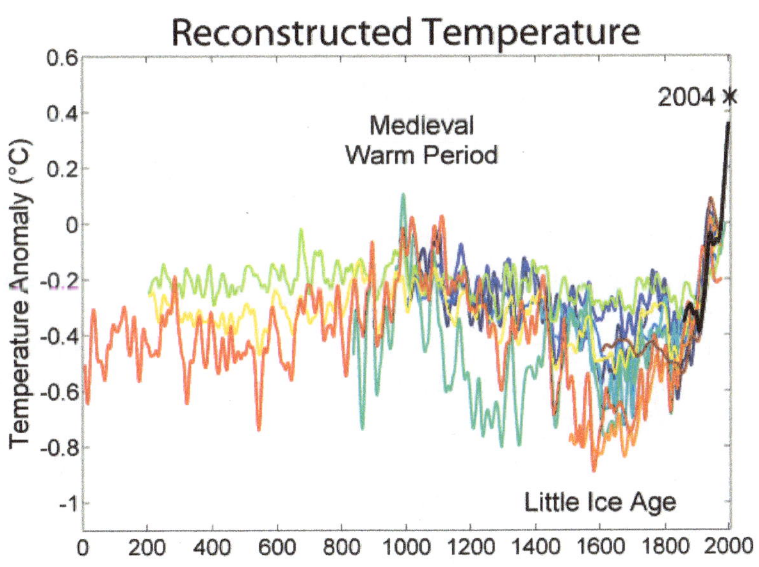

Two millennia of mean surface temperatures according to different reconstructions from climate proxies, each smoothed on a decadal scale, with the instrumental temperature record overlaid in black.

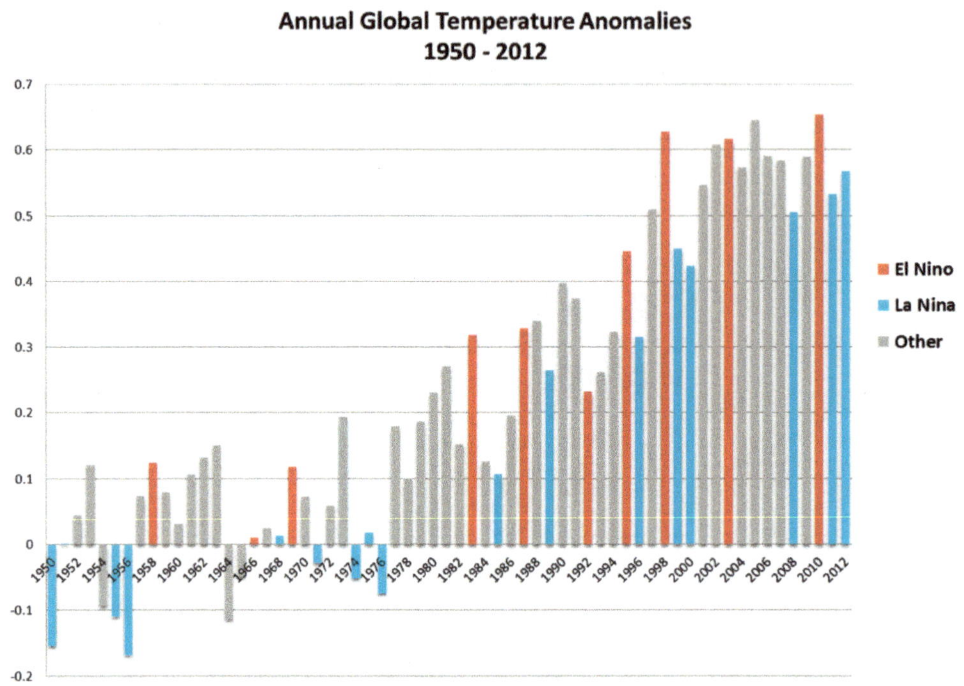

NOAA graph of Global Annual Temperature Anomalies 1950–2012, showing the El Niño Southern Oscillation.

The global average (land and ocean) surface temperature shows a warming of 0.85 [0.65 to 1.06] °C in the period 1880 to 2012, based on multiple independently produced datasets. Earth›s average surface temperature rose by 0.74±0.18 °C over the period 1906–2005. The rate of warming almost doubled for the last half of that period (0.13±0.03 °C per decade, versus 0.07±0.02 °C per decade).

The average temperature of the lower troposphere has increased between 0.13 and 0.22 °C (0.23 and 0.40 °F) per decade since 1979, according to satellite temperature measurements. Climate proxies show the temperature to have been relatively stable over the one or two thousand years before 1850, with regionally varying fluctuations such as the Medieval Warm Period and the Little Ice Age.

The warming that is evident in the instrumental temperature record is consistent with a wide range of observations, as documented by many independent scientific groups. Examples include sea level rise, widespread melting of snow and land ice, increased heat content of the oceans, increased humidity, and the earlier timing of spring events, e.g., the flowering of plants. The probability that these changes could have occurred by chance is virtually zero.

Trends

Temperature changes vary over the globe. Since 1979, land temperatures have increased about twice as fast as ocean temperatures (0.25 °C per decade against 0.13 °C per decade). Ocean temperatures increase more slowly than land temperatures because of the larger effective heat capacity of the oceans and because the ocean loses more heat by evaporation. Since the beginning of industrialisation the temperature difference between the hemispheres has increased due to melting of sea ice and snow in the North.Average arctic temperatures have been increasing at almost twice the rate of the rest of the world in the past 100 years; however arctic temperatures are also highly variable.Although more greenhouse gases are emitted in the Northern than Southern Hemisphere this does not contribute to the difference in warming because the major greenhouse gases persist long enough to mix between hemispheres.

The thermal inertia of the oceans and slow responses of other indirect effects mean that climate can take centuries or longer to adjust to changes in forcing. One climate commitment study concluded that if greenhouse gases were stabilized at year 2000 levels, surface temperatures would still increase by about one-half degree Celsius, and another found that if they were stabilized at 2005 levels surface warming could exceed a whole degree Celsius. Some of this surface warming will be driven by past natural forcings which are still seeking equilibrium in the climate system. One study using a highly simplified climate model indicates these past natural forcings may account for as much as 64% of the committed 2050 surface warming and their influence will fade with time compared to the human contribution.

Global temperature is subject to short-term fluctuations that overlay long-term trends and can temporarily mask them. The relative stability in surface temperature from 2002 to 2009, which has been dubbed the global warming hiatus by the media and some scientists, is consistent with such an episode.2015 updates to account for differing methods of measuring ocean surface temperature measurements show a positive trend over the recent decade.

Warmest Years

Fifteen of the top 16 warmest years have occurred since 2000. While record-breaking years can attract considerable public interest, individual years are less significant than the overall trend. Some climatologists have criticized the attention that the popular press gives to "warmest year" statistics; for example, Gavin Schmidt stated "the long-term trends or the expected sequence of records are far more important than whether any single year is a record or not." Ocean oscillations such as the El Niño Southern Oscillation (ENSO) can affect global average temperatures, causing temperatures of a given year to be abnormally warm or cold for reasons not directly related to the overall trend of climate change. For example, 1998 and 2015 temperatures were significantly warmed by strong El Niño conditions.

Initial causes of temperature changes (external forcings)

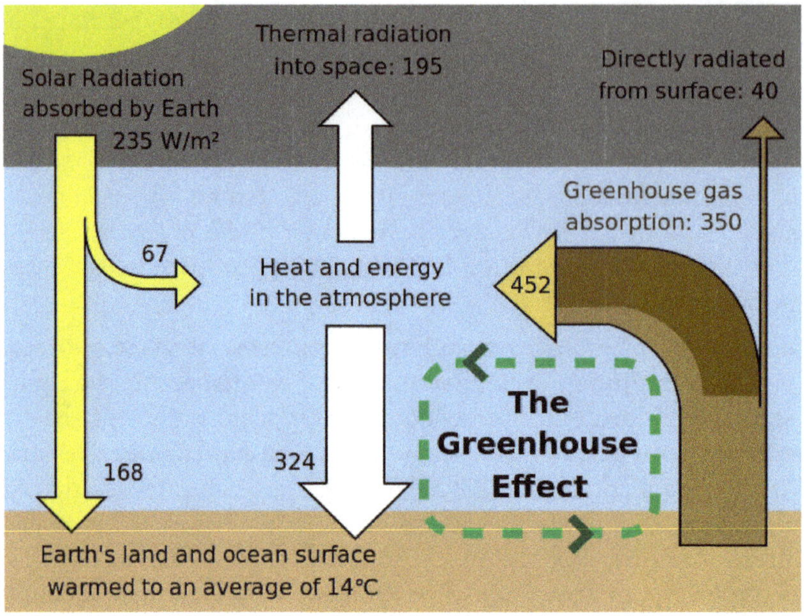

Greenhouse effect schematic showing energy flows between space, the atmosphere, and Earth's surface. Energy exchanges are expressed in watts per square metre (W/m^2).

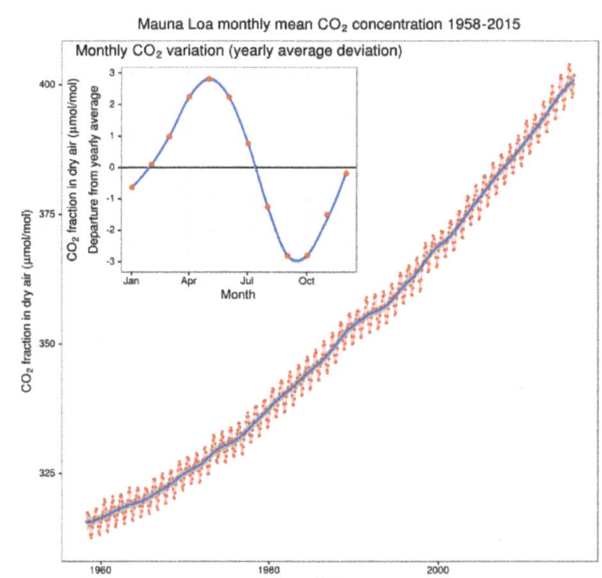

This graph, known as the Keeling Curve, documents the increase of atmospheric carbon dioxide concentrations from 1958–2015. Monthly CO_2 measurements display seasonal oscillations in an upward trend; each year's maximum occurs during the Northern Hemisphere›s late spring, and declines during its growing season as plants remove some atmospheric CO_2.

The climate system can spontaneously generate changes in global temperature for years to decades at a time but long-term changes in global temperature require *external forcings.* These forcings are "external" to the climate system but not necessarily external to Earth. Examples of external forcings include changes in atmospheric composition (e.g., increased concentrations of greenhouse gases), solar luminosity, volcanic eruptions, and variations in Earth›s orbit around the Sun.

Greenhouse Gases

The greenhouse effect is the process by which absorption and emission of infrared radiation by gases in a planet's atmosphere warm its lower atmosphere and surface. It was proposed by Joseph Fourier in 1824, discovered in 1860 by John Tyndall,was first investigated quantitatively by Svante Arrhenius in 1896,and was developed in the 1930s through 1960s by Guy Stewart Callendar.

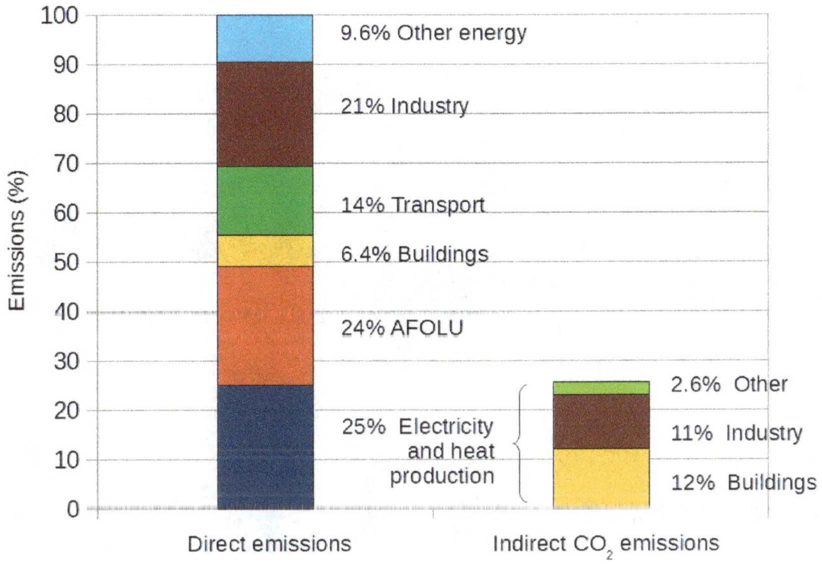

Annual world greenhouse gas emissions, in 2010, by sector.

Percentage share of global cumulative energy-related CO_2 emissions between 1751 and 2012 across different regions.

On Earth, naturally occurring amounts of greenhouse gases cause air temperature near the surface to be about 33 °C (59 °F) warmer than it would be in their absence.Without the Earth's atmosphere, the Earth's average temperature would be well below the freezing temperature of water. The major greenhouse gases are water vapour, which causes about 36–70% of the greenhouse effect; carbon dioxide (CO_2), which causes 9–26%; methane (CH_4), which causes 4–9%; and ozone (O_3), which causes 3–7%. Clouds also affect the radiation balance through cloud forcings similar to greenhouse gases.

Human activity since the Industrial Revolution has increased the amount of greenhouse gases in the atmosphere, leading to increased radiative forcing from CO_2, methane, tropospheric ozone, CFCs and nitrous oxide. According to work published in 2007, the concentrations of CO_2 and methane have increased by 36% and 148% respectively since 1750. These levels are much higher than at any time during the last 800,000 years, the period for which reliable data has been extracted from ice cores. Less direct geological evidence indicates that CO_2 values higher than this were last seen about 20 million years ago.

Fossil fuel burning has produced about three-quarters of the increase in CO_2 from human activity over the past 20 years. The rest of this increase is caused mostly by changes in land-use, particularly deforestation. Another significant non-fuel source of anthropogenic CO_2 emissions is the calcination of limestone for clinker production, a chemical process which releases CO_2. Estimates of global CO_2emissions in 2011 from fossil fuel combustion, including cement production and gas flaring, was 34.8 billion tonnes (9.5 ± 0.5 PgC), an increase of 54% above emissions in 1990. Coal burning was responsible for 43% of the total emissions, oil 34%, gas 18%, cement 4.9% and gas flaring 0.7%

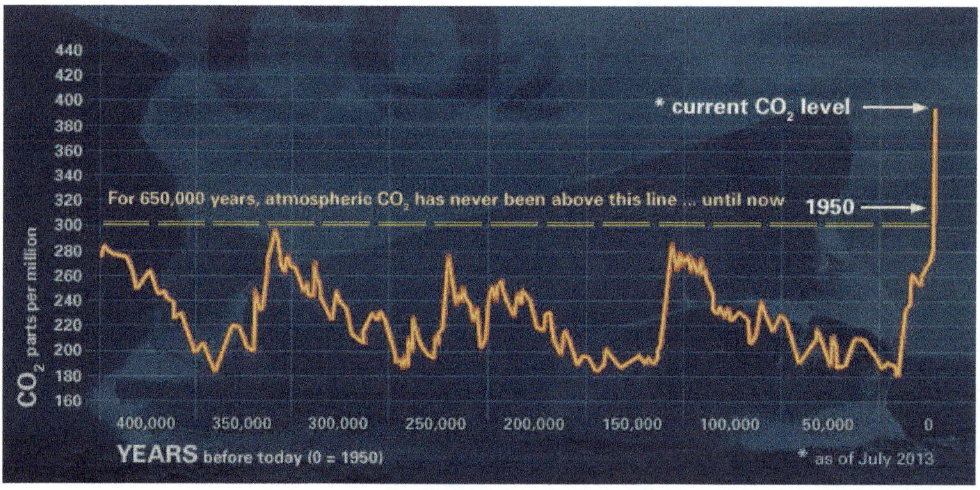

Atmospheric CO_2 concentration from 650,000 years ago to near present, using ice core proxy data and direct measurements.

In May 2013, it was reported that readings for CO_2 taken at the world's primary benchmark site in Mauna Loa surpassed 400 ppm. According to professor Brian Hoskins, this is likely the first time CO_2 levels have been this high for about 4.5 million years. Monthly global CO_2 concentrations exceeded 400 ppm in March 2015, probably for the first time in several million years. On 12 November 2015, NASA scientists reported that human-made carbon dioxide continues to increase above levels not seen in hundreds of thousands of years: currently, about half of the carbon dioxide released from the burning of fossil fuels is not absorbed by vegetation and the oceans and remains in the atmosphere.

Over the last three decades of the twentieth century, gross domestic product per capita and population growth were the main drivers of increases in greenhouse gas emissions. CO_2 emissions are continuing to rise due to the burning of fossil fuels and land-use change. Emissions can be attributed to different regions. Attributions of emissions due to land-use change are subject to considerable uncertainty.

Emissions scenarios, estimates of changes in future emission levels of greenhouse gases, have been projected that depend upon uncertain economic, sociological, technological, and natural developments. In most scenarios, emissions continue to rise over the century, while in a few, emissions are reduced. Fossil fuel reserves are abundant, and will not limit carbon emissions in the 21st century. Emission scenarios, combined with modelling of the carbon cycle, have been used to produce estimates of how atmospheric concentrations of greenhouse gases might change in the future. Using the six IPCC SRES "marker" scenarios, models suggest that by the year 2100, the atmospheric concentration of CO_2 could range between 541 and 970 ppm. This is 90–250% above the concentration in the year 1750.

The popular media and the public often confuse global warming with ozone depletion, i.e., the destruction of stratospheric ozone (e.g., the ozone layer) by chlorofluorocarbons. Although there are a few areas of linkage, the relationship between the two is not strong. Reduced stratospheric ozone has had a slight cooling influence on surface temperatures, while increased tropospheric ozone has had a somewhat larger warming effect.

Aerosols and Soot

Global dimming, a gradual reduction in the amount of global direct irradiance at the Earth's surface, was observed from 1961 until at least 1990. Solid and liquid particles known as *aerosols*, produced by volcanoes and human-made pollutants, are thought to be the main cause of this dimming. They exert a cooling effect by increasing the reflection of incoming sunlight. The effects of the products of fossil fuel combustion – CO_2 and aerosols – have partially offset one another in recent decades, so that net warming has been due to the increase in non-CO_2greenhouse gases such as methane. Radiative forcing due to aerosols is temporally limited due to the processes that remove aerosols from the atmosphere. Removal by clouds and precipitation gives tropospheric aerosols an atmospheric lifetime of only about a week, while stratospheric aerosols can remain for a few years. Carbon dioxide has a lifetime of a century or more, and as such, changes in aerosols will only delay climate changes due to carbon dioxide. Black carbon is second only to carbon dioxide for its contribution to global warming.

In addition to their direct effect by scattering and absorbing solar radiation, aerosols have indirect effects on the Earth›s radiation budget. Sulfate aerosols act as cloud condensation nuclei and thus lead to clouds that have more and smaller cloud droplets. These clouds reflect solar radiation more efficiently than clouds with fewer and larger droplets, a phenomenon known as the Twomey effect. This effect also causes droplets to be of more uniform size, which reduces growth of raindrops and makes the cloud more reflective to incoming sunlight, known as the Albrecht effect. Indirect effects are most noticeable in marine stratiform clouds, and have very little radiative effect on convective clouds. Indirect effects of aerosols represent the largest uncertainty in radiative forcing.

Soot may either cool or warm Earth's climate system, depending on whether it is airborne or deposited. Atmospheric soot directly absorbs solar radiation, which heats the atmosphere and cools the surface. In isolated areas with high soot production, such as rural India, as much as 50% of surface warming due to greenhouse gases may be masked by atmospheric brown clouds. When deposited, especially on glaciers or on ice in arctic regions, the lower surface albedo can also directly heat the surface. The influences of atmospheric particles, including black carbon, are most pronounced in the tropics and sub-tropics, particularly in Asia, while the effects of greenhouse gases are dominant in the extratropics and southern hemisphere.

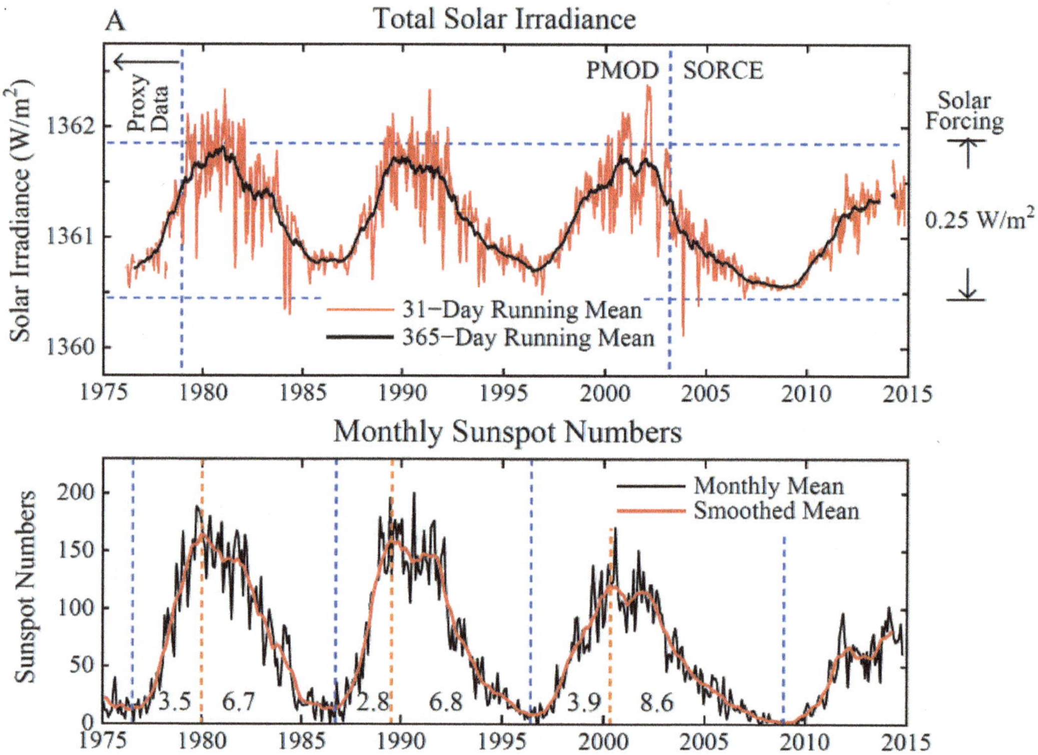

Changes in Total Solar Irradiance (TSI) and monthly sunspot numbers since the mid-1970s.

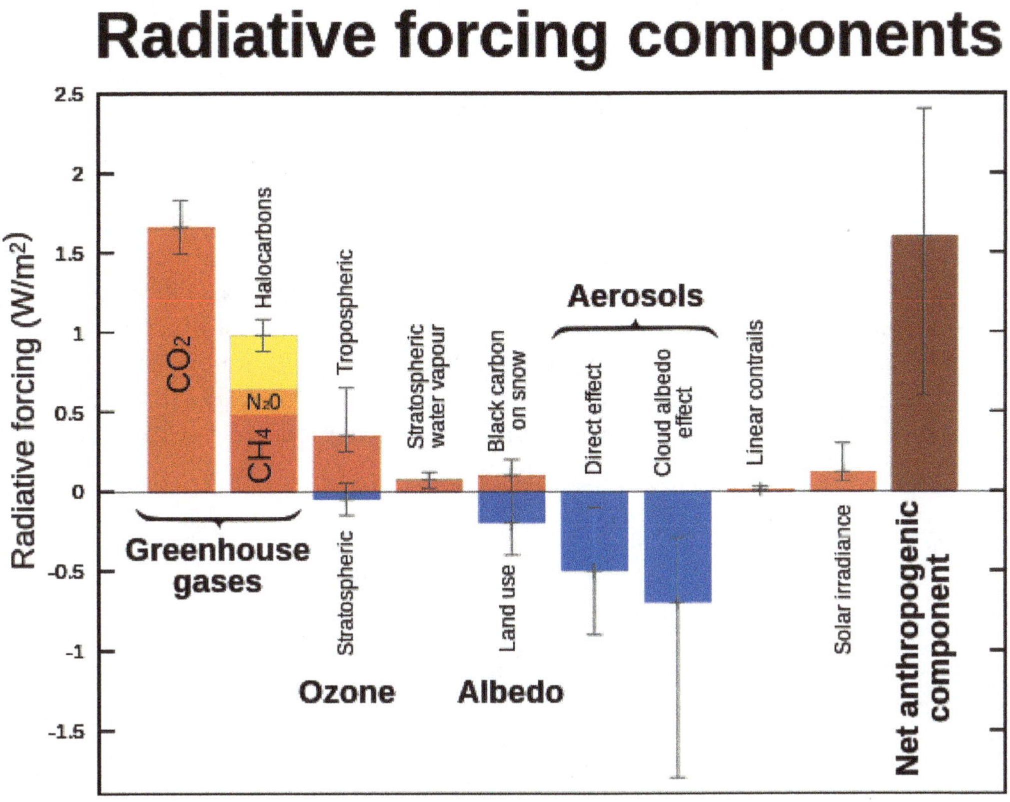

Contribution of natural factors and human activities to radiative forcing of climate change. Radiative forcing values are for the year 2005, relative to the pre-industrial era (1750). The contribution of solar irradiance to radiative forcing is 5% the value of the combined radiative forcing due to increases in the atmospheric concentrations of carbon dioxide, methane and nitrous oxide.

Solar Activity

Since 1978, solar irradiance has been measured by satellites. These measurements indicate that the Sun›s radiative output has not increased during that time, so the warming during the past 40 years cannot be attributed to an increase in solar energy reaching the Earth.

Climate models have been used to examine the role of the Sun in recent climate change. Models are unable to reproduce the rapid warming observed in recent decades when they only take into account variations in solar output and volcanic activity. Models are, however, able to simulate the observed 20th century changes in temperature when they include all of the most important external forcings, including human influences and natural forcings.

Another line of evidence is differing temperature changes at different levels in the Earth›s atmosphere. Basic physical principles require that the greenhouse effect produces warming of the lower atmosphere (the troposphere) but cooling of the upper atmosphere (the stratosphere). Depletion of the ozone layer by chemical refrigerants has also resulted in a strong cooling effect in the stratosphere. If solar variations were responsible for observed warming, warming of both the troposphere and stratosphere would be expected.

Variations in Earth›s orbit

The tilt of the Earth's axis and the shape of its orbit around the Sun vary slowly over tens of thousands of years. This changes climate by changing the seasonal and latitudinal distribution of incoming solar energy at Earth's surface. During the last few thousand years, this phenomenon contributed to a slow cooling trend at high latitudes of the Northern Hemisphere during summer, a trend that was reversed by greenhouse-gas-induced warming during the 20th century. Orbital cycles favorable for glaciation are not expected within the next 50,000 years.

FEEDBACK

Sea ice, shown here in Nunavut, in northern Canada, reflects more sunshine, while open ocean absorbs more, accelerating melting.

The climate system includes a range of *feedbacks*, which alter the response of the system to changes in external forcings. Positive feedbacks increase the response of the climate system to an initial forcing, while negative feedbacks reduce it.

There are a range of feedbacks in the climate system, including water vapour, changes in ice-albedo (snow and ice cover affect how much the Earth's surface absorbs or reflects incoming sunlight), clouds, and changes in the Earth's carbon cycle (e.g., the release of carbon from soil). The main negative feedback is the energy the Earth›s surface radiates into space as infrared radiation. According to the Stefan-Boltzmann law, if the absolute temperature (as measured in kelvin) doubles, radiated energy increases by a factor of 16 (2 to the 4th power).

Feedbacks are an important factor in determining the sensitivity of the climate system to increased atmospheric greenhouse gas concentrations. Other factors being equal, a higher *climate sensitivity* means that more warming will occur for a given increase in greenhouse gas forcing. Uncertainty over the effect of feedbacks is a major reason why different climate models project different magnitudes of warming for a given forcing scenario. More research is needed to understand the role of clouds and carbon cycle feedbacks in climate projections.

The IPCC projections previously mentioned span the "likely" range (greater than 66% probability, based on expert judgement) for the selected emissions scenarios. However, the IPCC›s projections do not reflect the full range of uncertainty. The lower end of the "likely" range appears to be better constrained than the upper end.

Climate models

Global climate model

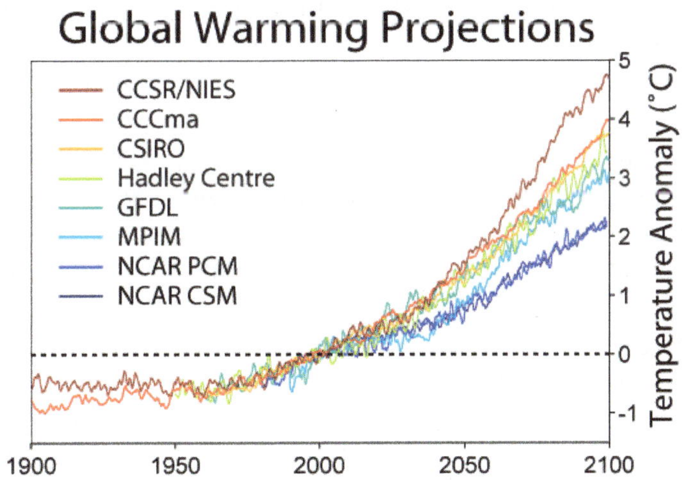

Calculations of global warming prepared in or before 2001 from a range of climate models under the SRES A2 emissions scenario, which assumes no action is taken to reduce emissions and regionally divided economic development.

Projected change in annual mean surface air temperature from the late 20th century to the middle 21st century, based on a medium emissions scenario (SRES A1B). This scenario assumes that no future policies are adopted to limit greenhouse gas emissions. Image credit: NOAA GFDL.

A climate model is a representation of the physical, chemical and biological processes that affect the climate system.Such models are based on scientific disciplines such as fluid dynamics and thermodynamics as well as physical processes such as radiative transfer. The models may be used to predict a range of variables such as local air movement, temperature, clouds, and other atmospheric properties; ocean temperature, salt content, and circulation; ice cover on land and sea; the transfer of heat and moisture from soil and vegetation to the atmosphere; and chemical and biological processes, among others.

Although researchers attempt to include as many processes as possible, simplifications of the actual climate system are inevitable because of the constraints of available computer power and limitations in knowledge of the climate system. Results from models can also vary due to different greenhouse gas inputs and the model›s climate sensitivity. For example, the uncertainty in IPCC›s 2007 projections is caused by (1) the use of multiple models with differing sensitivity to greenhouse gas concentrations, (2) the use of differing estimates of humanity›s future greenhouse gas emissions, (3) any additional emissions from climate feedbacks that were not included in the models IPCC used to prepare its report, i.e., greenhouse gas releases from permafrost.

The models do not assume the climate will warm due to increasing levels of greenhouse gases. Instead the models predict how greenhouse gases will interact with radiative transfer and other physical processes. Warming or cooling is thus a result, not an assumption, of the models.

Clouds and their effects are especially difficult to predict. Improving the models› representation of clouds is therefore an important topic in current research. Another prominent research topic is expanding and improving representations of the carbon cycle.

Models are also used to help investigate the causes of recent climate change by comparing the observed changes to those that the models project from various natural and human causes. Although these models do not unambiguously attribute the warming that occurred from approximately 1910 to 1945 to either natural variation or human effects, they do indicate that the warming since 1970 is dominated by anthropogenic greenhouse gas emissions.

The physical realism of models is tested by examining their ability to simulate contemporary or past climates. Climate models produce a good match to observations of global temperature changes over the last century, but do not simulate all aspects of climate. Not all effects of global warming are accurately predicted by the climate models used by the IPCC. Observed Arctic shrinkage has been faster than that predicted. Precipitation increased proportionally to atmospheric humidity, and hence significantly faster than global climate models predict. Since 1990, sea level has also risen considerably faster than models predicted it would.

Observed and expected environmental effects

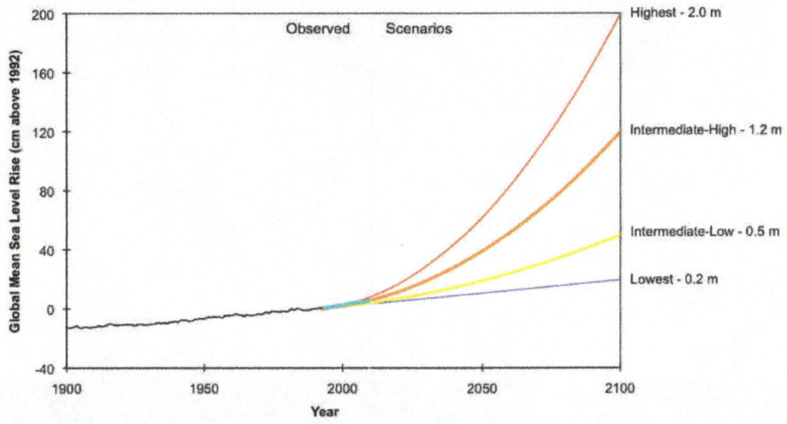

Projections of global mean sea level rise by Parris and others. Probabilities have not been assigned to these projections. Therefore, none of these projections should be interpreted as a "best estimate" of future sea level rise. Image credit: NOAA.

Anthropogenic forcing has likely contributed to some of the observed changes, including sea level rise, changes in climate extremes (such as the number of warm and cold days), declines in Arctic sea ice extent, glacier retreat, and greening of the Sahara.

During the 21st century, glaciers and snow cover are projected to continue their widespread retreat. Projections of declines in Arctic sea ice vary.Recent projections suggest that Arctic summers could be ice-free (defined as ice extent less than 1 million square km) as early as 2025-2030.

"Detection" is the process of demonstrating that climate has changed in some defined statistical sense, without providing a reason for that change. Detection does not imply attribution of the detected change to a particular cause. "Attribution" of causes of climate change is the process of establishing the most likely causes for the detected change with some defined level of confidence. Detection and attribution may also be applied to observed changes in physical, ecological and social systems.

Extreme Weather

Changes in regional climate are expected to include greater warming over land, with most warming at high northern latitudes, and least warming over the Southern Ocean and parts of the North Atlantic Ocean.

Future changes in precipitation are expected to follow existing trends, with reduced precipitation over subtropical land areas, and increased precipitation at subpolar latitudes and some equatorial regions. Projections suggest a probable increase in the frequency and severity of some extreme weather events, such as heat waves.

A 2015 study published in *Nature Climate Change*, states:

"About 18% of the moderate daily precipitation extremes over land are attributable to the observed temperature increase since pre-industrial times, which in turn primarily results from human influence. For 2 °C of warming the fraction of precipitation extremes attributable to human influence rises to about 40%. Likewise, today about 75% of the moderate daily hot extremes over land are attributable to warming. It is the most rare and extreme events for which the largest fraction is anthropogenic, and that contribution increases nonlinearly with further warming.

Data analysis of extreme events from 1960 till 2010 suggests that droughts and heat waves appear simultaneously with increased frequency. Extremely wet or dry events within the monsoon period have increased since 1980.

Sea Level Rise

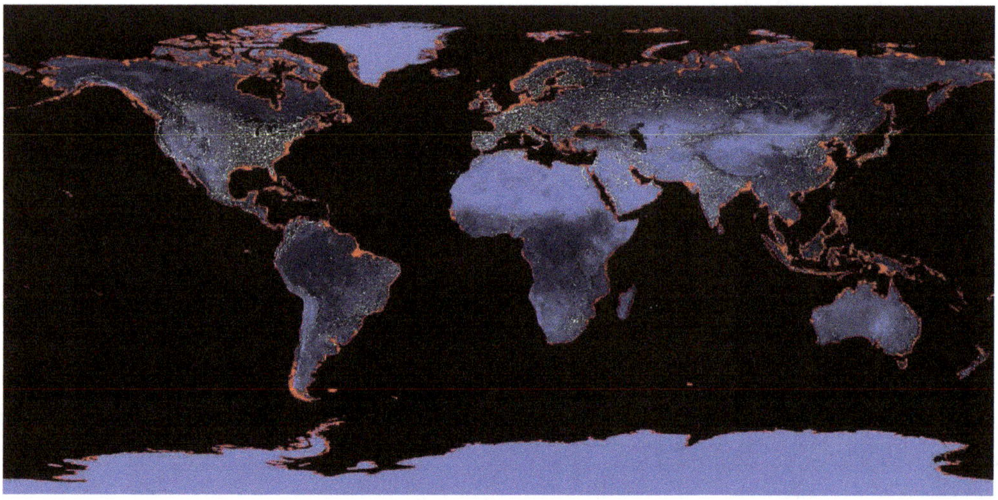

Map of the Earth with a six-metre sea level rise represented in red.

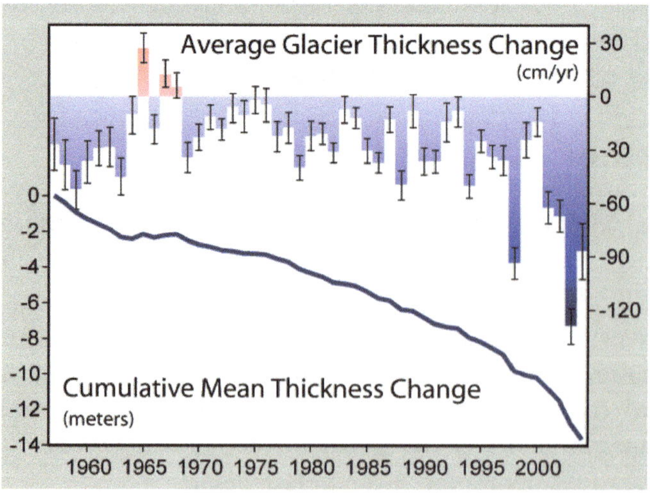

Sparse records indicate that glaciers have been retreating since the early 1800s. In the 1950s measurements began that allow the monitoring of glacial mass balance, reported to the World Glacier Monitoring Service (WGMS) and the National Snow and Ice Data Center (NSIDC).

The sea level rise since 1993 has been estimated to have been on average 2.6 mm and 2.9 mm per year ± 0.4 mm. Additionally, sea level rise has accelerated from 1995 to 2015. Over the 21st century, the IPCC projects for a high emissions scenario, that global mean sea level could rise by 52–98 cm. The IPCC's projections are conservative, and may underestimate future sea level rise. Other estimates suggest that for the same period, global mean sea level could rise by 0.2 to 2.0 m (0.7–6.6 ft), relative to mean sea level in 1992.

Widespread coastal flooding would be expected if several degrees of warming is sustained for millennia. For example, sustained global warming of more than 2 °C (relative to pre-industrial levels) could lead to eventual sea level rise of around 1 to 4 m due to thermal expansion of sea water and the melting of glaciers and small ice caps. Melting of the Greenland ice sheet could contribute an additional 4 to 7.5 m over many thousands of years. It has been estimated that we are already committed to a sea-level rise of approximately 2.3 metres for each degree of temperature rise within the next 2,000 years.

Warming beyond the 2 °C target would potentially lead to rates of sea-level rise dominated by ice loss from Antarctica. Continued CO_2 emissions from fossil sources could cause additional tens of metres of sea level rise, over the next millennia and eventually ultimately eliminate the entire Antarctic ice sheet, causing about 58 metres of sea level rise.

Ecological Systems

In terrestrial ecosystems, the earlier timing of spring events, as well as poleward and upward shifts in plant and animal ranges, have been linked with high confidence to recent warming. Future climate change is expected to affect particular ecosystems, including tundra, mangroves, and coral reefs. It is expected that most ecosystems will be affected by higher atmospheric CO_2 levels, combined with higher global temperatures. Overall, it is expected that climate change will result in the extinction of many species and reduced diversity of ecosystems.

Increases in atmospheric CO_2 concentrations have led to an increase in ocean acidity. Dissolved CO_2 increases ocean acidity, measured by lower pH values. Between 1750 and 2000, surface-ocean pH has decreased by ≈0.1, from ≈8.2 to ≈8.1. Surface-ocean pH has probably not been below ≈8.1 during the past 2 million years. Projections suggest that surface-ocean pH could decrease by an additional 0.3–0.4 units by 2100. Future ocean acidification could threaten coral reefs, fisheries, protected species, and other natural resources of value to society.

Ocean deoxygenation is projected to increase hypoxia by 10%, and triple suboxic waters (oxygen concentrations 98% less than the mean surface concentrations), for each 1 °C of upper ocean warming.

Long-term Effects

On the timescale of centuries to millennia, the magnitude of global warming will be determined primarily by anthropogenic CO_2 emissions. This is due to carbon dioxide's very long lifetime in the atmosphere.

Stabilizing the global average temperature would require large reductions in CO_2 emissions, as well as reductions in emissions of other greenhouse gases such as methane and nitrous oxide.Emissions of CO_2 would need to be reduced by more than 80% relative to their peak level. Even if this were achieved, global average temperatures would remain close to their highest level for many centuries. As of 2016, emissions of CO_2 from burning fossil fuels had stopped increasing, but the Guardian reports they need to be "reduced to have a real impact on climate change". Meanwhile, this greenhouse gas continues to accumulate in the atmosphere. Also, CO_2 is not the only factor driving climate change. Concentrations of atmospheric methane, another greenhouse gas, rose dramatically between 2006-2016 for unknown reasons. This undermines efforts to combat global warming and there is a risk of an uncontrollable runaway greenhouse effect.

Long-term effects also include a response from the Earth›s crust, due to ice melting and deglaciation, in a process called post-glacial rebound, when land masses are no longer depressed by the weight of ice. This could lead to landslides and increased seismic and volcanic activities. Tsunamis could be generated by submarine landslides caused by warmer ocean water thawing ocean-floor permafrost or releasing gas hydrates. Some world regions, such as the French Alps, already show signs of an increase in landslide frequency.

Large-scale and Abrupt Impacts

Climate change could result in global, large-scale changes in natural and social systems. Examples include the possibility for the Atlantic Meridional Overturning Circulation to slow- or shutdown, which in the instance of a shutdown would change weather in Europe and North America considerably, ocean acidification caused by increased atmospheric concentrations of carbon dioxide, and the long-term melting of ice sheets, which contributes to sea level rise.

Some large-scale changes could occur abruptly, i.e., over a short time period, and might also be irreversible. Examples of abrupt climate change are the rapid release of methane and carbon dioxide from permafrost, which would lead to amplified global warming, or the shutdown of thermohaline circulation. Scientific understanding of abrupt climate change is generally poor. The probability of abrupt change for some climate related feedbacks may be low. Factors that may increase the probability of abrupt climate change include higher magnitudes of global warming, warming that occurs more rapidly, and warming that is sustained over longer time periods.

OBSERVED AND EXPECTED EFFECTS ON SOCIAL SYSTEMS

The effects of climate change on human systems, mostly due to warming or shifts in precipitation patterns, or both, have been detected worldwide. Production of wheat and maize globally has been impacted by climate change. While crop production has increased in some mid-latitude regions such as the UK and Northeast China, economic losses due to extreme weather events have increased globally. There has been a shift from cold- to heat-related mortality in some regions as a result of warming. Livelihoods of indigenous peoples of the Arctic have been altered by climate change, and there is emerging evidence of climate change impacts on livelihoods of indigenous peoples in other regions. Regional impacts of climate change are now observable at more locations than before, on all continents and across ocean regions.

The future social impacts of climate change will be uneven. Many risks are expected to increase with higher magnitudes of global warming. All regions are at risk of experiencing negative impacts.Low-latitude, less developed areas face the greatest risk. A study from 2015 concluded that economic growth (gross domestic product) of poorer countries is much more impaired with projected future climate warming, than previously thought.

A meta-analysis of 56 studies concluded in 2014 that each degree of temperature rise will increase violence by up to 20%, which includes fist fights, violent crimes, civil unrest or wars.

Examples of impacts include:

- *Food*: Crop production will probably be negatively affected in low latitude countries, while effects at northern latitudes may be positive or negative. Global warming of around 4.6 °C relative to pre-industrial levels could pose a large risk to global and regional food security.
- *Health*: Generally impacts will be more negative than positive. Impacts include: the effects of extreme weather, leading to injury and loss of life; and indirect effects, such as undernutrition brought on by crop failures.

Habitat inundation

In small islands and mega deltas, inundation as a result of sea level rise is expected to threaten vital infrastructure and human settlements. This could lead to issues of homelessness in countries with low-lying areas such as Bangladesh, as well as statelessness for populations in countries such as the Maldives and Tuvalu.

Economy

Estimates based on the IPCC A1B emission scenario from additional CO_2 and CH_4 greenhouse gases released from permafrost, estimate associated impact damages by US$43 trillion.

Infrastructure

Continued permafrost degradation will likely result in unstable infrastructure in Arctic regions, or Alaska before 2100. Thus, impacting roads, pipelines and buildings, as well as water distribution, and cause slope failures.

Possible responses to global warming

See also: Climate action

Mitigation

The graph on the right shows three "pathways" to meet the UNFCCC's 2 °C target, labelled "global technology", "decentralized solutions", and "consumption change". Each pathway shows how various measures (e.g., improved energy efficiency, increased use of renewable energy) could contribute to emissions reductions. Image credit: PBL Netherlands Environmental Assessment Agency.

Mitigation of climate change are actions to reduce greenhouse gas emissions, or enhance the capacity of carbon sinks to absorb GHGs from the atmosphere. There is a large potential for future reductions in emissions by a combination of activities, including: energy conservation and increased energy efficiency; the use of low-carbon energy technologies, such as renewable energy, nuclear energy, and carbon capture and storage; and enhancing carbon sinks through, for example, reforestation and preventing deforestation. A 2015 report by Citibank concluded that transitioning to a low carbon economy would yield positive return on investments.

Near- and long-term trends in the global energy system are inconsistent with limiting global warming at below 1.5 or 2 °C, relative to pre-industrial levels. Pledges made as part of the Cancún agreements are broadly consistent with having a likely chance (66 to 100% probability) of limiting global warming (in the 21st century) at below 3 °C, relative to pre-industrial levels.

In limiting warming at below 2 °C, more stringent emission reductions in the near-term would allow for less rapid reductions after 2030. Many integrated models are unable to meet the 2 °C target if pessimistic assumptions are made about the availability of mitigation technologies.

Adaptation

Other policy responses include adaptation to climate change. Adaptation to climate change may be planned, either in reaction to or anticipation of climate change, or spontaneous, i.e., without government intervention. Planned adaptation is already occurring on a limited basis. The barriers, limits, and costs of future adaptation are not fully understood.

A concept related to adaptation is *adaptive capacity*, which is the ability of a system (human, natural or managed) to adjust to climate change (including climate variability and extremes) to moderate potential damages, to take advantage of opportunities, or to cope with consequences. Unmitigated climate change (i.e., future climate change without efforts to limit greenhouse gas emissions) would, in the long term, be likely to exceed the capacity of natural, managed and human systems to adapt.

Environmental organizations and public figures have emphasized changes in the climate and the risks they entail, while promoting adaptation to changes in infrastructural needs and emissions reductions.

Climate Engineering

Climate engineering (sometimes called *geoengineering* or *climate intervention*) is the deliberate modification of the climate. It has been investigated as a possible response to global warming, e.g. by NASA and the Royal Society. Techniques under research fall generally into the categories solar radiation management and carbon dioxide removal, although various other schemes have been suggested. A study from 2014 investigated the most common climate engineering methods and concluded they are either ineffective or have potentially severe side effects and cannot be stopped without causing rapid climate change.

DISCOURSE ABOUT GLOBAL WARMING

Political Discussion

Most countries in the world are parties to the United Nations Framework Convention on Climate Change (UNFCCC). The ultimate objective of the Convention is to prevent dangerous human interference of the climate system. As stated in the Convention, this requires that GHG concentrations are stabilized in the atmosphere at a level where ecosystems can adapt naturally to climate change, food production is not threatened, and economic development can proceed in a sustainable fashion. The Framework Convention was agreed in 1992, but since then, global emissions have risen.

During negotiations, the G77 (a lobbying group in the United Nations representing 133 developing nations) pushed for a mandate requiring developed countries to "[take] the lead" in reducing their emissions. This was justified on the basis that: the developed world's emissions had contributed most to the cumulation of GHGs in the atmosphere; per-capita emissions (i.e., emissions per head of population) were still relatively low in developing countries; and the emissions of developing countries would grow to meet their development needs.

This mandate was sustained in the Kyoto Protocol to the Framework Convention, which entered into legal effect in 2005. In ratifying the Kyoto Protocol, most developed countries accepted legally binding commitments to limit their emissions. These first-round commitments expired in 2012. United States President George W. Bush rejected the treaty on the basis that "it exempts 80% of the world, including major population centres such as China and India, from compliance, and would cause serious harm to the US economy."

At the 15th UNFCCC Conference of the Parties, held in 2009 at Copenhagen, several UNFCCC Parties produced the Copenhagen Accord. Parties associated with the Accord (140 countries, as of November 2010) aim to limit the future increase in global mean temperature to below 2 °C. The 16th Conference of the Parties (COP16) was held at Cancún in 2010. It produced an agreement, not a binding treaty, that the Parties should take urgent action to reduce greenhouse gas emissions to meet a goal of limiting global warming to 2 °C above pre-industrial temperatures. It also recognized the need to consider strengthening the goal to a global average rise of 1.5 °C.

Scientific Discussion

There is continuing discussion through published peer-reviewed scientific papers, which are assessed by scientists working in the relevant fields taking part in the Intergovernmental Panel on Climate Change. The scientific consensus as of 2013 stated in the IPCC Fifth Assessment Report is that it "is extremely likely that human influence has been the dominant cause of the observed warming since the mid-20th century". A 2008 report by the U.S. National Academy of Sciences stated that most scientists by then agreed that observed warming in recent decades was primarily caused by human activities increasing the amount of greenhouse gases in the atmosphere. In 2005 the Royal Society stated that while the overwhelming majority of scientists were in agreement on the main points, some individuals and organizations opposed to the consensus on urgent action needed to reduce greenhouse gas emissions have tried to undermine the science and work of the IPCC. National science academies have called on world leaders for policies to cut global emissions.

In the scientific literature, there is a strong consensus that global surface temperatures have increased in recent decades and that the trend is caused mainly by human-induced emissions of greenhouse gases. No scientific body of national or international standing disagrees with this view.

Discussion by the Public and in Popular Media

Global warming was the cover story in this 2007 issue of *Ms. magazine.*

The global warming controversy refers to a variety of disputes, substantially more pronounced in the popular media than in the scientific literature, regarding the nature, causes, and consequences of global warming. The disputed issues include the causes of increased global average air temperature, especially since the mid-20th century, whether this warming trend is unprecedented or within normal climatic variations, whether humankind has contributed significantly to it, and whether the increase is completely or partially an artefact of poor measurements. Additional disputes concern estimates of climate sensitivity, predictions of additional warming, and what the consequences of global warming will be.

By 1990, American conservative think tanks had begun challenging the legitimacy of global warming as a social problem. They challenged the scientific evidence, argued that global warming will have benefits, and asserted that proposed solutions would do more harm than good. Some people dispute aspects of climate change science. Organizations such as the libertarian Competitive Enterprise Institute, conservative commentators, and some companies such as ExxonMobil have challenged IPCC climate change scenarios, funded scientists who disagree with the scientific consensus, and provided their own projections of the economic cost of stricter controls.On the other hand, some fossil fuel companies have scaled back their efforts in recent years, or even called for policies to reduce global warming. Global oil companies have begun to acknowledge climate change exists and is caused by human activities and the burning of fossil fuels.

Surveys of Public Opinion

The world public, or at least people in economically advanced regions, became broadly aware of the global warming problem in the late 1980s. Polling groups began to track opinions on the subject, at first mainly in the United States. The longest consistent polling, by Gallup in the US, found relatively small deviations of 10% or so from 1998 to 2015 in opinion on the seriousness of global warming, but with increasing polarization between those concerned and those unconcerned.

The first major worldwide poll, conducted by Gallup in 2008-2009 in 127 countries, found that some 62% of people worldwide said they knew about global warming. In the advanced countries of North America, Europe and Japan, 90% or more knew about it (97% in the U.S., 99% in Japan); in less developed countries, especially in Africa, fewer than a quarter knew about it, although many had noticed local weather changes. Among those who knew about global warming, there was a wide variation between nations in belief that the warming was a result of human activities.

By 2010, with 111 countries surveyed, Gallup determined that there was a substantial decrease since 2007–08 in the number of Americans and Europeans who viewed global warming as a serious threat. In the US, just a little over half the population (53%) now viewed it as a serious concern for either themselves or their families; this was 10 points below the 2008 poll (63%). Latin America had the biggest rise in concern: 73% said global warming is a serious threat to their families. This global poll also found that people are more likely to attribute global warming to human activities than to natural causes, except in the US where nearly half (47%) of the population attributed global warming to natural causes.

A March–May 2013 survey by Pew Research Center for the People & the Press polled 39 countries about global threats. According to 54% of those questioned, global warming featured top of the perceived global threats. In a January 2013 survey, Pew found that 69% of Americans say there is solid evidence that the Earth's average temperature has got warmer over the past few decades, up six points since November 2011 and 12 points since 2009.

A 2010 survey of 14 industrialized countries found that skepticism about the danger of global warming was highest in Australia, Norway, New Zealand and the United States, in that order, correlating positively with per capita emissions of carbon dioxide.

ETYMOLOGY

In the 1950s, research suggested increasing temperatures, and a 1952 newspaper reported "climate change". This phrase next appeared in a November 1957 report in *The Hammond Times* which described Roger Revelle›s research into the effects of increasing human-caused CO_2 emissions on the greenhouse effect, "a large scale global warming, with radical climate changes may result". Both phrases were only used occasionally until 1975, when Wallace Smith Broecker published a scientific paper on the topic; "Climatic Change: Are We on the Brink of a Pronounced Global Warming?" The phrase began to come into common use, and in 1976 Mikhail Budyko›s statement that "a global warming up has started" was widely reported. Other studies, such as a 1971 MIT report, referred to the human impact as "inadvertent climate modification", but an influential 1979 National Academy of Sciences study headed by Jule Charney followed Broecker in using *global warming* for rising surface temperatures, while describing the wider effects of increased CO_2 as *climate change*.

In 1986 and November 1987, NASA climate scientist James Hansen gave testimony to Congress on global warming. There were increasing heatwaves and drought problems in the summer of 1988, and when Hansen testified in the Senate on 23 June he sparked worldwide interest. He said: "global warming has reached a level such that we can ascribe with a high degree of confidence a cause and effect relationship between the greenhouse effect and the observed warming." Public attention increased over the summer, and *global warming* became the dominant popular term, commonly used both by the press and in public discourse.

In a 2008 NASA article on usage, Erik M. Conway defined *Global warming* as "the increase in Earth's average surface temperature due to rising levels of greenhouse gases", while *Climate change* was "a long-term change in the Earth's climate, or of a region on Earth." As effects such as changing patterns of rainfall and rising sea levels would probably have more impact than temperatures alone, he considered *global climate change* a more scientifically accurate term, and like the Intergovernmental Panel on Climate Change, the NASA website would emphasize this wider context.

LONG-TERM EFFECTS OF GLOBAL WARMING

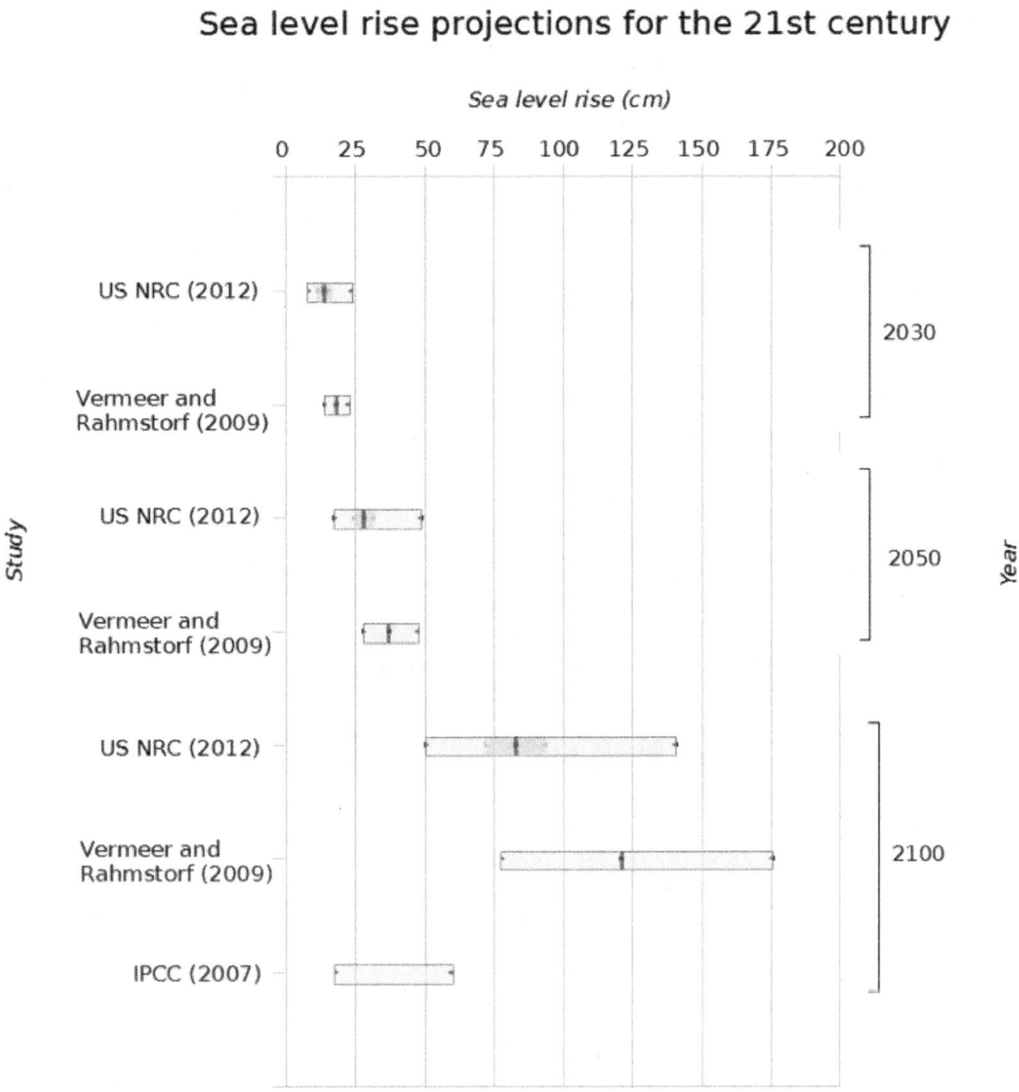

Sea level rise projections for the 21st century.

There are expected to be various **long-term effects of global warming**. Most discussion and research, including that by the Intergovernmental Panel on Climate Change (IPCC) reports, concentrates on the effects of global warming up to 2100, with only an outline of the effects beyond this.

Ice Loss and sea Level Rise

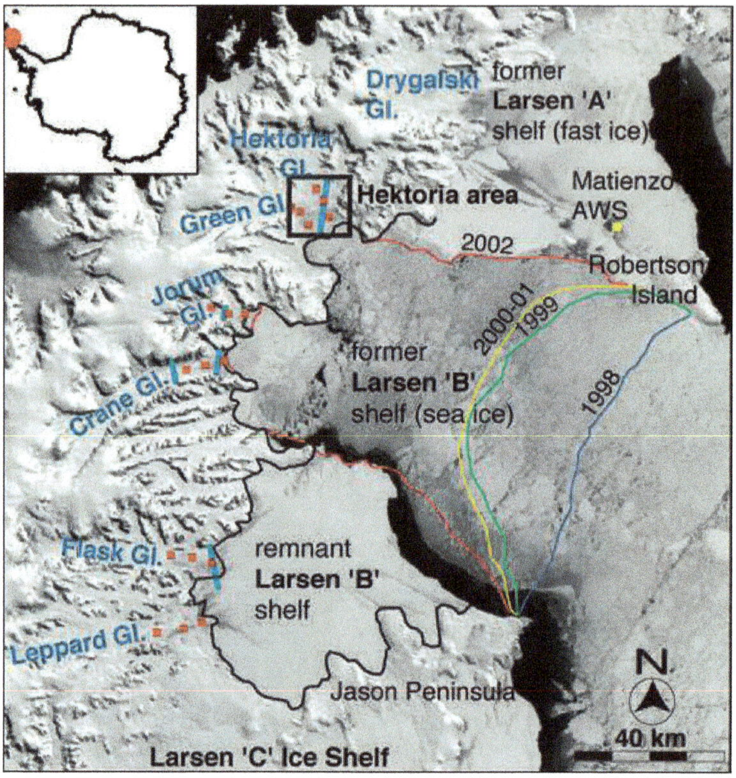

The collapse of Larsen B Shelf, showing the diminishing extent of the shelf from 1998 to 2002

Meltwater from melting ice sheets and glacier retreat contributes to a rise in the future sea level.

Antarctica

Concern has been expressed about the stability of the West Antarctic ice sheet (WAIS). In 2002 Vaughan & Spouge argued that the "WAIS most likely will not collapse in the next few centuries." In an inaugural article for members of the National Academy of Sciences elected in 2005, Timothy Lenton and others suggest that a collapse of the West Antarctic ice sheet could occur within a millennium. Specifically they state "Although the timescale is highly uncertain, a qualitative WAIS change could occur within this millennium, with collapse within 300 years being a worst-case scenario. Rapid sea-level rise (greater than 1 m per century) is more likely to come from the WAIS than from the Greenland ice sheet."

Greenland

Greenland's ice sheet contains enough fresh water as ice to raise sea level worldwide by 7 metres (23 ft). Greenland may become warm enough by 2100 to begin an almost complete melt over more than 1,000 years. James E. Hansen suggests that inadequate attention is being given to this issue.

One study suggests it would take 3,000 years to completely melt the Greenland ice sheet. This figure was derived from the assumed levels of greenhouse gases over the duration of the experiment.

As the Greenland ice sheet loses mass from calving of icebergs as well as by melting of ice, any such processes tend to accelerate the loss of the ice sheet.

Millennial-timescale events

Some long-term effects happen over thousands, not hundreds, of years.

Disruption to Thermohaline Circulation

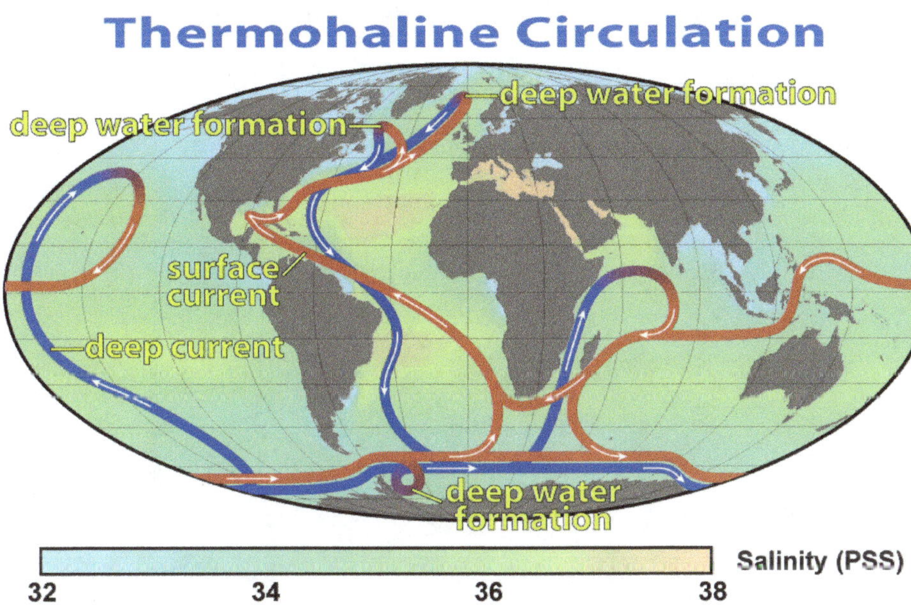

A schematic of the path of the thermohaline circulation. Blue paths represent deep-water currents, while red paths represent surface currents

Early work with simplified models suggested that global warming could cause a shutdown of the thermohaline circulation. This effect is not replicated in more sophisticated coupled ocean-atmosphere global climate models, which do not shutdown but do display varying degrees of slowdown.

This may take a considerable time to occur, as Knutti and Stocker found, again from a simplified model, when :"...thermohaline shutdowns can occur thousands of years after the warming has stopped."

Ocean Anoxia

One study suggests that the amount of oxygen dissolved in the oceans may decline, with adverse consequences for ocean life. This effect was determined using a model run of 100,000 years. The researchers predicted:

...severe, long-term ocean oxygen depletion, as well as a great expansion of ocean oxygen-minimum zones for scenarios with high emissions or high climate sensitivity. We find that climate feedbacks within the Earth system amplify the strength and duration of global warming, ocean heating and oxygen depletion. Decreased oxygen solubility from surface-layer warming accounts for most of the enhanced oxygen depletion in the upper 500 m of the ocean. Possible weakening of ocean overturning and convection lead to further oxygen depletion, also in the deep ocean.

Other research by the Leibniz Institute for Marine Sciences predicts that minimum oxygen zones, particularly in the pacific are expanding, with an estimated total reduction of ocean oxygen levels from 1% to 7% by the end of the century.

Clathrate Decomposition

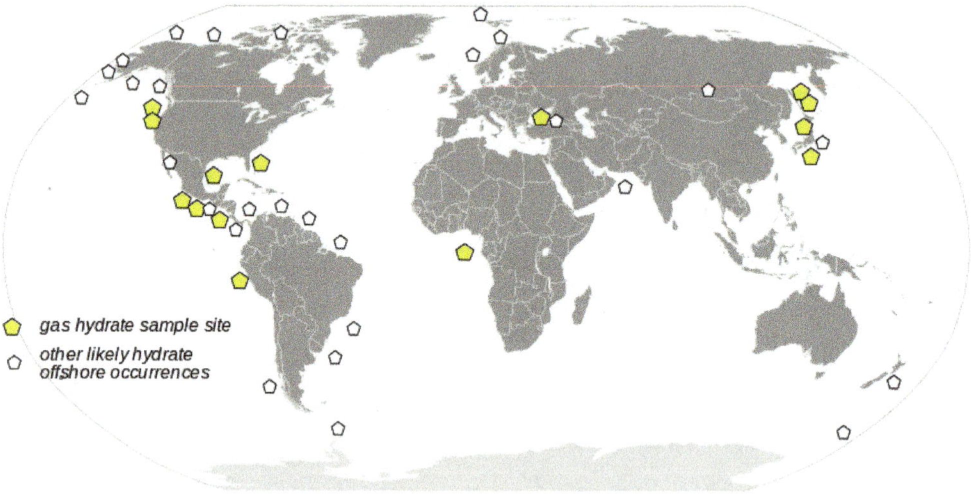

Worldwide distribution of confirmed or inferred offshore gas hydrate-bearing sediments, 1996.
Source: USGS

Methane clathrate, also called methane hydrate, is a form of water ice that contains a large amount of methane within its crystal structure. Extremely large deposits of methane clathrate have been found under sediments on the ocean floors of Earth (estimated at 3000–11,000 Gton C).

MacDonald suggests clathrate volumes are "about 11,000 Gt of carbon for ocean sediments, and about 400 Gt for sediments under permafrost regions". Buffett and

Archer predict "eventual releases of 2000–4000 Gton C in response to a ~2000 Gton C anthropogenic carbon release", for which they suggest no timescale.

Archer considered the timescale and determined that "...on longer time scales of 1–10 ky, there may be a positive feedback with ocean temperature, amplifying the long-term climate impact of anthropogenic CO2 release."

Long-term Return to Equilibrium

After the PETM peak there was an extended period of cooling or "climate whiplash".

While the warm surface waters of the oceans have limited ability to absorb anthropogenic carbon dioxide, the coldest surface waters near the poles (2–3% of ocean surfaces) can transfer significant amounts of carbon dioxide to deep-ocean reserves. Over a period of many centuries, this process and the process of calcium carbonate absorption of carbon dioxide on land and in the oceans will remove 60–80% of the excess carbon dioxide.

Igneous rock when exposed to a near surface environment absorbs carbon dioxide through a very slow weathering rate, but weathering increases in a warmer, higher rainfall climate, speeding the process. This geological weathering will absorb the remaining 20–40% of anthropogenic carbon dioxide over the period of tens of thousands to hundreds of thousands of years.

REFERENCES

1. 16 January 2015: NASA GISS: NASA GISS: NASA, NOAA Find 2014 Warmest Year in Modern Record, in: Research News. NASA Goddard Institute for Space Studies, New York, US. Retrieved 20 February 2015
2. Gillis, Justin (28 November 2015). "Short Answers to Hard Questions About Climate Change". The New York Times. Retrieved 29 November 2015.
3. Simpson, John. Oxford English Dictionary. (Global warming is a) gradual increase in the overall temperature of the earth's atmosphere generally attributed to the greenhouse effect caused by increased levels of carbon dioxide, CFC's, and other pollutants.
4. Hartmann, D. L.; Klein Tank, A. M. G.; Rusticucci, M. (2013). "2: Observations: Atmosphere and Surface" (PDF). IPCC WGI AR5 (Report). p. 198. Evidence for a warming world comes from multiple independent climate indicators, from high up in the atmosphere to the depths of the oceans. They include changes in surface, atmospheric and oceanic temperatures; glaciers; snow cover; sea ice; sea level and atmospheric water vapour. Scientists from all over the world have independently verified this evidence many times.
5. Myth vs Facts....". EPA (US). 2013.The U.S. Global Change Research Program, the National Academy of Sciences, and the Intergovernmental Panel on Climate Change (IPCC) have each independently concluded that warming of the climate system in recent decades is ‹unequivocal›. This conclusion is not drawn from any one source of data but is based on multiple lines of evidence, including three worldwide temperature datasets showing nearly identical warming trends

as well as numerous other independent indicators of global warming (e.g., rising sea levels, shrinking Arctic sea ice).

6. Borenstein, Seth (29 November 2015). "Earth is a wilder, warmer place since last climate deal made". Retrieved 29 November 2015.

7. Rhein, M.; Rintoul, S.R. (2013). "3: Observations: Ocean" (PDF). IPCC WGI AR5 (Report). p. 257. Ocean warming dominates the global energy change inventory. Warming of the ocean accounts for about 93% of the increase in the Earth›s energy inventory between 1971 and 2010 (high confidence), with warming of the upper (0 to 700 m) ocean accounting for about 64% of the total. Melting ice (including Arctic sea ice, ice sheets and glaciers) and warming of the continents and atmosphere account for the remainder of the change in energy.

8. IPCC, Climate Change 2013: The Physical Science Basis - Summary for Policymakers, Observed Changes in the Climate System, p. 2, in IPCC AR5 WG1 2013. "Warming of the climate system is unequivocal, and since the 1950s, many of the observed changes are unprecedented over decades to millennia."

9. ^ Jump up to:[a] [b] CLIMATE CHANGE 2014: Synthesis Report. Summary for Policymakers" (PDF). IPCC. Retrieved 1 November 2015. The following terms have been used to indicate the assessed likelihood of an outcome or a result: virtually certain 99–100% probability, very likely 90–100%, likely 66–100%, about as likely as not 33–66%, unlikely 0–33%, very unlikely 0–10%, exceptionally unlikely 0–1%. Additional terms (extremely likely: 95–100%, more likely than not >50–100%, more unlikely than likely 0–<50% and extremely unlikely 0–5%) may also be used when appropriate.

10. CLIMATE CHANGE 2014: Synthesis Report. Summary for Policymakers" (PDF). IPCC. Retrieved 7 March 2015. The evidence for human influence on the climate system has grown since the Fourth Assessment Report (AR4). It is extremely likely that more than half of the observed increase in global average surface temperature from 1951 to 2010 was caused by the anthropogenic increase in greenhouse gas concentrations and other anthropogenic forcings together

11. America›s Climate Choices: Panel on Advancing the Science of Climate Change; National Research Council (2010). Advancing the Science of Climate Change. Washington, D.C.: The National Academies Press. ISBN 0-309-14588-0. (p1) ... there is a strong, credible body of evidence, based on multiple lines of research, documenting that climate is changing and that these changes are in large part caused by human activities. While much remains to be learned, the core phenomenon, scientific questions, and hypotheses have been examined thoroughly and have stood firm in the face of serious scientific debate and careful evaluation of alternative explanations. * * * (p21-22) Some scientific conclusions or theories have been so thoroughly examined and tested, and supported by so many independent observations and results, that their likelihood of subsequently being found to be wrong is vanishingly small. Such conclusions and theories are then regarded as settled facts. This is the case for the conclusions that the Earth system is warming and that much of this warming is very likely due to human activities.

12. EPA". EPA. EPA.
13. ^ Jump up to:[a] [b] Buis, Alan; Ramsayer, Kate; Rasmussen, Carol (12 November 2015). "A Breathing Planet, Off Balance". NASA. Retrieved 13 November 2015.
14. Stocker et al., Technical Summary, in IPCC AR5 WG1 2013.
15. Joint Science Academies› Statement" (PDF). Retrieved 6 January 2014.
16. Kirby, Alex (17 May 2001). "Science academies back Kyoto". BBC News. Retrieved 27 July 2011.
17. DiMento, Joseph F. C.; Doughman, Pamela M. (2007). Climate Change: What It Means for Us, Our Children, and Our Grandchildren. The MIT Press. p. 68. ISBN 978-0-262-54193-0.
18. Parry, M.L.; et al., "Technical summary", Box TS.6. The main projected impacts for regions, in IPCC AR4 WG2 2007, pp. 59–63
19. Solomon et al., Technical Summary, Section TS.5.3: Regional-Scale Projections, in IPCC AR4 WG1 2007.
20. Lu, Jian; Vechhi, Gabriel A.; Reichler, Thomas (2007). "Expansion of the Hadley cell under global warming" (PDF). Geophysical Research Letters. **34** *(6): L06805. Bibcode:2007GeoRL..3406805L. doi:10.1029/2006GL028443.*
21. (2009). "Historical warnings of future food insecurity with unprecedented seasonal heat". Science. **323** *(5911): 240–4. doi:10.1126/science.1164363. PMID 19131626. Retrieved 13 April 2012.*
22. Peter, U.; et al. "Clark et al. 2016 Consequences of twenty-first-century policy for multi-millennial climate and sea-level change". Nature Climate Change. **6**: 360–369. doi:10.1038/NCLIMATE2923.
23. United Nations Framework Convention on Climate Change (UNFCCC) (2011). "Status of Ratification of the Convention". UNFCCC Secretariat: Bonn, Germany: UNFCCC.. Most countries in the world are Parties to the United Nations Framework Convention on Climate Change (UNFCCC), which has adopted the 2 °C limit. As of 25 November 2011, there are 195 parties (194 states and 1 regional economic integration organization (the European Union)) to the UNFCCC.
24. Article 2". The United Nations Framework Convention on Climate Change. The ultimate objective of this Convention and any related legal instruments that the Conference of the Parties may adopt is to achieve, in accordance with the relevant provisions of the Convention, stabilization of greenhouse gas concentrations in the atmosphere at a level that would prevent dangerous anthropogenic interference with the climate system. Such a level should be achieved within a time-frame sufficient to allow ecosystems to adapt naturally to climate change, to ensure that food production is not threatened and to enable economic development to proceed in a sustainable manner. Such a level should be achieved within a time-frame sufficient to allow ecosystems to adapt naturally to climate change, to ensure that food production is not threatened and to enable economic development to proceed in a sustainable manner, excerpt from the founding international treaty that took force on 21 March 1994.

25. United Nations Framework Convention on Climate Change (UNFCCC) (2011). "Conference of the Parties – Sixteenth Session: Decision 1/CP.16: The Cancun Agreements: Outcome of the work of the Ad Hoc Working Group on Long-term Cooperative Action under the Convention (English): Paragraph 4" (PDF). UNFCCC Secretariat: Bonn, Germany: UNFCCC: 3. "(...) deep cuts in global greenhouse gas emissions are required according to science, and as documented in the Fourth Assessment Report of the Intergovernmental Panel on Climate Change, with a view to reducing global greenhouse gas emissions so as to hold the increase in global average temperature below 2 °C above preindustrial levels"

26. America›s Climate Choices. Washington, D.C.: The National Academies Press. 2011. p. 15. ISBN 978-0-309-14585-5. The average temperature of the Earth›s surface increased by about 1.4 °F (0.8 °C) over the past 100 years, with about 1.0 °F (0.6 °C) of this warming occurring over just the past three decades.

27. Sutter, John D.; Berlinger, Joshua (12 December 2015). "Final draft of climate deal formally accepted in Paris". CNN. Cable News Network, Turner Broadcasting System, Inc. Retrieved 12 December 2015.

28. Vaughan, A. (12 December 2015). "Paris climate deal: key points at a glance". The Guardian. London and Manchester, UK. Retrieved 12 December 2015.. Archived 1 August 2016.

29. Stokes, Bruce; Wike, Richard; Carle, Jill (5 November 2015). "Global Concern about Climate Change, Broad Support for Limiting Emissions: U.S., China Less Worried; Partisan Divides in Key Countries". Pew Research Center. Retrieved 18 June 2016.

30. Brown, Dwayne; Cabbage, Michael; McCarthy, Leslie; Norton, Karen (20 January 2016). "NASA, NOAA Analyses Reveal Record-Shattering Global Warm Temperatures in 2015". NASA. Retrieved 21 January 2016.

31. Rhein, M., et al. (June 7, 2013): Box 3.1, in: Chapter 3: Observations: Ocean (final draft accepted by IPCC Working Group I), pp.11-12 (pp.14-15 of PDF chapter), in: IPCC AR5 WG1 2013

32. IPCC (November 11, 2013): D.3 Detection and Attribution of Climate Change, in: Summary for Policymakers (finalized version), p.15, in: IPCC AR5 WG1 2013

33. Climate Change 2013: The Physical Science Basis, IPCC Fifth Assessment Report (WGI AR5)" (PDF). WGI AR5. IPCC AR5. 2013. p. 5.

34. Climate Change 2007: Working Group I: The Physical Science Basis". IPCC AR4. 2007.

35. Jansen et al., Ch. 6, Palaeoclimate, Section 6.6.1.1: What Do Reconstructions Based on Palaeoclimatic Proxies Show?, pp. 466–478, in IPCC AR4 WG1 2007.

36. Kennedy, J.J.; et al. (2010). "How do we know the world has warmed? in: 2. Global Climate, in: State of the Climate in 2009". Bull. Amer. Meteor. Soc. **91** *(7): 26.*

37. Kennedy, C. (10 July 2012). "ClimateWatch Magazine >> State of the Climate: 2011 Global Sea Level". NOAA Climate Services Portal.

38. Summary for Policymakers". Direct Observations of Recent Climate Change., in IPCC AR4 WG1 2007

39. Summary for Policymakers". B. Current knowledge about observed impacts of climate change on the natural and human environment., in IPCC AR4 WG2 2007

40. Rosenzweig, C.; et al. "Ch 1: Assessment of Observed Changes and Responses in Natural and Managed Systems". Sec 1.3.5.1 Changes in phenology., in IPCC AR4 WG2 2007, p. 99

41. Trenberth et al., Chap 3, Observations: Atmospheric Surface and Climate Change, Executive Summary, p. 237, in IPCC AR4 WG1 2007.

42. Rowan T. Sutton; Buwen Dong; Jonathan M. Gregory (2007). "Land/sea warming ratio in response to climate change: IPCC AR4 model results and comparison with observations". Geophysical Research Letters. **34** *(2): L02701. Bibcode:2007GeoRL..3402701S. doi:10.1029/2006GL028164. Retrieved 19 September 2007.*

43. Feulner, Georg; Rahmstorf, Stefan; Levermann, Anders; Volkwardt, Silvia (March 2013). "On the Origin of the Surface Air Temperature Difference Between the Hemispheres in Earth›s Present-Day Climate". Journal of Climate. **26**: 130325101629005. doi:10.1175/JCLI-D-12-00636.1. Retrieved 25 April 2013.

44. TS.3.1.2 Spatial Distribution of Changes in Temperature, Circulation and Related Variables - AR4 WGI Technical Summary

45. Ehhalt et al., Chapter 4: Atmospheric Chemistry and Greenhouse Gases, Section 4.2.3.1: Carbon monoxide (CO) and hydrogen (H_2), p. 256, in IPCC TAR WG1 2001.

46. Meehl, Gerald A.; Washington, Warren M.; Collins, William D.; Arblaster, Julie M.; Hu, Aixue; Buja, Lawrence E.; Strand, Warren G.; Teng, Haiyan (18 March 2005). "How Much More Global Warming and Sea Level Rise" (PDF). Science. **307** *(5716): 1769–1772. Bibcode:2005Sci...307.1769M. doi:10.1126/science.1106663. PMID 15774757. Retrieved 11 February 2007.*

47. T. M. L. Wigley (2005). "The Climate Change Commitment" (PDF). doi:10.1126/science.1103934. Even if atmospheric composition were fixed today, global-mean temperature and sea level rise would continue due to oceanic thermal inertia. These constant-composition (CC) commitments and their uncertainties are quantified. Constant-emissions (CE) commitments are also considered. The CC warming commitment could exceed 1C. The CE warming commitment is 2 to 6C by the year 2400." (...) "A breakdown of the natural and anthropogenic components of the CC commitment, together with uncertainties arising from ocean mixing (Kz) uncertainties, is given in table S1. Past natural forcing (inclusion of which is the default case here) has a marked effect. The natural forcing component is surprisingly large, 64% of the total commitment in 2050, reducing to 52% by 2400.

48. England, Matthew (February 2014). "Recent intensification of wind-driven circulation in the Pacific and the ongoing warming hiatus". Nature Climate Change. **4**: 222–227. Bibcode:2014NatCC...4..222E. doi:10.1038/nclimate2106.

49. Knight, J.; Kenney, J.J.; Folland, C.; Harris, G.; Jones, G.S.; Palmer, M.; Parker, D.; Scaife, A.; Stott, P. (August 2009). "Do Global Temperature Trends Over the Last Decade Falsify Climate Predictions? [in "State of the Climate in 2008"]" (PDF). Bull. Amer. Meteor. Soc. **90** *(8): S75–S79. Retrieved 13 August 2011.*

50. Global temperature slowdown – not an end to climate change. UK Met Office. Retrieved 20 March 2011.

51. Gavin Schmidt (4 June 2015). "NOAA temperature record updates and the ‹hiatus›".

52. NOAA (4 June 2015). "Science publishes new NOAA analysis: Data show no recent slowdown in global warming".

53. Miller, Brandon (20 January 2016). "2015 is warmest year on record, NOAA and NASA say". CNN. Retrieved 27 March 2016.

54. Schmidt, Gavin (22 January 2015). "Thoughts on 2014 and ongoing temperature trends". RealClimate. Retrieved 4 September 2015.

55. Group (28 November 2004). "Forcings (filed under: Glossary)". RealClimate.

56. Pew Center on Global Climate Change / Center for Climate and Energy Solutions (September 2006). "Science Brief 1: The Causes of Global Climate Change" (PDF). Arlington, Virginia, USA: Center for Climate and Energy Solutions., p.2

57. Brown, Patrick T.; Li, Wenhong; Jiang, Jonathan H.; Su, Hui (2015-12-07). "Unforced Surface Air Temperature Variability and Its Contrasting Relationship with the Anomalous TOA Energy Flux at Local and Global Spatial Scales". Journal of Climate. **29** *(3): 925–940. doi:10.1175/JCLI-D-15-0384.1. ISSN 0894-8755.*

58. Hegerl et al., Chapter 9: Understanding and Attributing Climate Change, Section 9.4.1.5: The Influence of Other Anthropogenic and Natural Forcings, in IPCC AR4 WG1 2007, pp. 690–691. "Recent estimates indicate a relatively small combined effect of natural forcings on the global mean temperature evolution of the second half of the 20th century, with a small net cooling from the combined effects of solar and volcanic forcings." p. 690

59. Tyndall, John (1861). "On the Absorption and Radiation of Heat by Gases and Vapours, and on the Physical Connection of Radiation, Absorption, and Conduction" (PDF). Philosophical Magazine. 4. **22**: 169–94, 273–85. Retrieved 8 May 2013.

60. Weart, Spencer (2008). "The Carbon Dioxide Greenhouse Effect". The Discovery of Global Warming. American Institute of Physics. Retrieved 21 April 2009.

61. The Callendar Effect: the life and work of Guy Stewart Callendar (1898–1964) Amer Meteor Soc., Boston. ISBN 978-1-878220-76-9

62. Le Treut; et al. "Chapter 1: Historical Overview of Climate Change Science". FAQ 1.1., p. 97, in IPCC AR4 WG1 2007: "To emit 240 W m–2, a surface would have to have a temperature of around −19 °C. This is much colder than the conditions that actually exist at the Earth›s surface (the global mean surface temperature is about 14 °C). Instead, the necessary −19 °C is found at an altitude about 5 km above the surface."

63. Blue, Jessica. "What is the Natural Greenhouse Effect?". National Geographic. Retrieved 1 Jan 2015.

64. Kiehl, J.T.; Trenberth, K.E. (1997). "Earth›s Annual Global Mean Energy Budget" (PDF). Bulletin of the American Meteorological Society. **78** *(2): 197–208. Bibcode:1997BAMS...78..197K. doi:10.1175/1520-0477(1997)078<0197:EAGMEB>2.0.CO;2. ISSN 1520-0477. Archived from the original (PDF) on 24 June 2008. Retrieved 21 April 2009.*

65. Schmidt, Gavin (6 April 2005). "Water vapour: feedback or forcing?". RealClimate. Retrieved 21 April 2009.

66. Russell, Randy (16 May 2007). "The Greenhouse Effect & Greenhouse Gases". University Corporation for Atmospheric Research Windows to the Universe. Retrieved 27 December 2009.

67. EPA (2007). "Recent Climate Change: Atmosphere Changes". Climate Change Science Program. United States Environmental Protection Agency. Retrieved 21 April 2009.

68. Spahni, Renato; Jérôme Chappellaz; Thomas F. Stocker; Laetitia Loulergue; Gregor Hausammann; Kenji Kawamura; Jacqueline Flückiger; Jakob Schwander; Dominique Raynaud; Valérie Masson-Delmotte; Jean Jouzel (November 2005). "Atmospheric Methane and Nitrous Oxide of the Late Pleistocene from Antarctic Ice Cores". Science. **310** *(5752): 1317–1321. Bibcode:2005Sci...310.1317S. doi:10.1126/science.1120132. PMID 16311333.*

69. Siegenthaler, Urs; et al. (November 2005). "Stable Carbon Cycle–Climate Relationship During the Late Pleistocene" (PDF). Science. **310** *(5752): 1313–1317. Bibcode:2005Sci...310.1313S. doi:10.1126/science.1120130. PMID 16311332. Retrieved 25 August 2010.*

70. Petit, J. R.; et al. (3 June 1999). "Climate and atmospheric history of the past 420,000 years from the Vostok ice core, Antarctica" (PDF). Nature. **399** *(6735): 429–436. Bibcode:1999Natur.399..429P. doi:10.1038/20859. Retrieved 27 December 2009.*

71. Lüthi, D.; Le Floch, M.; Bereiter, B.; Blunier, T.; Barnola, J. M.; Siegenthaler, U.; Raynaud, D.; Jouzel, J.; Fischer, H.; Kawamura, K.; Stocker, T. F. (2008). "High-resolution carbon dioxide concentration record 650,000–800,000 years before present". Nature. **453** *(7193): 379–382. Bibcode:2008Natur.453..379L. doi:10.1038/nature06949. PMID 18480821.*

72. Pearson, PN; Palmer, MR (2000). "Atmospheric carbon dioxide concentrations over the past 60 million years". Nature. **406** *(6797): 695–699. doi:10.1038/35021000. PMID 10963587.*

73. IPCC, Summary for Policymakers, Concentrations of atmospheric greenhouse gases ..., p. 7, in IPCC TAR WG1 2001.

74. IPCC (2007) AR4. Climate Change 2007: Working Group III: Mitigation of Climate Change, section 7.4.5.1. https://www.ipcc.ch/publications_and_data/ar4/wg3/en/ch7s7-4-5.html

75. Le Quéré, C.;; Andres, R.J.; Boden, T.; Conway, T.; Houghton, R.A.; House,

J.I.; Marland, G.; Peters, G.P.; van der Werf, G.; Ahlström, A.; Andrew, R.M.; Bopp, L.; Canadell, J.G.; Ciais, P.; Doney, S.C.; Enright, C.; Friedlingstein, P.; Huntingford, C.; Jain, A.K.; Jourdain, C.; Kato, E.; Keeling, R.F.; Klein Goldewijk, K.; Levis, S.; Levy, P.; Lomas, M.; Poulter, B.; Raupach, M.R.; Schwinger, J.; Sitch, S.; Stocker, B.D.; Viovy, N.; Zaehle, S.; Zeng, N. (2 December 2012). "The global carbon budget 1959–2011". Earth System Science Data Discussions. **5** *(2): 1107–1157. Bibcode:2012ESSDD...5.1107L. doi:10.5194/essdd-5-1107-2012.*

76. Carbon dioxide passes symbolic mark". BBC. 10 May 2013. Retrieved 27 May 2013.
77. Pilita Clark (10 May 2013). "CO_2 *at highest level for millions of years". Financial Times. Retrieved 27 May 2013. (registration required (help)).*
78. Climate scientists discuss future of their field". 7 July 2015.
79. Rogner, H.-H., et al., Chap. 1, Introduction, Section 1.3.1.2: Intensities, in IPCC AR4 WG3 2007.
80. NRC (2008). "Understanding and Responding to Climate Change" (PDF). Board on Atmospheric Sciences and Climate, US National Academy of Sciences. p. 2. Retrieved 9 November 2010.
81. World Bank (2010). World Development Report 2010: Development and Climate Change. The International Bank for Reconstruction and Development / The World Bank, 1818 H Street NW, Washington, D.C. 20433. doi:10.1596/978-0-8213-7987-5. ISBN 978-0-8213-7987-5. Archived from the original on 5 March 2010. Retrieved 6 April 2010.
82. Banuri et al., Chapter 3: Equity and Social Considerations, Section 3.3.3: Patterns of greenhouse gas emissions, and Box 3.1, pp. 92–93 in IPCC SAR WG3 1996.
83. Liverman, D.M. (2008). "Conventions of climate change: constructions of danger and the dispossession of the atmosphere" (PDF). Journal of Historical Geography. **35** *(2): 279–296. doi:10.1016/j.jhg.2008.08.008. Retrieved 10 May 2011.*
84. Fisher et al., Chapter 3: Issues related to mitigation in the long-term context, Section 3.1: Emissions scenarios: Issues related to mitigation in the long term context in IPCC AR4 WG3 2007.
85. Morita, Chapter 2: Greenhouse Gas Emission Mitigation Scenarios and Implications, Section 2.5.1.4: Emissions and Other Results of the SRES Scenarios, in IPCC TAR WG3 2001.
86. Rogner et al., Ch. 1: Introduction, Figure 1.7, in IPCC AR4 WG3 2007.
87. IPCC, Summary for Policymakers, Introduction, paragraph 6, in IPCC TAR WG3 2001.
88. Prentence et al., Chapter 3: The Carbon Cycle and Atmospheric Carbon Dioxide Executive Summary, in IPCC TAR WG1 2001.
89. Newell, P.J., 2000: Climate for change: non-state actors and the global politics of greenhouse. Cambridge University Press, ISBN 0-521-63250-1.
90. Talk of the Nation. "Americans Fail the Climate Quiz". NPR. Retrieved 27

December 2011.

91. Shindell, Drew; Faluvegi, Greg; Lacis, Andrew; Hansen, James; Ruedy, Reto; Aguilar, Elliot (2006). "Role of tropospheric ozone increases in 20th-century climate change". Journal of Geophysical Research. **111** *(D8): D08302. Bibcode:2006JGRD..11108302S. doi:10.1029/2005JD006348.*

92. Solomon, S; D. Qin; M. Manning; Z. Chen; M. Marquis; K.B. Averyt; M. Tignor; H.L. Miller, eds. (2007). "3.4.4.2 Surface Radiation". Climate Change 2007: Working Group I: The Physical Science Basis. ISBN 978-0-521-88009-1.

93. Hansen, J; Sato, M; Ruedy, R; Lacis, A; Oinas, V (2000). "Global warming in the twenty-first century: an alternative scenario". Proc. Natl. Acad. Sci. U.S.A. **97** *(18): 9875–80. Bibcode:2000PNAS...97.9875H. doi:10.1073/pnas.170278997. PMC 27611* PMID 10944197.

94. Ramanathan, V.; Carmichael, G. (2008). "Global and regional climate changes due to black carbon". Nature Geoscience. **1** *(4): 221–227. Bibcode:2008NatGe...1..221R. doi:10.1038/ngeo156.*

95. V. Ramanathan and G. Carmichael, supra note 1, at 221 (". . . emissions of black carbon are the second strongest contribution to current global warming, after carbon dioxide emissions.") Numerous scientists also calculate that black carbon may be second only to CO_2 in its contribution to climate change, including Tami C. Bond & Haolin Sun, Can Reducing Black Carbon Emissions Counteract Global Warming, ENVIRON. SCI. TECHN. (2005), at 5921 ("BC is the second or third largest individual warming agent, following carbon dioxide and methane."); and J. Hansen, A Brighter Future, 53 CLIMATE CHANGE 435 (2002), available athttp://pubs.giss.nasa.gov/docs/2002/2002_Hansen_1.pdf (calculating the climate forcing of BC at 1.0±0.5 W/m^2).

96. Twomey, S. (1977). "Influence of pollution on shortwave albedo of clouds". J. Atmos. Sci. **34** *(7): 1149–1152. Bibcode:1977JAtS...34.1149T. doi:10.1175/1520-0469(1977)034<1149:TIOPOT>2.0.CO;2. ISSN 1520-0469.*

97. Albrecht, B. (1989). "Aerosols, cloud microphysics, and fractional cloudiness". Science. **245** *(4923): 1227–1239. Bibcode:1989Sci...245.1227A. doi:10.1126/science.245.4923.1227. PMID 17747885.*

98. IPCC, Aerosols, their Direct and Indirect Effects», pp. 291–292 in IPCC TAR WG1 2001.

99. Ramanathan, V.; Chung, C.; Kim, D.; Bettge, T.; Buja, L.; Kiehl, J. T.; Washington, W. M.; Fu, Q.; Sikka, D. R.; Wild, M. (2005). "Atmospheric brown clouds: Impacts on South Asian climate and hydrological cycle"(Full free text). Proceedings of the National Academy of Sciences. **102** *(15): 5326–5333. Bibcode:2005PNAS..102.5326R. doi:10.1073/pnas.0500656102. PMC 552786* . PMID 15749818.

100. Ramanathan, V.; et al. (2008). "Report Summary" (PDF). Atmospheric Brown Clouds: Regional Assessment Report with Focus on Asia. United Nations Environment Programme.

101. Ramanathan, V.; et al. (2008). "Part III: Global and Future Implications" (PDF).

Atmospheric Brown Clouds: Regional Assessment Report with Focus on Asia. United Nations Environment Programme.

102. IPCC, Summary for Policymakers, Human and Natural Drivers of Climate Change, Figure SPM.2, in IPCC AR4 WG1 2007.

103. US Environmental Protection Agency (2009). "3.2.2 Solar Irradiance". Volume 3: Attribution of Observed Climate Change. Endangerment and Cause or Contribute Findings for Greenhouse Gases under Section 202(a) of the Clean Air Act. EPA›s Response to Public Comments. US Environmental Protection Agency. Archived from the original on 16 June 2011. Retrieved June 23, 2011.

104. Hegerl, et al., Chapter 9: Understanding and Attributing Climate Change, Frequently Asked Question 9.2: Can the Warming of the 20th century be Explained by Natural Variability?, in IPCC AR4 WG1 2007.

105. Simmon, R.; D. Herring (November 2009). "Notes for slide number 7, titled "Satellite evidence also suggests greenhouse gas warming," in presentation, "Human contributions to global climate change"". Presentation library on the U.S. National Oceanic and Atmospheric Administration›s Climate Services website. Archived from the original on 3 July 2011. Retrieved 23 June 2011.

106. Hegerl et al., Chapter 9: Understanding and Attributing Climate Change, Frequently Asked Question 9.2: Can the Warming of the 20th century be Explained by Natural Variability?, in IPCC AR4 WG1 2007.

107. Randel, William J.; Shine, Keith P.; Austin, John; et al. (2009). "An update of observed stratospheric temperature trends". Journal of Geophysical Research. **114** *(D2): D02107. Bibcode:2009JGRD..11402107R. doi:10.1029/2008JD010421.*

108. USGCRP 2009, p. 20

109. R.S. Bradley; K.R. Briffa; J. Cole; M.K. Hughes; T.J. Osborn (2003). "The climate of the last millennium". In K.D. Alverson; R.S. Bradley; T.F. Pederson. Paleoclimate, global change and the future. Springer. pp. 105–141. ISBN 3-540-42402-4.

110. Kaufman, D. S.; Schneider, D. P.; McKay, N. P.; Ammann, C. M.; Bradley, R. S.; Briffa, K. R.; Miller, G. H.; Otto-Bliesner, B. L.; Overpeck, J. T.; Vinther, B. M.; Abbott, M.; Axford, M.; Bird, Y.; Birks, B.; Bjune, H. J. B.; Briner, A. E.; Cook, J.; Chipman, T.; Francus, M.; Gajewski, P.; Geirsdottir, K.; Hu, A.; Kutchko, F. S.; Lamoureux, B.; Loso, S.; MacDonald, M.; Peros, G.; Porinchu, M.; Schiff, D.; Seppa, C.; Seppa, H.; Arctic Lakes 2k Project Members (2009). "Recent Warming Reverses Long-Term Arctic Cooling". Science. **325** *(5945): 1236–1239. Bibcode:2009Sci...325.1236K. doi:10.1126/science.1173983. PMID 19729653.*

111. Arctic Warming Overtakes 2,000 Years of Natural Cooling". UCAR. 3 September 2009. Retrieved 8 June 2011.

112. Bello, David (4 September 2009). "Global Warming Reverses Long-Term Arctic Cooling". Scientific American. Retrieved 8 June 2011.

113. Mann, M. E.; Zhang, Z.; Hughes, M. K.; Bradley, R. S.; Miller, S. K.; Rutherford, S.; Ni, F. (2008). "Proxy-based reconstructions of hemispheric and global surface

temperature variations over the past two millennia". Proceedings of the National Academy of Sciences. **105** *(36): 13252–7. Bibcode:2008PNAS..10513252M. doi:10.1073/pnas.0805721105. PMC 2527990*. PMID 18765811.

114. Berger, A. (2002). "CLIMATE: An Exceptionally Long Interglacial Ahead?". Science. **297** *(5585): 1287–8. doi:10.1126/science.1076120. PMID 12193773.*

115. Masson-Delmotte V.M.; et al. (2013). "Information from paleoclimate archives". In Stocker T.F. et al. Climate Change 2013: The Physical Science Basis. Contribution of Working Group I to the Fifth Assessment Report of the Intergovernmental Panel on Climate Change. Cambridge University Press. pp. 383–464. ISBN 978-1-107-66182-0.

116. Jackson, R.; A. Jenkins (17 November 2012). "Vital signs of the planet: global climate change and global warming: uncertainties". Earth Science Communications Team at NASA›s Jet Propulsion Laboratory / California Institute of Technology.

117. Riebeek, H. (16 June 2011). "The Carbon Cycle: Feature Articles: Effects of Changing the Carbon Cycle". Earth Observatory, part of the EOS Project Science Office located at NASA Goddard Space Flight Center.

118. US National Research Council (2003). "Ch. 1 Introduction". Understanding Climate Change Feedbacks. Washington, D.C., USA: National Academies Press., p.19

119. Lindsey, R. (14 January 2009). "Earth›s Energy Budget (p.4), in: Climate and Earth›s Energy Budget: Feature Articles". Earth Observatory, part of the EOS Project Science Office, located at NASA Goddard Space Flight Center.

120. US National Research Council (2006). "Ch. 1 Introduction to Technical Chapters". Surface Temperature Reconstructions for the Last 2,000 Years. Washington, D.C., USA: National Academies Press., pp.26-27

121. AMS Council (20 August 2012). "2012 American Meteorological Society (AMS) Information Statement on Climate Change". Boston, Massachusetts, USA: AMS.

122. Meehl, G.A.; et al. "Ch 10: Global Climate Projections". Sec 10.5.4.6 Synthesis of Projected Global Temperature at Year 2100]., in IPCC AR4 WG1 2007

123. NOAA (January 2007). "Patterns of greenhouse warming" (PDF). GFDL Climate Modeling Research Highlights. Princeton, New Jersey, USA: The National Oceanic and Atmospheric Administration (NOAA) Geophysical Fluid Dynamics Laboratory (GFDL). **1** *(6).*, revision 2 February 2007, 8:50.08 AM.

124. NOAA Geophysical Fluid Dynamics Laboratory (GFDL) (9 October 2012). "NOAA GFDL Climate Research Highlights Image Gallery: Patterns of Greenhouse Warming". NOAA GFDL.

125. IPCC, Glossary A-D: "Climate Model", in IPCC AR4 SYR 2007.

126. Karl, TR; et al., eds. (2009). "Global Climate Change". Global Climate Change Impacts in the United States. Cambridge University Press. ISBN 978-0-521-14407-0.

127. KEVIN SCHAEFER; TINGJUN ZHANG; LORI BRUHWILER; ANDREW P. BARRETT (2011). "Amount and timing of permafrost carbon release in response to

climate warming". Tellus Series B. **63** *(2): 165–180. Bibcode:2011TellB..63..165S. doi:10.1111/j.1600-0889.2011.00527.x.*

128. Hansen, James (2000). "Climatic Change: Understanding Global Warming". In Robert Lanza. One World: The Health & Survival of the Human Species in the 21st century. Health Press (New Mexico). pp. 173–190. ISBN 0-929173-33-3. Retrieved 18 August 2007.

129. Stocker et al., Chapter 7: Physical Climate Processes and Feedbacks, Section 7.2.2: Cloud Processes and Feedbacks, in IPCC TAR WG1 2001.

130. Torn, Margaret; Harte, John (2006). "Missing feedbacks, asymmetric uncertainties, and the underestimation of future warming" (PDF). Geophysical Research Letters. **33** *(10): L10703. Bibcode:2006GeoRL..3310703T. doi:10.1029/2005GL025540. Retrieved 4 March 2007.*

131. Harte, John; Saleska, Scott; Shih, Tiffany (2006). "Shifts in plant dominance control carbon-cycle responses to experimental warming and widespread drought". Environmental Research Letters. **1** *(1): 014001. Bibcode:2006ERL.....1a4001H. doi:10.1088/1748-9326/1/1/014001. Retrieved 2 May 2007.*

132. Scheffer, Marten; Brovkin, Victor; Cox, Peter (2006). "Positive feedback between global warming and atmospheric CO_2 *concentration inferred from past climate change" (PDF). Geophysical Research Letters.* **33** *(10): L10702. Bibcode:2006GeoRL..3310702S. doi:10.1029/2005gl025044. Retrieved 4 May 2007.*

133. Randall et al., Chapter 8, Climate Models and Their Evaluation, Sec. FAQ 8.1 in IPCC AR4 WG1 2007.

134. IPCC, Technical Summary, p. 54, in IPCC TAR WG1 2001.

135. Stroeve, J.; et al. (2007). "Arctic sea ice decline: Faster than forecast". Geophysical Research Letters. **34** *(9): L09501. Bibcode:2007GeoRL..3409501S. doi:10.1029/2007GL029703.*

136. Wentz,FJ; et al. (2007). "How Much More Rain Will Global Warming Bring?". Science. **317** *(5835): 233–5. Bibcode:2007Sci...317..233W. doi:10.1126/science.1140746. PMID 17540863.*

137. Liepert, Beate G.; Previdi, Michael (2009). "Do Models and Observations Disagree on the Rainfall Response to Global Warming?". Journal of Climate. **22** *(11): 3156–3166. Bibcode:2009JCli...22.3156L. doi:10.1175/2008JCLI2472.1. Recently analyzed satellite-derived global precipitation datasets from 1987 to 2006 indicate an increase in global-mean precipitation of 1.1%–1.4% decade−1. This trend corresponds to a hydrological sensitivity (HS) of 7% K−1 of global warming, which is close to the Clausius–Clapeyron (CC) rate expected from the increase in saturation water vapor pressure with temperature. Analysis of two available global ocean evaporation datasets confirms this observed intensification of the atmospheric water cycle. The observed hydrological sensitivity over the past 20-yr period is higher by a factor of 5 than the average HS of 1.4% K−1 simulated in state-of-the-art coupled atmosphere–ocean climate models for the twentieth and twenty-first centuries.*

138. Rahmstorf, S.; Cazenave, A.; Church, J. A.; Hansen, J. E.; Keeling, R. F.; Parker, D. E.; Somerville, R. C. J. (4 May 2007). "Recent Climate Observations Compared to Projections". Science. **316** *(5825): 709–709. Bibcode:2007Sci...316..709R. doi:10.1126/science.1136843. PMID 17272686.*

139. 4. Global Mean Sea Level Rise Scenarios, in: Main Report, in Parris & others 2012, p. 12

140. Executive Summary, in Parris & others 2012, p. 1

141. Hegerl, G.C.; et al. "Ch 9: Understanding and Attributing Climate Change". Executive Summary., in IPCC AR4 WG1 2007

142. Sahara Desert Greening Due to Climate Change?". National Geographic. Retrieved 12 June 2010.

143. Meehl, G.A.; et al. "Ch 10: Global Climate Projections". Box 10.1: Future Abrupt Climate Change, 'Climate Surprises', and Irreversible Changes: Glaciers and ice caps., in IPCC AR4 WG1 2007, p. 776

144. Meehl, G.A.; et al. "Ch 10: Global Climate Projections". Sec 10.3.3.2 Changes in Snow Cover and Frozen Ground., in IPCC AR4 WG1 2007, pp. 770, 772

145. Meehl, G.A.; et al. "Ch 10: Global Climate Projections". Sec 10.3.3.1 Changes in Sea Ice Cover., in IPCC AR4 WG1 2007, p. 770

146. Wang, M.; Overland, J. E. (2009). "A sea ice free summer Arctic within 30 years?". Geophys. Res. Lett. **36** *(7). Bibcode:2009GeoRL..3607502W. doi:10.1029/2009GL037820. Retrieved 2 May 2011.*

147. Met Office. "Arctic sea ice 2012". Exeter, UK: Met Office.

148. IPCC, Glossary A-D: "Detection and attribution", in IPCC AR4 WG1 2007. See also Hegerl et al., Section 9.1.2: What are Climate Change Detection and Attribution?, in IPCC AR4 WG1 2007.

149. Rosenzweig et al., Chapter 1: Assessment of Observed Changes and Responses in Natural and Managed Systems Section 1.2 Methods of detection and attribution of observed changes, in IPCC AR4 WG2 2007 .

150. IPCC, Synthesis Report Summary for Policymakers, Section 3: Projected climate change and its impacts, in IPCC AR4 SYR 2007.

151. NOAA (February 2007). "Will the wet get wetter and the dry drier?" (PDF). GFDL Climate Modeling Research Highlights. Princeton, New Jersey, USA: National Oceanic and Atmospheric Administration (NOAA) Geophysical Fluid Dynamics Laboratory (GFDL). **1** *(5).*, p.1. Revision 15 October 2008, 4:47:16 PM.

152. D. Future Climate Extremes, Impacts, and Disaster Losses, in: Summary for policymakers". MANAGING THE RISKS OF EXTREME EVENTS AND DISASTERS TO ADVANCE CLIMATE CHANGE ADAPTATION., in IPCC SREX 2012, pp. 9–13

153. Justin Gillis (27 April 2015). "New Study Links Weather Extremes to Global Warming". The New York Times. Retrieved 27 April 2015. "The bottom line is that things are not that complicated," Dr. Knutti said. "You make the world a degree or two warmer, and there will be more hot days. There will be more

moisture in the atmosphere, so that must come down somewhere."

154. E. M. Fischer; R. Knutti (27 April 2015). "Anthropogenic contribution to global occurrence of heavy-precipitation and high-temperature extremes" (online). Nature Climate Change. **5**: 560–564. Bibcode:2015NatCC...5..560F. doi:10.1038/nclimate2617. Retrieved 27 April 2015. We show that at the present-day warming of 0.85 °C about 18% of the moderate daily precipitation extremes over land are attributable to the observed temperature increase since pre-industrial times, which in turn primarily results from human influence. ... Likewise, today about 75% of the moderate daily hot extremes over land are attributable to warming.
155. UCI study finds dramatic increase in concurrent droughts, heat waves". UCI. 2015.
156. Indian Monsoons Are Becoming More Extreme". Scientific American. 2014.
157. Christopher S. Watson; Neil J. White; John A. Church; Matt A. King; Reed J. Burgette; Benoit Legresy (11 May 2015). "Unabated global mean sea-level rise over the satellite altimeter era". Nature Climate Change. **5**: 565–568. Bibcode:2015NatCC...5..565W. doi:10.1038/nclimate2635.
158. Churchs, John; Clark, Peter. "Chapter 13: Sea Level Change - Final Draft Underlying Scientific-Technical Assessment" (PDF). IPCC Working Group I. Retrieved 21 January 2015.
159. PROJECTIONS OF FUTURE SEA LEVEL RISE, pp.243-244, in: Ch. 7. Sea Level Rise and the Coastal Environment, in National Research Council 2010
160. BOX SYN-1: SUSTAINED WARMING COULD LEAD TO SEVERE IMPACTS, p.5, in: Synopsis, in National Research Council 2011
161. Anders Levermann; Peter U. Clark; Ben Marzeion; Glenn A. Milne; David Pollard; Valentina Radic; Alexander Robinson (13 June 2013). "The multimillennial sea-level commitment of global warming". PNAS. **110**: 13745–13750. Bibcode:2013PNAS..11013745L. doi:10.1073/pnas.1219414110.
162. Ricarda Winkelmann; Anders Levermann; Andy Ridgwell; Ken Caldeira (11 September 2015). "Combustion of available fossil fuel resources sufficient to eliminate the Antarctic Ice Sheet". doi:10.1126/sciadv.1500589.
163. IPCC, Synthesis Report Summary for Policymakers, Section 1: Observed changes in climate and their effects, in IPCC AR4 SYR 2007.
164. Fischlin, et al., Chapter 4: Ecosystems, their Properties, Goods and Services, Executive Summary, p. 213, in IPCC AR4 WG2 2007. Executive summary not present in on-line text; see pdf.
165. Schneider et al., Chapter 19: Assessing Key Vulnerabilities and the Risk from Climate Change, Section 19.3.4: Ecosystems and biodiversity, in IPCC AR4 WG2 2007.
166. Ocean Acidification, in: Ch. 2. Our Changing Climate, in NCADAC 2013, pp. 69–70
167. Introduction, in Zeebe 2012, p. 142
168. Ocean acidification, in: Executive summary, in Good & others 2010, p. 14

169. UNEP 2010

170. 5. Ocean acidification, in Good & others 2010, pp. 73–81

171. IAP 2009

172. Deutsch; et al. (2011). "Climate-Forced Variability of Ocean Hypoxia" (PDF). Science. **333**: 336–339. Bibcode:2011Sci...333..336D. doi:10.1126/science.1202422.

173. Summary, pp.14-19, in National Research Council 2011

174. FAQ 12.3, in: Chapter 12: Long-term Climate Change: Projections, Commitments and Irreversibility, in IPCC AR5 WG1 2013, pp.88-89 (pp.90-91 of PDF chapter)

175. BOX 2.1: STABILIZATION AND NON-CO_2 GREENHOUSE GASES (p.65), in: Chapter 2: Emissions, Concentrations, and Related Factors, in National Research Council 2011

176. The world›s carbon-dioxide emissions have stabilised". The Economist. 16 March 2016. Retrieved 12 December 2016.

177. Rapid rise in methane emissions in 10 years surprises scientists The Guardian

178. Bill McGuire. "Climate forcing of geological and geomorphological hazards". Philosophical Transactions A. Royal Society. **368**: 2311–2315. Bibcode:2010RSPTA.368.2311M. doi:10.1098/rsta.2010.0077.

179. Jérôme Lopez Saez; Christophe Corona; Markus Stoffel; Frédéric Berger. "Climate change increases frequency of shallow spring landslides in the French Alps". Geology. **41**: 619–622. doi:10.1130/G34098.1.

180. Smith, J.B.; et al. "Ch. 19. Vulnerability to Climate Change and Reasons for Concern: A Synthesis". Sec 19.6. Extreme and Irreversible Effects., in IPCC TAR WG2 2001

181. Smith, J. B.; Schneider, S. H.; Oppenheimer, M.; Yohe, G. W.; Hare, W.; Mastrandrea, M. D.; Patwardhan, A.; Burton, I.; Corfee-Morlot, J.; Magadza, C. H. D.; Füssel, H.-M.; Pittock, A. B.; Rahman, A.; Suarez, A.; van Ypersele, J.-P. (17 March 2009). "Assessing dangerous climate change through an update of the Intergovernmental Panel on Climate Change (IPCC) ‹reasons for concern›". Proceedings of the National Academy of Sciences. **106** *(11): 4133–7. Bibcode:2009PNAS..106.4133S. doi:10.1073/pnas.0812355106. PMC 2648893.* PMID 19251662.

182. Clark, P.U.; et al. (December 2008). "Executive Summary". Abrupt Climate Change. A Report by the U.S. Climate Change Science Program and the Subcommittee on Global Change Research. Reston, Virginia, USA: U.S. Geological Survey., pp. 1–7. Report website

183. Siberian permafrost thaw warning sparked by cave data". BBC. 22 February 2013. Retrieved 24 February 2013.

184. US National Research Council (2010). "Advancing the Science of Climate Change: Report in Brief". Washington, D.C., USA: National Academies Press., p.3. PDF of Report

185. IPCC. "Summary for Policymakers". Sec. 2.6. The Potential for Large-Scale

and Possibly Irreversible Impacts Poses Risks that have yet to be Reliably Quantified., in IPCC TAR WG2 2001

186. Cramer, W., et al., Executive summary, in: Chapter 18: Detection and attribution of observed impacts(archived 8 July 2014), pp.3-4, in IPCC AR5 WG2 A 2014
187. FAQ 7 and 8, in: Volume-wide Frequently Asked Questions (FAQs) (archived 8 July 2014), pp.2-3, in IPCC AR5 WG2 A 2014
188. Oppenheimer, M., et al., Section 19.6.3: Updating Reasons for Concern, in: Chapter 19: Emergent risks and key vulnerabilities (archived 8 July 2014), pp.39-46, in IPCC AR5 WG2 A 2014
189. Field, C., et al., B-3: Regional Risks and Potential for Adaptation, in: Technical Summary (archived 8 July 2014), pp.27-30, in IPCC AR5 WG2 A 2014
190. Oppenheimer, M., et al., Section 19.6.3: Updating Reasons for Concern, in: Chapter 19: Emergent risks and key vulnerabilities (archived 8 July 2014), pp.42-43, in IPCC AR5 WG2 A 2014
191. Dana Nuccitelli (26 January 2015). "Climate change could impact the poor much more than previously thought". The Guardian.
192. Chris Mooney (22 October 2014). "There›s a surprisingly strong link between climate change and violence". The Washington Post.
193. Porter, J.R., et al., Executive summary, in: Chapter 7: Food security and food production systems(archived 8 July 2014), p.3, in IPCC AR5 WG2 A 2014
194. Reference temperature period converted from late-20th century to pre-industrial times (approximated in the source as 1850-1900).
195. Assessment Box SPM-1 (p.14) and B-2. Sectoral Risks and Potential for Adaptation: Food security and food production systems (p.18), in: Summary for Policymakers (archived 8 July 2014), in IPCC AR5 WG2 A 2014
196. Smith, K.R., et al., FAQ 11.2, in: Chapter 11: Human health: impacts, adaptation, and co-benefits(archived 8 July 2014), p.37, in IPCC AR5 WG2 A 2014
197. Costello, Anthony; Abbas, Mustafa; Allen, Adriana; Ball, Sarah; Bell, Sarah; Bellamy, Richard; Friel, Sharon; Groce, Nora; Johnson, Anne; Kett, Maria; Lee, Maria; Levy, Caren; Maslin, Mark; McCoy, David; McGuire, Bill; Montgomery, Hugh; Napier, David; Pagel, Christina; Patel, Jinesh; de Oliveira, Jose Antonio Puppim; Redclift, Nanneke; Rees, Hannah; Rogger, Daniel; Scott, Joanne; Stephenson, Judith; Twigg, John; Wolff, Jonathan; Patterson, Craig (May 2009). "Managing the health effects of climate change". The Lancet. **373** *(9676): 1693–1733. doi:10.1016/S0140-6736(09)60935-1.*
198. Watts, Nick; Adger, W Neil; Agnolucci, Paolo; Blackstock, Jason; Byass, Peter; Cai, Wenjia; Chaytor, Sarah; Colbourn, Tim; Collins, Mat; Cooper, Adam; Cox, Peter M; Depledge, Joanna; Drummond, Paul; Ekins, Paul; Galaz, Victor; Grace, Delia; Graham, Hilary; Grubb, Michael; Haines, Andy; Hamilton, Ian; Hunter, Alasdair; Jiang, Xujia; Li, Moxuan; Kelman, Ilan; Liang, Lu; Lott, Melissa; Lowe, Robert; Luo, Yong; Mace, Georgina; Maslin, Mark; Nilsson, Maria; Oreszczyn, Tadj; Pye, Steve; Quinn, Tara; Svensdotter, My; Venevsky, Sergey;

Warner, Koko; Xu, Bing; Yang, Jun; Yin, Yongyuan; Yu, Chaoqing; Zhang, Qiang; Gong, Peng; Montgomery, Hugh; Costello, Anthony (November 2015). "Health and climate change: policy responses to protect public health". The Lancet. **386** *(10006): 1861–1914. doi:10.1016/S0140-6736(15)60854-6. PMID 26111439. Retrieved 4 January 2016.*

199. Smith, K.R., et al., Section 11.4: Direct Impacts of Climate and Weather on Health, in: Chapter 11: Human health: impacts, adaptation, and co-benefits (archived 8 July 2014), pp.10-13, in IPCC AR5 WG2 A 2014
200. Smith, K.R., et al., Section 11.6.1. Nutrition, in: Chapter 11: Human health: impacts, adaptation, and co-benefits (archived 8 July 2014), pp.10-13, in IPCC AR5 WG2 A 2014
201. IPCC AR4 SYR 2007. 3.3.3 Especially affected systems, sectors and regions. Synthesis report.
202. Mimura, N.; et al. (2007). "Executive summary". In Parry, M.L. et al. Chapter 16: Small Islands. Climate change 2007: impacts, adaptation and vulnerability: contribution of Working Group II to the fourth assessment report of the Intergovernmental Panel on Climate Change (IPCC). Cambridge University Press (CUP): Cambridge, UK: Print version: CUP. This version: IPCC website. ISBN 0521880106. Retrieved 15 September 2011.
203. Climate change and the risk of statelessness" (PDF). May 2011. Retrieved 13 April 2012.
204. Chris Hope; Kevin Schaefer (2015). "Economic impacts of carbon dioxide and methane released from thawing permafrost". Nature. **6**: 56–59. Bibcode:2016NatCC...6...56H. doi:10.1038/nclimate2807.
205. North Slope permafrost thawing sooner than expected". University of Alaska Fairbanks. 2015.
206. PBL Netherlands Environment Agency (June 15, 2012). "Figure 6.14, in: Chapter 6: The energy and climate challenge". In van Vuuren, D.; M. Kok. Roads from Rio+20 (PDF). ISBN 978-90-78645-98-6., p.177, Report no. 500062001. Report website.
207. Mitigation, in USGCRP 2015
208. IPCC, Synthesis Report Summary for Policymakers, Section 4: Adaptation and mitigation options, in IPCC AR4 SYR 2007.
209. Edenhofer, O., et al., Table TS.3, in: Technical summary (archived 30 December 2014), in: IPCC AR5 WG3 2014, p. 68
210. Citi report: slowing global warming would save tens of trillions of dollars". The Guardian. 2015.
211. Clarke, L., et al., Executive summary, in: Chapter 6: Assessing Transformation Pathways (archived 30 December 2014), in: IPCC AR5 WG3 2014, p. 418
212. SPM4.1: Long-term mitigation pathways, in: Summary for Policymakers (archived 27 December 2014), in: IPCC AR5 WG3 2014, pp. 10–13
213. Edenhofer, O., et al., TS.3.1.2: Short- and long-term requirements of mitigation

pathways, in: Technical summary (archived 30 December 2014), in: IPCC AR5 WG3 2014, pp. 55–56

214. Edenhofer, O., et al., TS.3.1.3: Costs, investments and burden sharing, in: Technical summary (archived 30 December 2014), in: IPCC AR5 WG3 2014, p. 58

215. Smit et al., Chapter 18: Adaptation to Climate Change in the Context of Sustainable Development and Equity, Section 18.2.3: Adaptation Types and Forms, in IPCC TAR WG2 2001.

216. Appendix I. Glossary". Adaptive capacity., in IPCC AR4 WG2 2007

217. Synthesis report". Sec 6.3 Responses to climate change: Robust findings]., in IPCC AR4 SYR 2007

218. New Report Provides Authoritative Assessment of National, Regional Impacts of Global Climate Change" (Press release). U.S. Global Change Research Program. 16 June 2009. Retrieved 14 January 2016.

219. Workshop on managing solar radiation" (PDF). NASA. April 2007. Retrieved 23 May 2009.

220. Stop emitting CO2 or geoengineering could be our only hope" (Press release). The Royal Society. 28 August 2009. Retrieved 14 June 2011.

221. P. Keller, David; Feng, Ellias Y.; Oschlies, Andreas (January 2014). "Potential climate engineering effectiveness and side effects during a high carbon dioxide-emission scenario". Nature. **5**: 3304. Bibcode:2014NatCo...5E3304K. doi:10.1038/ncomms4304. Retrieved 31 March 2014. We find that even when applied continuously and at scales as large as currently deemed possible, all methods are, individually, either relatively ineffective with limited (<8%) warming reductions, or they have potentially severe side effects and cannot be stopped without causing rapid climate change.

222. Quoted in IPCC SAR SYR 1996, "Synthesis of Scientific-Technical Information Relevant to Interpreting Article 2 of the UN Framework Convention on Climate Change", paragraph 4.1, p. 8 (pdf p. 18.)

223. Granger Morgan, M. (Lead Author), H. Dowlatabadi, M. Henrion, D. Keith, R. Lempert, S. McBride, M. Small and T. Wilbanks (Contributing Authors) (2009). "Non-Technical Summary: BOX NT.1 Summary of Climate Change Basics". Synthesis and Assessment Product 5.2: Best practice approaches for characterizing, communicating, and incorporating scientific uncertainty in decisionmaking. A Report by the U.S. Climate Change Science Program and the Subcommittee on Global Change Research (PDF). Washington, D.C., USA.: National Oceanic and Atmospheric Administration. p. 11. Retrieved June 1, 2011.

224. UNFCCC (n.d.). "Essential Background". UNFCCC website. Retrieved 18 May 2010.

225. UNFCCC (n.d.). "Full text of the Convention, Article 2". UNFCCC website. Retrieved 18 May 2010.

226. Rogner et al., Chapter 1: Introduction, Executive summary, in IPCC AR4

WG3 2007.

227. Raupach, R.; Marland, G.; Ciais, P.; Le Quere, C.; Canadell, G.; Klepper, G.; Field, B. (Jun 2007). "Global and regional drivers of accelerating CO2 emissions" (Free full text). Proceedings of the National Academy of Sciences. **104** *(24): 10288–10293. Bibcode:2007PNAS..10410288R. doi:10.1073/pnas.0700609104. ISSN 0027-8424. PMC 1876160.* PMID 17519334.

228. Dessai, S. (2001). "The climate regime from The Hague to Marrakech: Saving or sinking the Kyoto Protocol?" (PDF). Tyndall Centre Working Paper 12. Tyndall Centre website. Retrieved 5 May 2010.

229. Grubb, M. (July–September 2003). "The Economics of the Kyoto Protocol" (PDF). World Economics. **4** *(3): 144–145. Retrieved 25 March 2010.*

230. UNFCCC (n.d.). "Kyoto Protocol". UNFCCC website. Retrieved 21 May 2011.

231. Müller, Benito (February 2010). Copenhagen 2009: Failure or final wake-up call for our leaders? EV 49(PDF). Oxford Institute for Energy Studies. p. i. ISBN 978-1-907555-04-6. Retrieved 18 May 2010.

232. Rudd, Kevin (25 May 2015). "Paris Can›t Be Another Copenhagen". The New York Times. Retrieved 26 May 2015.

233. United Nations Environment Programme (November 2010). "Technical summary". The Emissions Gap Report: Are the Copenhagen Accord pledges sufficient to limit global warming to 2 °C or 1.5 °C? A preliminary assessment (advance copy) (PDF). UNEP website. Retrieved 11 May 2011. This publication is also available in e-book format

234. UNFCCC (30 March 2010). "Decision 2/CP. 15 Copenhagen Accord. In: Report of the Conference of the Parties on its fifteenth session, held in Copenhagen from 7 to 19 December 2009. Addendum. Part Two: Action taken by the Conference of the Parties at its fifteenth session" (PDF). United Nations Office at Geneva, Switzerland. p. 5. Retrieved 17 May 2010.

235. Outcome of the work of the Ad Hoc Working Group on long-term Cooperative Action under the Convention" (PDF). PRESIDENCIA DE LA REPÚBLICA, MÉXICO. 11 December 2010. p. 2. Retrieved 12 January 2011.

236. IPCC, "Summary for Policymakers", Detection and Attribution of Climate Change, "It is extremely likely that human influence has been the dominant cause of the observed warming since the mid-20th century" (page 15) and "In this Summary for Policymakers, the following terms have been used to indicate the assessed likelihood of an outcome or a result: (...) extremely likely: 95–100%" (page 2)., in IPCC AR5 WG1 2013.

237. Royal Society (13 April 2005). "Letter from The Royal Society: A GUIDE TO FACTS AND FICTIONS ABOUT CLIMATE CHANGE: Misleading arguments: Many scientists do not think that climate change is a problem. Some scientists have signed petitions stating that climate change is not a problem. ... There are some individuals and organisations, some of which are funded by the US oil industry, that seek to undermine the science of climate change and the work of the IPCC. They appear motivated in their arguments by opposition to

the United Nations Framework Convention on Climate Change and the Kyoto Protocol, which seek urgent action to tackle climate change through a reduction in greenhouse gas emissions.". Economic Affairs – Written Evidence. The Economics of Climate Change, the Second Report of the 2005–2006 session, produced by the UK Parliament House of Lords Economics Affairs Select Committee. UK Parliament website. Retrieved 9 July 2011. This document is also available in PDF format

238. Academia Brasileira de Ciéncias (Brazil), Royal Society of Canada, Chinese Academy of Sciences, Académie des Sciences (France), Deutsche Akademie der Naturforscher Leopoldina (Germany), Indian National Science Academy, Accademia Nazionale dei Lincei (Italy), Science Council of Japan, Academia Mexicana de Ciencias, Russian Academy of Sciences, Academy of Science of South Africa, Royal Society (United Kingdom), National Academy of Sciences (United States of America) (May 2009). "G8+5 Academies› joint statement: Climate change and the transformation of energy technologies for a low carbon future"(PDF). US National Academies website. Retrieved 5 May 2010.

239. John Cook; et al. (13 April 2016). "Consensus on consensus: a synthesis of consensus estimates on human-caused global warming". Environmental Research Letters. **11** *(4): 048002. doi:10.1088/1748-9326/11/4/048002. Retrieved 21 July 2016.*

240. Julie Brigham-Grette; et al. (September 2006). "Petroleum Geologists› Award to Novelist Crichton Is Inappropriate" (PDF). Eos. **87** *(36): 364. Bibcode:2006EOSTr..87..364B. doi:10.1029/2006EO360008. Retrieved 23 January 2007. The AAPG stands alone among scientific societies in its denial of human-induced effects on global warming.*

241. Boykoff, M.; Boykoff, J. (July 2004). "Balance as bias: global warming and the US prestige press". Global Environmental Change Part A. **14** *(2): 125–136. doi:10.1016/j.gloenvcha.2003.10.001.*

242. Oreskes, Naomi; Conway, Erik. Merchants of Doubt: How a Handful of Scientists Obscured the Truth on Issues from Tobacco Smoke to Global Warming (first ed.). Bloomsbury Press. ISBN 978-1-59691-610-4.

243. Aaron M. McCright and Riley E. Dunlap, "Challenging Global Warming as a Social Problem: An Analysis of the Conservative Movement's Counter-Claims", Social Problems, November 2000, Vol. 47 Issue 4, pp 499–522 in JSTOR

244. Weart, S. (July 2009). "The Public and Climate Change (cont. – since 1980). Section: After 1988". American Institute of Physics website. Retrieved 5 May 2010.

245. SEPP (n.d.). "Frequently Asked Questions About Climate Change". Science & Environmental Policy Project (SEPP) website. Archived from the original on 11 May 2008. Retrieved 5 May 2010.

246. Begley, Sharon (13 August 2007). "The Truth About Denial". Newsweek. Retrieved 13 August 2007.

247. Adams, David (20 September 2006). "Royal Society tells Exxon: stop funding climate change denial". The Guardian. London. Retrieved 9 August 2007.

248. Exxon cuts ties to global warming skeptics". MSNBC. 12 January 2007. Retrieved 2 May 2007.

249. Sandell, Clayton (3 January 2007). "Report: Big Money Confusing Public on Global Warming". ABC. Retrieved 27 April 2007.

250. Greenpeace: Exxon still funding climate skeptics". USA Today. Reuters. 18 May 2007. Retrieved 21 January 2010.

251. Global Warming Resolutions at U.S. Oil Companies Bring Policy Commitments from Leaders, and Record High Votes at Laggards" (Press release). Ceres. 13 May 2004. Retrieved 4 March 2010.

252. Oil Company Positions on the Reality and Risk of Climate Change". Environmental Studies. University of Oshkosh - Wisconsin. Retrieved 27 March 2016.

253. Weart, S. (February 2015). "The Public and Climate Change (cont. – since 1980). Section: after 1988". American Institute of Physics website. Retrieved 18 August 2015.

254. Environment". Gallup. 2015. Retrieved 18 August 2015.

255. Pelham, Brett (2009). "Awareness, Opinions about Global Warming Vary Worldwide". Gallup. Retrieved 18 August 2015.

256. Pugliese, Anita (20 April 2011). "Fewer Americans, Europeans View Global Warming as a Threat". Gallup. Retrieved 22 April 2011.

257. Ray, Julie; Anita Pugliese (22 April 2011). "Worldwide, Blame for Climate Change Falls on Humans". Gallup.Com. Retrieved 3 May 2011. People nearly everywhere, including majorities in developed Asia and Latin America, are more likely to attribute global warming to human activities rather than natural causes. The U.S. is the exception, with nearly half (47%) – and the largest percentage in the world – attributing global warming to natural causes.

258. Climate Change and Financial Instability Seen as Top Global Threats". Pew Research Center for the People & the Press.

259. Climate Change: Key Data Points from Pew Research | Pew Research Center

260. Tranter, Bruce; Booth, Kate (July 2015). "Scepticism in a Changing Climate: A Cross-national Study". Global Environmental Change. **33**: 54–164. doi:10.1016/j.gloenvcha.2015.05.003.

261. Weart, Spencer R. (February 2014). "The Discovery of Global Warming; The Public and Climate Change: Suspicions of a Human-Caused Greenhouse (1956-1969)". American Institute of Physics. Retrieved 12 May 2015., and footnote 27

262. Erik Conway. "What's in a Name? Global Warming vs. Climate Change", NASA, 5 December 2008

263. Weart, Spencer R. (February 2014). "The Discovery of Global Warming; The Public and Climate Change: The Summer of 1988". American Institute of Physics. Retrieved 12 May 2015.

264. U.S. Senate, Committee on Energy and Natural Resources, "Greenhouse Effect and Global Climate Change, part 2" 100th Cong., 1st sess., 23 June 1988, p. 44.

CHAPTER 4

Climate Change and Mitigation

Chapter Contents

- **Introduction**
- **Climate Change Mitigation**
- **Environmental Migrant**
- **Abrupt Climate Change**

Climate change occurs when the Earth's average temperature changes dramatically over time. As little as one or two degrees can be considered dramatic change because the Earth's ecosystem depends on a very delicate balance, and even small shifts can have a far-reaching impact. A drop in average temperature can also be considered climate change, but in modern times people using the term are usually talking about global warming.

One reason that climate change has become a popular and, at times, controversial topic is that many people believe it is mostly the result of human activity. Burning fossil fuels, such as oil and coal, produces byproducts such as carbon dioxide gas. Since there are not enough plants on Earth to quickly transform all this emitted carbon dioxide into oxygen, the gas remains in the atmosphere. Through a process known as the greenhouse effect, the carbon dioxide traps solar heat, which leads to the warming of the planet.

INTRODUCTION

Climate change is a change in the statistical distribution of weather patterns when that change lasts for an extended period of time (i.e., decades to millions of years). Climate change may refer to a change in average weather conditions, or in the time variation of weather around longer-term average conditions (i.e., more or fewer extreme weather events). Climate change is caused by factors such as biotic processes, variations in solar radiation received by Earth, plate tectonics, and volcanic eruptions. Certain human activities have also been identified as significant causes of recent climate change, often referred to as *global warming*. Scientists actively work to understand past and future climate by using observations and theoretical models. A climate record—extending deep into the Earth's past—has been assembled, and continues to be built up, based on geological evidence from borehole temperature profiles, cores removed from deep accumulations of ice, floral and faunal records, glacial and periglacial processes, stable-isotope and other analyses of sediment layers, and records of past sea levels. More recent data are provided by the instrumental record. General circulation models, based on the physical sciences, are often used in theoretical approaches to match past climate data, make future projections, and link causes and effects in climate change.

Terminology

The most general definition of *climate change* is a change in the statistical properties (principally its mean and spread) of the climate system when considered over long periods of time, regardless of cause. Accordingly, fluctuations over periods shorter than a few decades, such as El Niño, do not represent climate change.

The term sometimes is used to refer specifically to climate change caused by human activity, as opposed to changes in climate that may have resulted as part of Earth's natural processes. In this sense, especially in the context of environmental policy, the term *climate change* has become synonymous with *anthropogenic global warming*. Within scientific journals, *global warming* refers to surface temperature increases while *climate change* includes global warming and everything else that increasing greenhouse gas levels affect.

A related term is "climatic change". In 1966, the World Meteorological Organization (WMO) proposed the term "climatic change" to encompass all forms of climatic variability on time-scales longer than 10 years, regardless of cause. Change was a given and climatic was used as an adjective to describe this kind of change (as opposed to political or economic change). When it was realized that human activities had a potential to drastically alter the climate, the term climate change replaced climatic change as the dominant term to reflect an anthropogenic cause. Climate change was incorporated in the title of the Intergovernmental Panel on Climate Change (IPCC) and the UN Framework Convention on Climate Change (UNFCCC). Climate change, used as a noun, became an issue rather than the technical description of changing weather.

Causes

On the broadest scale, the rate at which energy is received from the Sun and the rate at which it is lost to space determine the equilibrium temperature and climate of Earth. This energy is distributed around the globe by winds, ocean currents, and other mechanisms to affect the climates of different regions.

Factors that can shape climate are called climate forcings or "forcing mechanisms". These include processes such as variations in solar radiation, variations in the Earth's orbit, variations in the albedo or reflectivity of the continents and oceans, mountain-building and continental drift and changes in greenhouse gas concentrations. There are a variety of climate change feedbacks that can either amplify or diminish the initial forcing.

Some parts of the climate system, such as the oceans and ice caps, respond more slowly in reaction to climate forcings, while others respond more quickly. There are also key threshold factors which when exceeded can produce rapid change.

Forcing mechanisms can be either "internal" or "external". Internal forcing mechanisms are natural processes within the climate system itself (e.g., the thermohaline circulation). External forcing mechanisms can be either natural (e.g., changes in solar output) or anthropogenic (e.g., increased emissions of greenhouse gases).

Whether the initial forcing mechanism is internal or external, the response of the climate system might be fast (e.g., a sudden cooling due to airborne volcanic ash reflecting sunlight), slow (e.g. thermal expansion of warming ocean water), or a combination (e.g., sudden loss of albedo in the arctic ocean as sea ice melts, followed by more gradual thermal expansion of the water). Therefore, the climate system can respond abruptly, but the full response to forcing mechanisms might not be fully developed for centuries or even longer.

Internal Forcing Mechanisms

Scientists generally define the five components of earth's climate system to include atmosphere, hydrosphere, cryosphere, lithosphere (restricted to the surface soils, rocks, and sediments), and biosphere. Natural changes in the climate system ("internal forcings") result in internal "climate variability". Examples include the type and distribution of species, and changes in ocean-atmosphere circulations.

Ocean-Atmosphere Variability

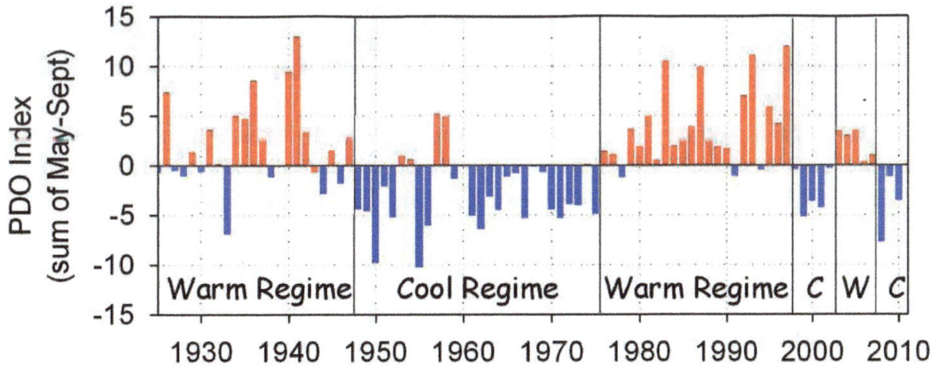

Pacific Decadal Oscillation 1925 to 2010.

The ocean and atmosphere can work together to spontaneously generate internal climate variability that can persist for years to decades at a time. Examples of this type of variability include the El Niño-Southern Oscillation, the Pacific decadal oscillation, and the Atlantic Multidecadal Oscillation. These variations can affect global average surface temperature by redistributing heat between the deep ocean and the atmopshere and/or by altering the cloud/water vapor/sea ice distribution which can affect the total energy budget of the earth.

The oceanic aspects of these circulations can generate variability on centennial timescales due to the ocean having hundreds of times more mass than in the atmosphere, and thus very high thermal inertia. For example, alterations to ocean processes such as thermohaline circulation play a key role in redistributing heat in the world's oceans. Due to the long timescales of this circulation, ocean temperature at depth is still adjusting to effects of the Little Ice Age which occurred between the 1600 and 1800s.

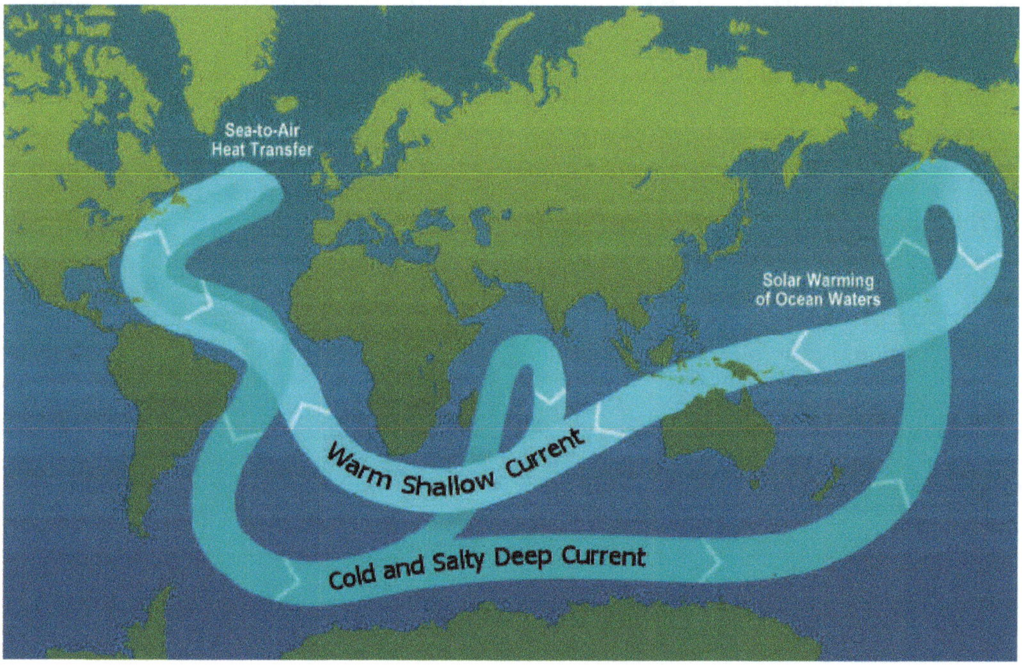

A schematic of modern thermohaline circulation. Tens of millions of years ago, continental-plate movement formed a land-free gap around Antarctica, allowing the formation of the ACC, which keeps warm waters away from Antarctica.

Life

Life affects climate through its role in the carbon and water cycles and through such mechanisms as albedo, evapotranspiration, cloud formation, and weathering.Examples of how life may have affected past climate include:

- glaciation 2.3 billion years ago triggered by the evolution of oxygenic photosynthesis, which depleted the atmosphere of the greenhouse gas carbon dioxide and introduced free oxygen.
- another glaciation 300 million years ago ushered in by long-term burial of decomposition-resistant detritus of vascular land-plants (creating a carbon sink and forming coal)

- termination of the Paleocene-Eocene Thermal Maximum 55 million years ago by flourishing marine phytoplankton
- reversal of global warming 49 million years ago by 800,000 years of arctic azolla blooms
- global cooling over the past 40 million years driven by the expansion of grass-grazer ecosystems

External Forcing Mechanisms

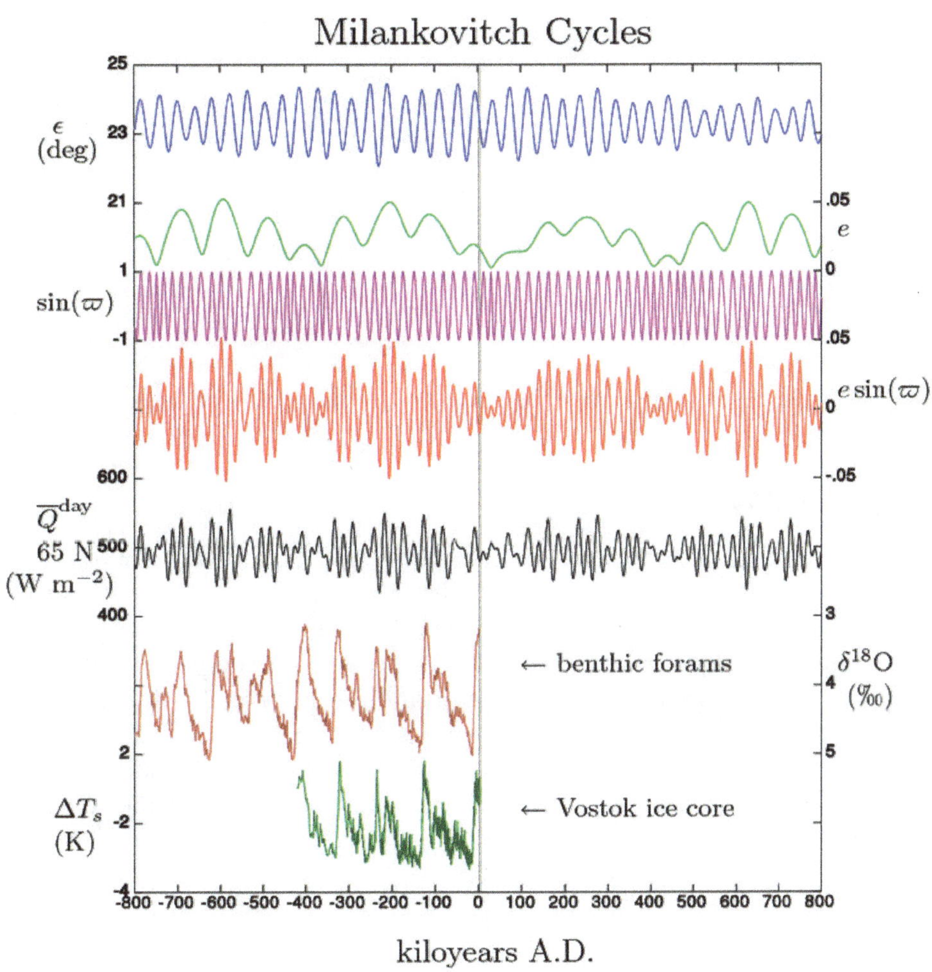

Milankovitch cycles from 800,000 years ago in the past to 800,000 years in the future.

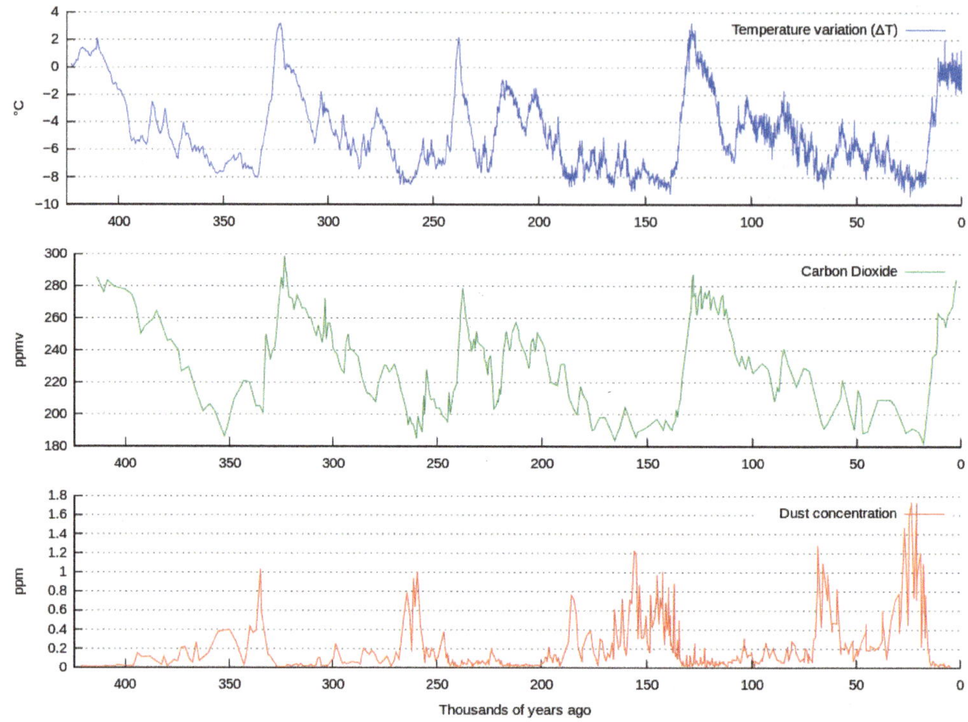

Variations in CO_2, temperature and dust from the Vostok ice core over the last 450,000 years.

Orbital Variations

Slight variations in Earth's orbit lead to changes in the seasonal distribution of sunlight reaching the Earth's surface and how it is distributed across the globe. There is very little change to the area-averaged annually averaged sunshine; but there can be strong changes in the geographical and seasonal distribution. The three types of orbital variations are variations in Earth's eccentricity, changes in the tilt angle of Earth's axis of rotation, and precession of Earth's axis. Combined together, these produce Milankovitch cycles which have a large impact on climate and are notable for their correlation to glacial and interglacial periods, their correlation with the advance and retreat of the Sahara, and for their appearance in the stratigraphic record.

The IPCC notes that Milankovitch cycles drove the ice age cycles, CO_2 followed temperature change "with a lag of some hundreds of years," and that as a feedback amplified temperature change. The depths of the ocean have a lag time in changing temperature (thermal inertia on such scale). Upon seawater temperature change, the solubility of CO_2 in the oceans changed, as well as other factors impacting air-sea CO_2exchange.

Solar Output

The Sun is the predominant source of energy input to the Earth. Other sources include geothermal energy from the Earth's core, and heat from the decay of radioactive compounds. Both long- and short-term variations in solar intensity are known to affect global climate.

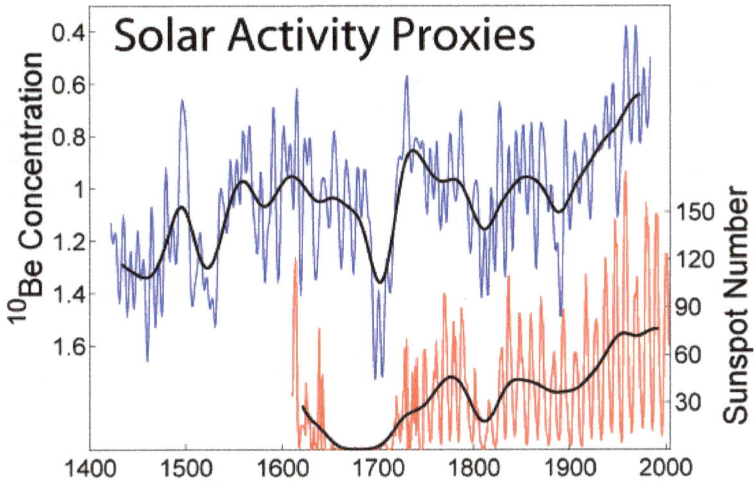

Variations in solar activity during the last several centuries based on observations of sunspots and beryllium isotopes. The period of extraordinarily few sunspots in the late 17th century was the Maunder minimum.

Three to four billion years ago, the Sun emitted only 70% as much power as it does today. If the atmospheric composition had been the same as today, liquid water should not have existed on Earth. However, there is evidence for the presence of water on the early Earth, in the Hadean and Archean eons, leading to what is known as the faint young Sun paradox. Hypothesized solutions to this paradox include a vastly different atmosphere, with much higher concentrations of greenhouse gases than currently exist. Over the following approximately 4 billion years, the energy output of the Sun increased and atmospheric composition changed. The Great Oxygenation Event – oxygenation of the atmosphere around 2.4 billion years ago – was the most notable alteration. Over the next five billion years, the Sun's ultimate death as it becomes a red giant and then a white dwarf will have large effects on climate, with the red giant phase possibly ending any life on Earth that survives until that time.

Solar output also varies on shorter time scales, including the 11-year solar cycle and longer-term modulations. Solar intensity variations possibly as a result of the Wolf, Spörer and Maunder Minimum are considered to have been influential in triggering the Little Ice Age, and some of the warming observed from 1900 to 1950. The cyclical nature of the Sun's energy output is not yet fully understood; it differs from the very slow change that is happening within the Sun as it ages and evolves. Research indicates that solar variability has had effects including the Maunder minimum from 1645 to 1715 A.D., part of the Little Ice Age from 1550 to 1850 A.D. that was marked by relative cooling and greater glacier extent than the centuries before and afterward. Some studies point toward solar radiation increases from cyclical sunspot activity affecting global warming, and climate may be influenced by the sum of all effects (solar variation, anthropogenic radiative forcings, etc.).

Interestingly, a 2010 study *suggests*, "that the effects of solar variability on temperature throughout the atmosphere may be contrary to current expectations."

In an Aug 2011 Press Release, CERN announced the publication in the Nature journal the initial results from its CLOUD experiment. The results indicate that ionisation from cosmic rays significantly enhances aerosol formation in the presence of sulfuric acid and water, but in the lower atmosphere where ammonia is also required, this is insufficient to account for aerosol formation and additional trace vapours must be involved. The next step is to find more about these trace vapours, including whether they are of natural or human origin.

Volcanism

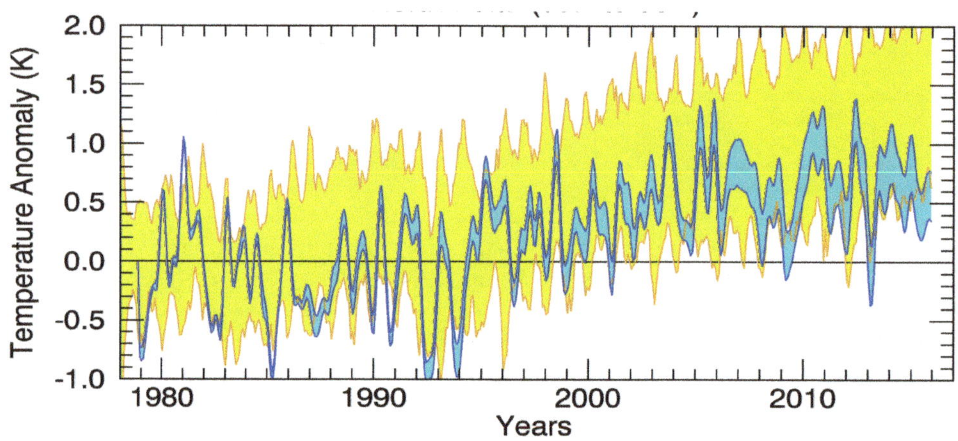

In atmospheric temperature from 1979 to 2010, determined by MSUNASA satellites, effects appear from aerosols released by major volcanic eruptions (El Chichón and Pinatubo). El Niño is a separate event, from ocean variability.

The eruptions considered to be large enough to affect the Earth's climate on a scale of more than 1 year are the ones that inject over 100,000 tons of SO_2 into the stratosphere. This is due to the optical properties of SO_2 and sulfate aerosols, which strongly absorb or scatter solar radiation, creating a global layer of sulfuric acid haze. On average, such eruptions occur several times per century, and cause cooling (by partially blocking the transmission of solar radiation to the Earth's surface) for a period of a few years.

The eruption of Mount Pinatubo in 1991, the second largest terrestrial eruption of the 20th century, affected the climate substantially, subsequently global temperatures decreased by about 0.5 °C (0.9 °F) for up to three years. Thus, the cooling over large parts of the Earth reduced surface temperatures in 1991-93, the equivalent to a reduction in net radiation of 4 watts per square meter. The Mount Tambora eruption in 1815 caused the Year Without a Summer. Much larger eruptions, known as large igneous provinces, occur only a few times every fifty - hundred million years - through flood basalt, and caused in Earth past global warming and mass extinctions.

Small eruptions, with injections of less than 0.1 Mt of sulfur dioxide into the stratosphere, impact the atmosphere only subtly, as temperature changes are comparable with natural variability. However, because smaller eruptions occur at a much higher frequency, they too have a significant impact on Earth's atmosphere.

Seismic monitoring maps current and future trends in volcanic activities, and tries to develop early warning systems. In climate modelling the aim is to study the physical mechanisms and feedbacks of volcanic forcing.

Volcanoes are also part of the extended carbon cycle. Over very long (geological) time periods, they release carbon dioxide from the Earth's crust and mantle, counteracting the uptake by sedimentary rocks and other geological carbon dioxide sinks. The US Geological Survey estimates are that volcanic emissions are at a much lower level than the effects of current human activities, which generate 100–300 times the amount of carbon dioxide emitted by volcanoes. A review of published studies indicates that annual volcanic emissions of carbon dioxide, including amounts released from mid-ocean ridges, volcanic arcs, and hot spot volcanoes, are only the equivalent of 3 to 5 days of human caused output. The annual amount put out by human activities may be greater than the amount released by supererruptions, the most recent of which was the Toba eruption in Indonesia 74,000 years ago.

Although volcanoes are technically part of the lithosphere, which itself is part of the climate system, the IPCC explicitly defines volcanism as an external forcing agent.

Plate Tectonics

Over the course of millions of years, the motion of tectonic plates reconfigures global land and ocean areas and generates topography. This can affect both global and local patterns of climate and atmosphere-ocean circulation.

The position of the continents determines the geometry of the oceans and therefore influences patterns of ocean circulation. The locations of the seas are important in controlling the transfer of heat and moisture across the globe, and therefore, in determining global climate. A recent example of tectonic control on ocean circulation is the formation of the Isthmus of Panama about 5 million years ago, which shut off direct mixing between the Atlantic and Pacific Oceans. This strongly affected the ocean dynamics of what is now the Gulf Stream and may have led to Northern Hemisphere ice cover.During the Carboniferous period, about 300 to 360 million years ago, plate tectonics may have triggered large-scale storage of carbon and increased glaciation. Geologic evidence points to a "megamonsoonal" circulation pattern during the time of the supercontinent Pangaea, and climate modeling suggests that the existence of the supercontinent was conducive to the establishment of monsoons.

The size of continents is also important. Because of the stabilizing effect of the oceans on temperature, yearly temperature variations are generally lower in coastal areas than they are inland. A larger supercontinent will therefore have more area in which climate is strongly seasonal than will several smaller continents or islands.

Human Influences

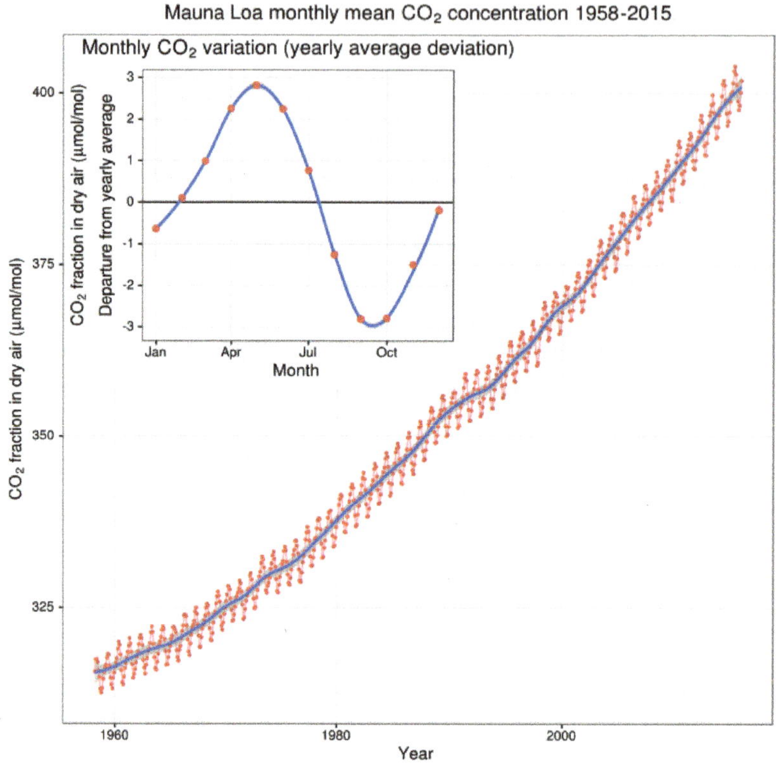

Increase in atmospheric CO_2 levels.

In the context of climate variation, anthropogenic factors are human activities which affect the climate. The scientific consensus on climate change is "that climate is changing and that these changes are in large part caused by human activities," and it "is largely irreversible."

"Science has made enormous inroads in understanding climate change and its causes, and is beginning to help develop a strong understanding of current and potential impacts that will affect people today and in coming decades. This understanding is crucial because it allows decision makers to place climate change in the context of other large challenges facing the nation and the world. There are still some uncertainties, and there always will be in understanding a complex system like Earth's climate. Nevertheless, there is a strong, credible body of evidence, based on multiple lines of research, documenting that climate is changing and that these changes are in large part caused by human activities. While much remains to be learned, the core phenomenon, scientific questions, and hypotheses have been examined thoroughly and have stood firm in the face of serious scientific debate and careful evaluation of alternative explanations."

—United States National Research Council, Advancing the Science of Climate Change

Of most concern in these anthropogenic factors is the increase in CO_2 levels due to emissions from fossil fuel combustion, followed by aerosols (particulate matter in the atmosphere) and the CO_2 released by cement manufacture. Other factors, including land use, ozone depletion, animal agriculture and deforestation, are also of concern in the roles they play – both separately and in conjunction with other factors – in affecting climate, microclimate, and measures of climate variables.

Physical Evidence

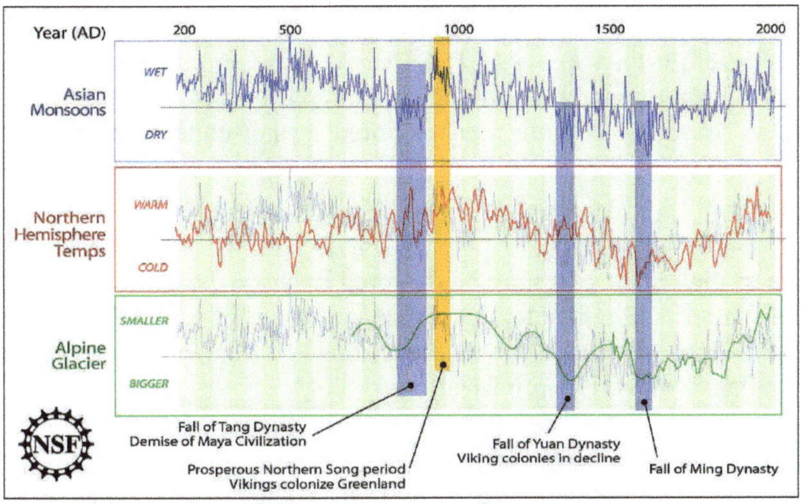

Comparisons between Asian Monsoons from 200 AD to 2000 AD (staying in the background on other plots), Northern Hemisphere temperature, Alpine glacier extent (vertically inverted as marked), and human history as noted by the U.S. NSF.

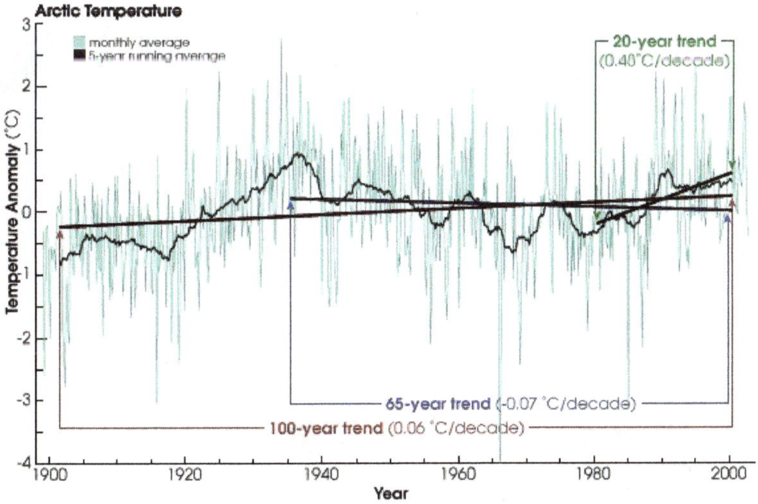

Arctic temperature anomalies over a 100-year period as estimated by NASA. Typical high monthly variance can be seen, while longer-term averages highlight trends.

Evidence for climatic change is taken from a variety of sources that can be used to reconstruct past climates. Reasonably complete global records of surface temperature are available beginning from the mid-late 19th century. For earlier periods, most of the evidence is indirect—climatic changes are inferred from changes in proxies, indicators that reflect climate, such as vegetation, ice cores, dendrochronology, sea level change, and glacial geology.

Temperature Measurements and Proxies

The instrumental temperature record from surface stations was supplemented by radiosonde balloons, extensive atmospheric monitoring by the mid-20th century, and, from the 1970s on, with global satellite data as well. The $^{18}O/^{16}O$ ratio in calcite and ice core samples used to deduce ocean temperature in the distant past is an example of a temperature proxy method, as are other climate metrics noted in subsequent categories.

Historical and Archaeological Evidence

Climate change in the recent past may be detected by corresponding changes in settlement and agricultural patterns. Archaeological evidence, oral history and historical documents can offer insights into past changes in the climate. Climate change effects have been linked to the collapse of various civilizations.

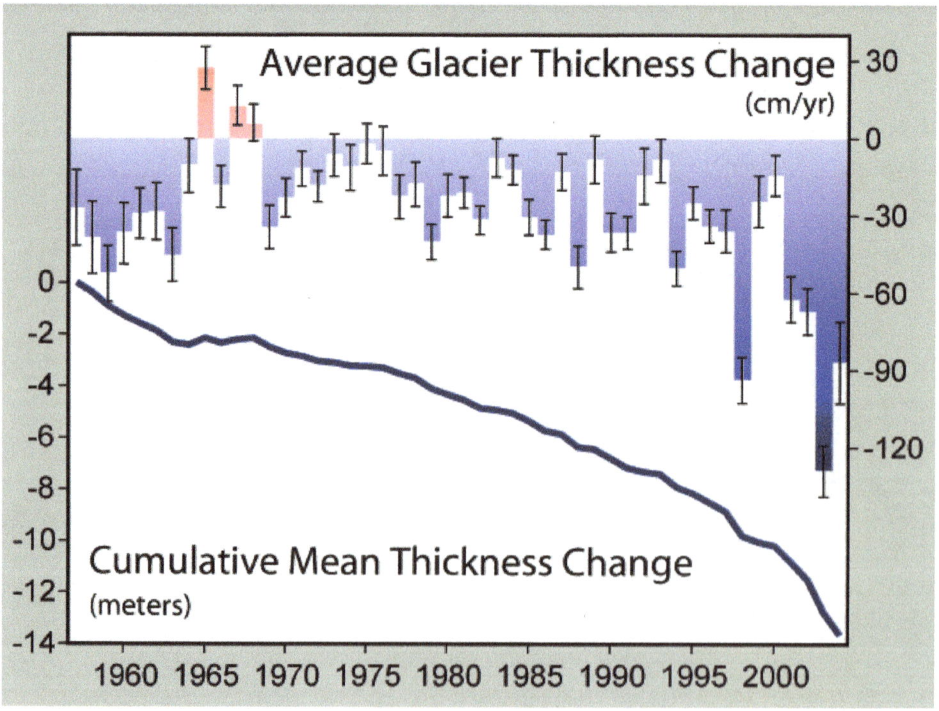

Decline in Thickness of Glaciers Worldwide over the Past Half-Century.

Glaciers

Glaciers are considered among the most sensitive indicators of climate change. Their size is determined by a mass balance between snow input and melt output. As temperatures warm, glaciers retreat unless snow precipitation increases to make up for the additional melt; the converse is also true.

Glaciers grow and shrink due both to natural variability and external forcings. Variability in temperature, precipitation, and englacial and subglacial hydrology can strongly determine the evolution of a glacier in a particular season. Therefore, one must average over a decadal or longer time-scale and/or over many individual glaciers to smooth out the local short-term variability and obtain a glacier history that is related to climate.

A world glacier inventory has been compiled since the 1970s, initially based mainly on aerial photographs and maps but now relying more on satellites. This compilation tracks more than 100,000 glaciers covering a total area of approximately 240,000 km^2, and preliminary estimates indicate that the remaining ice cover is around 445,000 km^2. The World Glacier Monitoring Service collects data annually on glacier retreat and glacier mass balance. From this data, glaciers worldwide have been found to be shrinking significantly, with strong glacier retreats in the 1940s, stable or growing conditions during the 1920s and 1970s, and again retreating from the mid-1980s to present.

The most significant climate processes since the middle to late Pliocene (approximately 3 million years ago) are the glacial and interglacial cycles. The present interglacial period (the Holocene) has lasted about 11,700 years. Shaped by orbital variations, responses such as the rise and fall of continental ice sheets and significant sea-level changes helped create the climate. Other changes, including Heinrich events, Dansgaard–Oeschger events and the Younger Dryas, however, illustrate how glacial variations may also influence climate without the orbital forcing.

Glaciers leave behind moraines that contain a wealth of material—including organic matter, quartz, and potassium that may be dated—recording the periods in which a glacier advanced and retreated. Similarly, by tephrochronological techniques, the lack of glacier cover can be identified by the presence of soil or volcanic tephra horizons whose date of deposit may also be ascertained.

Arctic Sea Ice Loss

Main articles: Arctic sea ice decline and Climate change in the Arctic

The decline in Arctic sea ice, both in extent and thickness, over the last several decades is further evidence for rapid climate change. Sea ice is frozen seawater that floats on the ocean surface. It covers millions of square miles in the polar regions, varying with the seasons. In the Arctic, some sea ice remains year after year, whereas almost all Southern Ocean or Antarctic sea ice melts away and reforms annually. Satellite observations show that Arctic sea ice is now declining at a rate of 13.3 percent per decade, relative to the 1981 to 2010 average.

This video summarizes how climate change, associated with increased carbon dioxide levels, has affected plant growth.

Vegetation

A change in the type, distribution and coverage of vegetation may occur given a change in the climate. Some changes in climate may result in increased precipitation and warmth, resulting in improved plant growth and the subsequent sequestration of airborne CO_2. A gradual increase in warmth in a region will lead to earlier flowering and fruiting times, driving a change in the timing of life cycles of dependent organisms. Conversely, cold will cause plant bio-cycles to lag. Larger, faster or more radical changes, however, may result in vegetation stress, rapid plant loss and desertification in certain circumstances. An example of this occurred during the Carboniferous Rainforest Collapse (CRC), an extinction event 300 million years ago. At this time vast rainforests covered the equatorial region of Europe and America. Climate change devastated these tropical rainforests, abruptly fragmenting the habitat into isolated ‹islands’ and causing the extinction of many plant and animal species.

Pollen Analysis

Palynology is the study of contemporary and fossil palynomorphs, including pollen. Palynology is used to infer the geographical distribution of plant species, which vary under different climate conditions. Different groups of plants have pollen with distinctive shapes and surface textures, and since the outer surface of pollen is composed of a very resilient material, they resist decay. Changes in the type of pollen found in different layers of sediment in lakes, bogs, or river deltas indicate changes in plant communities. These changes are often a sign of a changing climate. As an example, palynological studies have been used to track changing vegetation patterns throughout the Quaternary glaciationsand especially since the last glacial maximum.

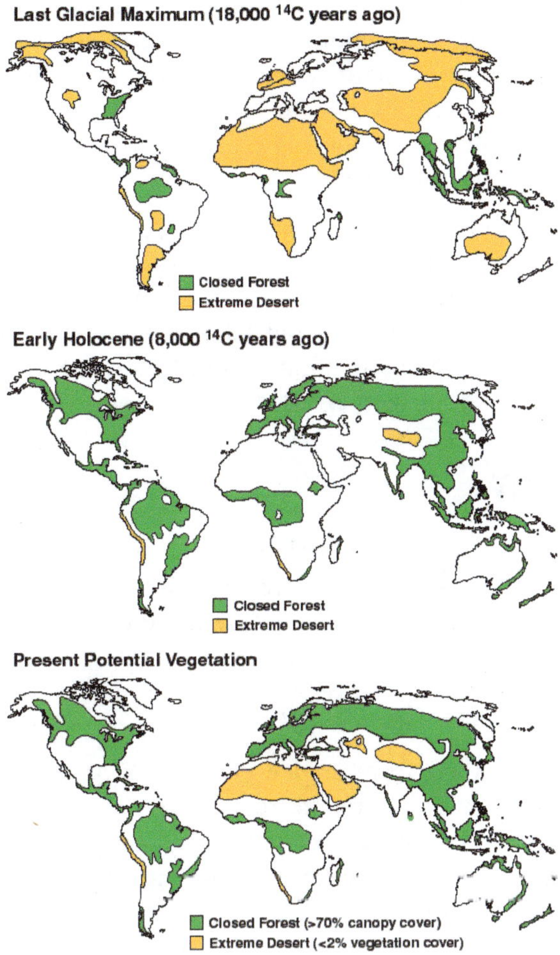

Top: Arid ice age climate

Middle: Atlantic Period, warm and wet

Bottom: Potential vegetation in climate now if not for human effects like agriculture.

Cloud cover and precipitation

Past precipitation can be estimated in the modern era with the global network of precipitation gauges. Surface coverage over oceans and remote areas is relatively sparse, but, reducing reliance on interpolation, satellite clouds and precipitation data has been available since the 1970s. Quantification of climatological variation of precipitation in prior centuries and epochs is less complete but approximated using proxies such as marine sediments, ice cores, cave stalagmites, and tree rings. In July 2016 scientists published evidence of increased cloud cover over polar regions, as predicted by climate models.

Climatological temperatures substantially affect cloud cover and precipitation. For instance, during the Last Glacial Maximum of 18,000 years ago, thermal-driven evaporation from the oceans onto continental landmasses was low, causing large areas of extreme desert, including polar deserts (cold but with low rates of cloud cover and precipitation). In contrast, the world's climate was cloudier and wetter than today near the start of the warm Atlantic Period of 8000 years ago.

Estimated global land precipitation increased by approximately 2% over the course of the 20th century, though the calculated trend varies if different time endpoints are chosen, complicated by ENSO and other oscillations, including greater global land cloud cover precipitation in the 1950s and 1970s than the later 1980s and 1990s despite the positive trend over the century overall. Similar slight overall increase in global river runoff and in average soil moisture has been perceived.

Dendroclimatology

Dendroclimatology is the analysis of tree ring growth patterns to determine past climate variations.Wide and thick rings indicate a fertile, well-watered growing period, whilst thin, narrow rings indicate a time of lower rainfall and less-than-ideal growing conditions.

Ice cores

Analysis of ice in a core drilled from an ice sheet such as the Antarctic ice sheet, can be used to show a link between temperature and global sea level variations. The air trapped in bubbles in the ice can also reveal the CO_2 variations of the atmosphere from the distant past, well before modern environmental influences. The study of these ice cores has been a significant indicator of the changes in CO_2 over many millennia, and continues to provide valuable information about the differences between ancient and modern atmospheric conditions.

Animals

Remains of beetles are common in freshwater and land sediments. Different species of beetles tend to be found under different climatic conditions. Given the extensive lineage of beetles whose genetic makeup has not altered significantly over the millennia, knowledge of the present climatic range of the different species, and the age of the sediments in which remains are found, past climatic conditions may be inferred.

Similarly, the historical abundance of various fish species has been found to have a substantial relationship with observed climatic conditions. Changes in the primary productivity of autotrophs in the oceans can affect marine food webs.

Sea level Change

Global sea level change for much of the last century has generally been estimated using tide gauge measurements collated over long periods of time to give a long-term average.

More recently, altimeter measurements — in combination with accurately determined satellite orbits — have provided an improved measurement of global sea level change. To measure sea levels prior to instrumental measurements, scientists have dated coral reefs that grow near the surface of the ocean, coastal sediments, marine terraces, ooids in limestones, and nearshore archaeological remains. The predominant dating methods used are uranium series and radiocarbon, with cosmogenic radionuclides being sometimes used to date terraces that have experienced relative sea level fall. In the early Pliocene, global temperatures were 1–2°C warmer than the present temperature, yet sea level was 15–25 meters higher than today.

CLIMATE CHANGE MITIGATION

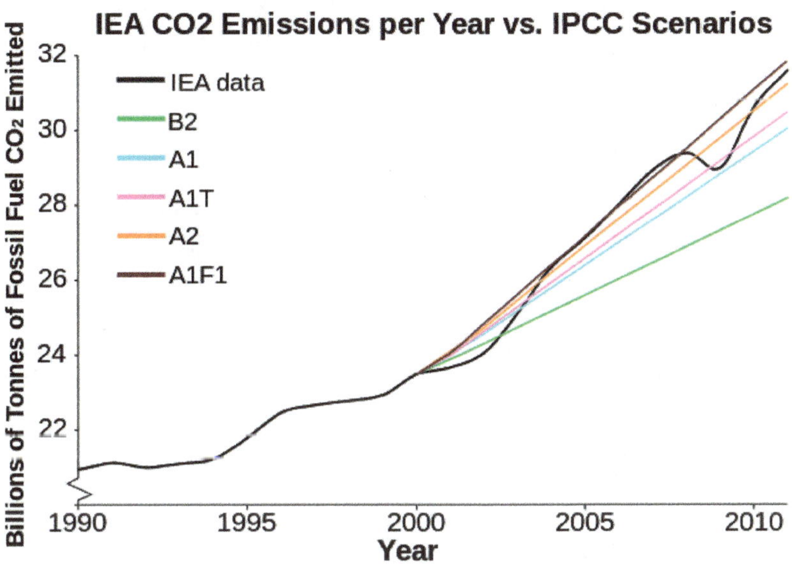

Fossil fuel related CO2 emissions compared to five of IPCC's emissions scenarios. The dips are related to global recessions. Data from IPCC SRES scenarios; Data spreadsheet included with International Energy Agency's "CO2 Emissions from Fuel Combustion 2010 – Highlights"; and Supplemental IEA data. Image source: Skeptical Science

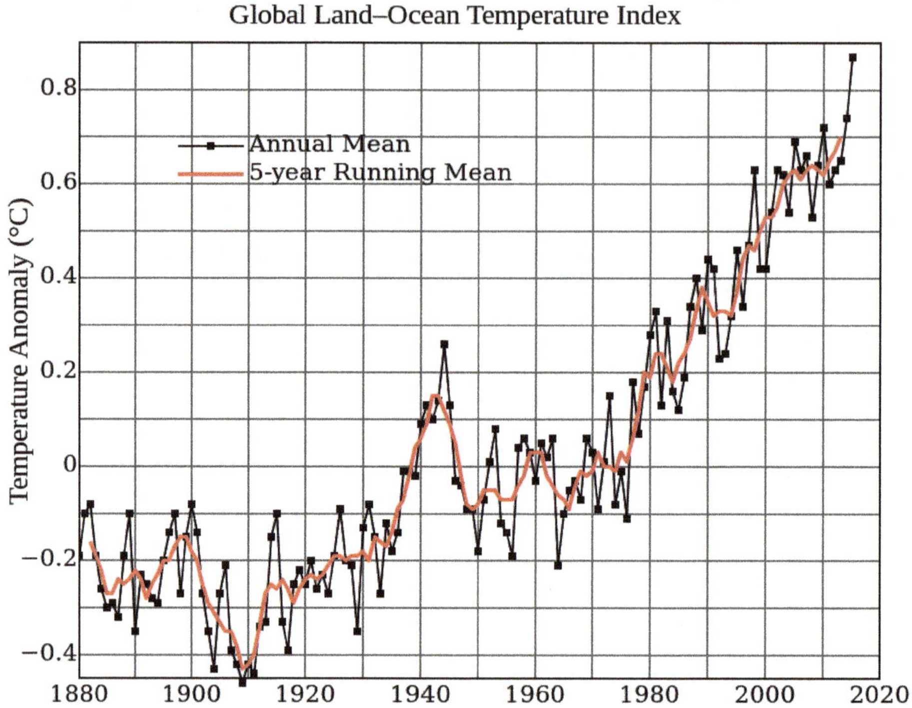

Global mean surface temperature change since 1880, relative to the 1951–1980 mean. The black line is the annual mean and the red line is the 5-year running mean. Source: NASA GISS. Global dimming, from sulfate aerosol air pollution, between 1950 and 1980 is believed to have mitigated global warming somewhat.

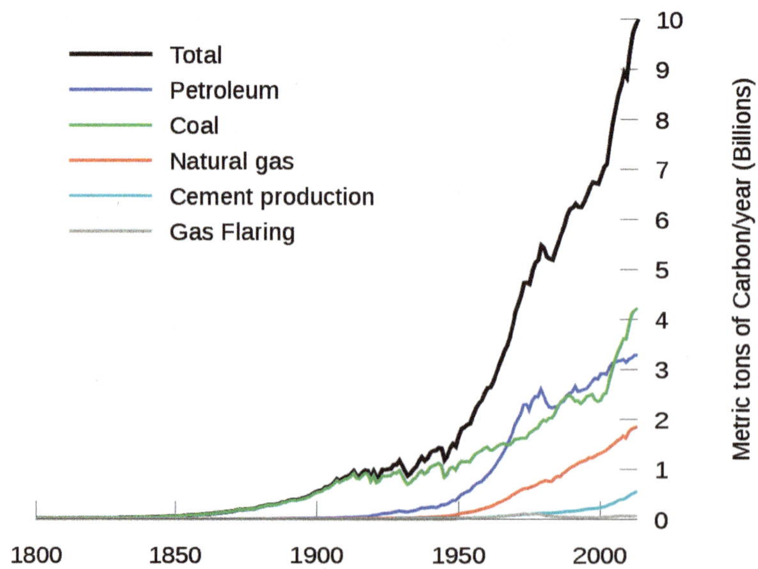

Global carbon dioxide emissions from human activities 1800–2007.

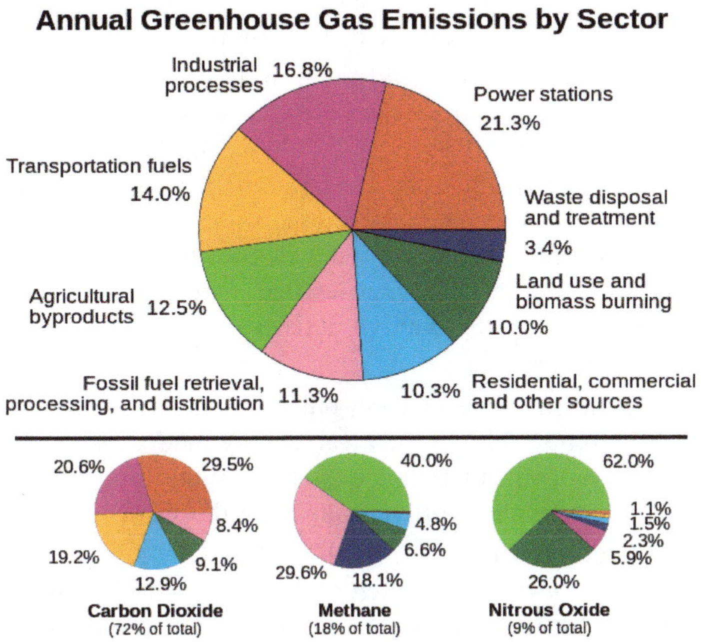

Greenhouse gas emissions by sector. See World Resources Institute for a detailed breakdown.

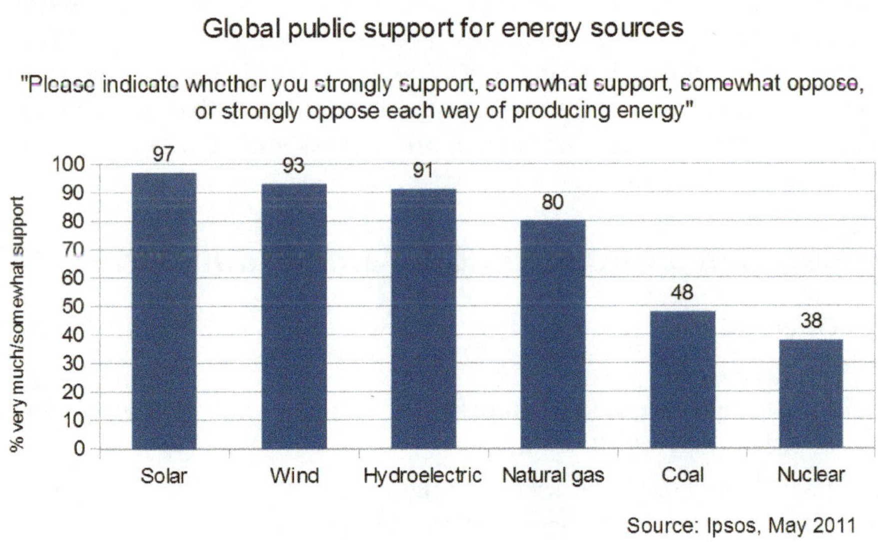

Global public support for energy sources, based on a survey by Ipsos (2011).

Climate change mitigation consists of actions to limit the magnitude or rate of long-term climate change. Climate change mitigation generally involves reductions in human (anthropogenic) emissions of greenhouse gases (GHGs).Mitigation may also be achieved by increasing the capacity of carbon sinks, e.g., through reforestation. Mitigation policies can substantially reduce the risks associated with human-induced global warming.

According to the IPCC's 2014 assessment report, "Mitigation is a public good; climate change is a case of the ‹tragedy of the commons. Effective climate change mitigation will not be achieved if each agent (individual, institution or country) acts independently in its own selfish interest, (See International Cooperation and Emissions Trading) suggesting the need for collective action. Some adaptation actions, on the other hand, have characteristics of a private good as benefits of actions may accrue more directly to the individuals, regions, or countries that undertake them, at least in the short term. Nevertheless, financing such adaptive activities remains an issue, particularly for poor individuals and countries."

Examples of mitigation include phasing out fossil fuels by switching to low-carbon energy sources, such as renewable and nuclear energy, and expanding forests and other "sinks" to remove greater amounts of carbon dioxide from the atmosphere. Energy efficiency may also play a role, for example, through improving the insulation of buildings.Another approach to climate change mitigation is climate engineering.

Most countries are parties to the United Nations Framework Convention on Climate Change (UNFCCC). The ultimate objective of the UNFCCC is to stabilize atmospheric concentrations of GHGs at a level that would prevent dangerous human interference of the climate system. Scientific analysis can provide information on the impacts of climate change, but deciding which impacts are dangerous requires value judgments.

In 2010, Parties to the UNFCCC agreed that future global warming should be limited to below 2.0 °C (3.6 °F) relative to the pre-industrial level. With the Paris Agreement of 2015 this was confirmed, but was revised with a new target laying down "parties will do the best" to achieve warming below 1.5 °C.The current trajectory of global greenhouse gas emissions does not appear to be consistent with limiting global warming to below 1.5 or 2 °C. Other mitigation policies have been proposed, some of which are more stringent or modest than the 2 °C limit.

Greenhouse gas concentrations and Stabilization

Stabilizing CO_2 emissions at their present level would not stabilize its concentration in the atmosphere.

Stabilizing the atmospheric concentration of CO_2 at a constant level would require emissions to be effectively eliminated.

One of the issues often discussed in relation to climate change mitigation is the stabilization of greenhouse gas concentrations in the atmosphere. The United Nations Framework Convention on Climate Change (UNFCCC) has the ultimate objective of preventing "dangerous" anthropogenic (i.e., human) interference of the climate system. As is stated in Article 2 of the Convention, this requires that greenhouse gas (GHG) concentrations are stabilized in the atmosphere at a level where ecosystems can adapt naturally to climate change, food production is not threatened, and economic development can proceed in a sustainable fashion. There are a number of anthropogenic greenhouse gases. These include carbon dioxide (chemical formula: CO_2), methane (CH_4), nitrous oxide (N_2O), and a group of gases referred to as halocarbons. The emissions reductions necessary to stabilize the atmospheric concentrations of these gases varies. CO_2 is the most important of the anthropogenic greenhouse gases (see radiative forcing).

There is a difference between stabilizing CO_2 emissions and stabilizing atmospheric concentrations of CO_2. Stabilizing emissions of CO_2 at current levels would not lead to a stabilization in the atmospheric concentration of CO_2. In fact, stabilizing emissions at current levels would result in the atmospheric concentration of CO_2 continuing to rise over the 21st century and beyond (see the graphs opposite).

The reason for this is that human activities are adding CO_2 to the atmosphere far faster than natural processes can remove it (see carbon dioxide in Earth's atmosphere for a complete explanation). This is analogous to a flow of water into a bathtub. So long as the tap runs water (analogous to the emission of carbon dioxide) into the tub faster than water escapes through the plughole (the natural removal of carbon dioxide from the atmosphere), then the level of water in the tub (analogous to the concentration of carbon dioxide in the atmosphere) will continue to rise.

According to some studies, stabilizing atmospheric CO_2 concentrations would require anthropogenic CO_2 emissions to be reduced by 80% relative to the peak emissions level. An 80% reduction in emissions would stabilize CO_2 concentrations for around a century, but even greater reductions would be required beyond this. Other research has found that, after leaving room for emissions for food production for 9 billion people and to keep the global temperature rise below 2 °C, emissions from energy production and transport will have to peak almost immediately in the developed world and decline at ca. 10% per annum until zero emissions are reached around 2030. In developing countries energy and transport emissions would have to peak by 2025 and then decline similarly.

Stabilizing the atmospheric concentration of the other greenhouse gasses humans emit also depends on how fast their emissions are added to the atmosphere, and how fast the GHGs are removed. Stabilization for these gases is described in the later section on non-CO_2 GHGs.

Projections

Projections of future greenhouse gas emissions are highly uncertain. In the absence of policies to mitigate climate change, GHG emissions could rise significantly over the 21st century.

Numerous assessments have considered how atmospheric GHG concentrations could be stabilized.The lower the desired stabilization level, the sooner global GHG emissions must peak and decline.GHG concentrations are unlikely to stabilize this century without major policy changcs.

Energy consumption by power source

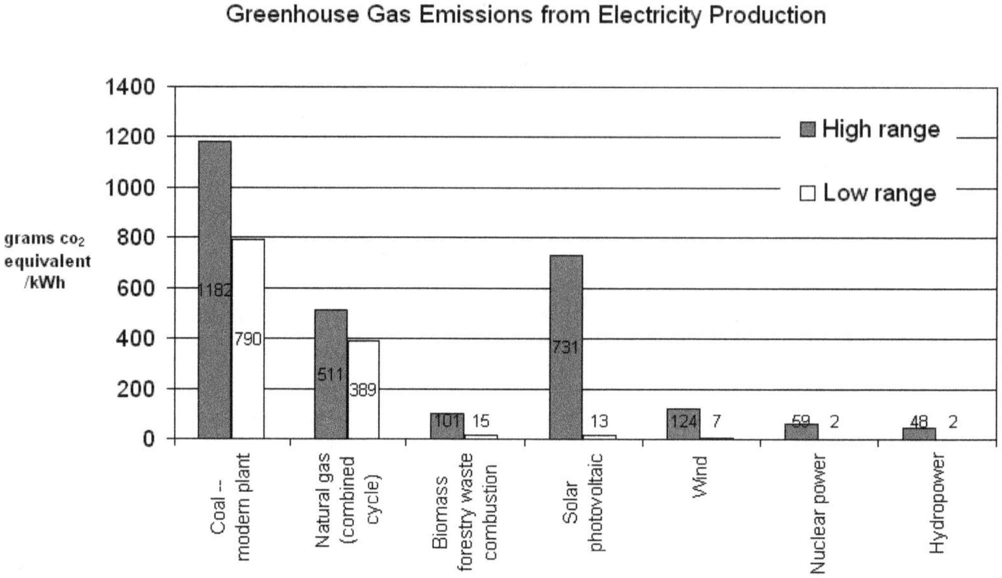

Source: IEA 2000

"Hydropower-Internalised Costs and Externalised Benefits"; Frans H. Koch; International Energy Agency (IEA)-Implementing Agreement for Hydropower Technologies and Programmes; 2000.

To create lasting climate change mitigation, the replacement of high carbon emission intensity power sources, such as conventional fossil fuels—oil, coal and natural gas—with low-carbon power sources is required. Fossil fuels supply humanity with the vast majority of our energy demands, and at a growing rate. In 2012 the IEA noted that coal accounted for half the increased energy use of the prior decade, growing faster than all renewable energy sources. Both hydroelectricity and nuclear power together provide the majority of the generated low-carbon power fraction of global total power consumption.

Fuel type	**Average total global power consumption in TW**		
	1980	**2004**	**2006**
Oil	4.38	5.58	5.74
Gas	1.80	3.45	3.61
Coal	2.34	3.87	4.27
Hydroelectric	0.60	0.93	1.00
Nuclear power	0.25	0.91	0.93
Geothermal, wind, solar energy, wood	0.02	0.13	0.16
Total	**9.48**	**15.0**	**15.8**
Source: The USA Energy Information Administration			

Change and use of energy, by source, in units of (PWh) in that year.				
	Fossil	**Nuclear**	**All renewables**	**Total**
1990	83.374	6.113	13.082	102.569
2000	94.493	7.857	15.337	117.687
2008	117.076	8.283	18.492	143.851
Change 2000–2008	22.583	0.426	3.155	26.164

Methods and Means

This graph shows the projected contribution of various energy sources to world primary electricity consumption (PEC). It is based on a climate change mitigation scenario, in which GHG emissions are substantially reduced over the 21st century. In the scenario, emission reductions are achieved using a portfolio of energy sources, as well as reductions in energy demand. *Also available in greyscale.*

Assessments often suggest that GHG emissions can be reduced using a portfolio of low-carbon technologies. At the core of most proposals is the reduction of greenhouse gas (GHG) emissions through reducing energy waste and switching to low-carbon power sources of energy. As the cost of reducing GHG emissions in the electricity sector appears to be lower than in other sectors, such as in the transportation sector, the electricity sector may deliver the largest proportional carbon reductions under an economically efficient climate policy.

"Economic tools can be useful in designing climate change mitigation policies." "While the limitations of economics and social welfare analysis, including cost–benefit analysis, are widely documented, economics nevertheless provides useful tools for assessing the pros and cons of taking, or not taking, action on climate change mitigation, as well as of adaptation measures, in achieving competing societal goals. Understanding these pros and cons can help in making policy decisions on climate change mitigation and can influence the actions taken by countries, institutions and individuals."

Other frequently discussed means include energy conservation, increasing fuel economy in automobiles (which includes the use of electric hybrids), charging plug-in hybrids and electric cars by low-carbon electricity, making individual-lifestyle changes (e.g., cycling instead of driving), and changing business practices. Many fossil fuel driven vehicles can be converted to use electricity, the U.S. has an estimated capacity of supporting 73% light duty vehicles (LDV). In terms of transportation, the net result would be a 27% total reduction in emissions of the greenhouse gases carbon dioxide, methane, and nitrous oxide, a 31% total reduction in nitrogen oxides, a slight reduction in nitrous oxide emissions, an increase in particulate matter emissions, the same sulfur dioxide emissions, and the near elimination of carbon monoxide and volatile organic compound emissions (a 98% decrease in carbon monoxide and a 93% decrease in volatile organic compounds). The emissions would be displaced away from street level, where they have "high human-health implications."

A range of energy technologies may contribute to climate change mitigation. These include nuclear power and renewable energy sources such as biomass, hydroelectricity, wind power, solar power, geothermal power, ocean energy, and; the use of carbon sinks, and carbon capture and storage. For example, Pacala and Socolow of Princeton have proposed a 15 part program to reduce CO_2emissions by 1 billion metric tons per year – or 25 billion tons over the 50-year period using today's technologies as a type of Global warming game.

Another consideration is how future socio-economic development proceeds. Development choices (or "pathways") can lead differences in GHG emissions. Political and social attitudes may affect how easy or difficult it is to implement effective policies to reduce emissions.

Alternative Energy Sources

Renewable Energy

The 22,500 MW nameplate capacityThree Gorges Dam in the Peoples Republic of China, the largest hydroelectric power station in the world.

The Shepherds Flat Wind Farm is an 845 megawatt (MW) nameplate capacity, wind farm in the U.S. state of Oregon, each turbine is a nameplate 2 or 2.5 MW electricity generator.

The 150 MW Andasol solar power station is a commercial parabolic trough solar thermal power plant, located in Spain. The Andasol plant uses tanks of molten salt to store solar energy so that it can continue generating electricity for 7.5 hours after the sun has stopped shining.

Solar cookers use sunlight as energy source for outdoor cooking.

Renewable energy flows involve natural phenomena such as sunlight, wind, rain, tides, plant growth, and geothermal heat, as the International Energy Agency explains:

Renewable energy is derived from natural processes that are replenished constantly. In its various forms, it derives directly from the sun, or from heat generated deep within the earth. Included in the definition is electricity and heat generated from solar, wind, ocean, hydropower, biomass, geothermal resources, and biofuels and hydrogen derived from renewable resources.

Climate change concerns and the need to reduce carbon emissions are driving increasing growth in the renewable energy industries. Low-carbon renewable energy replaces conventional fossil fuels in three main areas: power generation, hot water/ space heating, and transport fuels. In 2011, the share of renewables in electricity generation worldwide grew for the fourth year in a row to 20.2%. Based on REN21's 2014 report, renewables contributed 19% to supply global energy consumption. This energy consumption is divided as 9% coming from burning biomass, 4.2% as heat energy (non-biomass), 3.8% hydro electricity and 2% as electricity from wind, solar, geothermal, and biomass thermal power plants.

Renewable energy use has grown much faster than anyone anticipated. The Intergovernmental Panel on Climate Change (IPCC) has said that there are few fundamental technological limits to integrating a portfolio of renewable energy technologies to meet most of total global energy demand. At the national level, at least 30 nations around the world already have renewable energy contributing more than 20% of energy supply.

As of 2012, renewable energy accounts for almost half of new electricity capacity installed and costs are continuing to fall.Public policy and political leadership helps to "level the playing field" and drive the wider acceptance of renewable energy technologies. As of 2011, 118 countries have targets for their own renewable energy futures, and have enacted wide-ranging public policies to promote renewables. Leading renewable energy companies include BrightSource Energy, First Solar, Gamesa, GE Energy, Goldwind, Sinovel, Suntech, Trina Solar, Vestas and Yingli.

The incentive to use 100% renewable energy has been created by global warming and other ecological as well as economic concerns. Mark Z. Jacobson says producing all new energy with wind power, solar power, and hydropower by 2030 is feasible and existing energy supply arrangements could be replaced by 2050. Barriers to implementing the renewable energy plan are seen to be "primarily social and political, not technological or economic". Jacobson says that energy costs with a wind, solar, water system should be similar to today's energy costs. According to a 2011 projection by the (IEA)International Energy Agency, solar power generators may produce most of the world's electricity within 50 years, dramatically reducing harmful greenhouse gas emissions. Critics of the "100% renewable energy" approach include Vaclav Smil and James E. Hansen. Smil and Hansen are concerned about the variable output of solar and wind power, NIMBYism, and a lack of infrastructure.

Economic analysts expect market gains for renewable energy (and efficient energy use) following the 2011 Japanese nuclear accidents. In his 2012 State of the Union address, President Barack Obama restated his commitment to renewable energy and mentioned the long-standing Interior Department commitment to permit 10,000 MW of renewable energy projects on public land in 2012.Globally, there are an estimated 3 million direct jobs in renewable energy industries, with about half of them in the biofuels industry.

Some countries, with favorable geography, geology and weather well suited to an economical exploitation of renewable energy sources, already get most of their electricity from renewables, including from geothermal energy in Iceland (100 percent), and Hydroelectric power in Brazil (85 percent), Austria (62 percent), New Zealand (65 percent), and Sweden (54 percent). Renewable power generators are spread across many countries, with wind power providing a significant share of electricity in some regional areas: for example, 14 percent in the U.S. state of Iowa, 40 percent in the northern German state of Schleswig-Holstein, and 20 percent in Denmark. Solar water heating makes an important and growing contribution in many countries, most notably in China, which now has 70 percent of the global total (180 GWth). Worldwide, total installed solar water heating systems meet a portion of the water heating needs of over 70 million households. The use of biomass for heating continues to grow as well. In Sweden, national use of biomass energy has surpassed that of oil. Direct geothermal heating is also growing rapidly. Renewable biofuels for transportation, such as ethanol fuel and biodiesel, have contributed to a significant decline in oil consumption in the United States since 2006. The 93 billion liters of biofuels produced worldwide in 2009 displaced the equivalent of an estimated 68 billion liters of gasoline, equal to about 5 percent of world gasoline production.

Some of the world's largest solar power stations: Ivanpah (CSP) and Topaz (PV)

Nuclear Power

Since about 2001 the term "nuclear renaissance" has been used to refer to a possible nuclear power industry revival, driven by rising fossil fuel prices and new concerns about meeting greenhouse gas emission limits. However, in March 2011 the Fukushima nuclear disaster in Japan and associated shutdowns at other nuclear facilities raised questions among some commentators over the future of nuclear power. Platts has reported that "the crisis at Japan's Fukushima nuclear plants has prompted leading energy-consuming countries to review the safety of their existing reactors and cast doubt on the speed and scale of planned expansions around the world".

Blue Cherenkov light being produced near the core of the Fission powered Advanced Test Reactor.

The World Nuclear Association has reported that nuclear electricity generation in 2012 was at its lowest level since 1999. Several previous international studies and assessments,suggested that as part of the portfolio of other low-carbon energy technologies, nuclear power will continue to play a role in reducing greenhouse gas emissions. Historically, nuclear power usage is estimated to have prevented the atmospheric emission of 64 gigatonnes of CO2-equivalent as of 2013. Public concerns about nuclear power include the fate of spent nuclear fuel, nuclear accidents, security risks, nuclear proliferation, and a concern that nuclear power plants are very expensive. Of these concerns, nuclear accidents and disposal of long-lived radioactive fuel/"waste" have probably had the greatest public impact worldwide. Although generally unaware of it, both of these glaring public concerns are greatly diminished by present passive safety designs, the experimentally proven, "melt-down proof" EBR-II, future molten salt reactors, and the use of conventional and more advanced fuel/"waste" pyroprocessing, with the latter recycling or reprocessing not presently being commonplace as it is often considered to be cheaper to use a once-through nuclear fuel cycle in many countries, depending on the varying levels of intrinsic value given by a society in reducing the long-lived waste in their country, with France doing a considerable amount of reprocessing when compared to the US.

Nuclear power, with a 10.6% share of world electricity production as of 2013, is second only to hydroelectricity as the largest source of low-carbon power. Over 400 reactors generate electricity in 31 countries.

A Yale University review published in the Journal of Industrial Ecology analyzing CO_2 life cycle assessment(LCA) emissions from nuclear power(Light water reactors) determined that: "The collective LCA literature indicates that life cycle GHG emissions from nuclear power are only a fraction of traditional fossil sources and comparable to renewable technologies." While some have raised uncertainty surrounding the future GHG emissions of nuclear power as a result of an extreme potential decline in uranium ore grade without a corresponding increase in the efficiency of enrichment methods. In a scenario analysis of future global nuclear development, as it could be effected by a decreasing global uranium market of average ore grade, the analysis determined that depending on conditions, median life cycle nuclear power GHG emissions could be between 9 and 110 g CO_2-eq/kWh by 2050, with the latter high figure being derived from a "worst-case scenario" that is not "considered very robust" by the authors of the paper, as the "ore grade" in the scenario is lower than the uranium concentration in many lignite coal ashes.

Although this future analyses primarily deals with extrapolations for present Generation II reactor technology, the same paper also summarizes the literature on "FBRs"/Fast Breeder Reactors, of which two are in operation as of 2014 with the newest being the BN-800, for these reactors it states that the "median life cycle GHG emissions ... [are] similar to or lower than [present light water reactors] LWRs and purports to consume little or no uranium ore.

In their 2014 report, the IPCC comparison of energy sources global warming potential per unit of electricity generated, which notably included albedo effects, mirror the median emission value derived from the Warner and Heath Yale meta-analysis for the more common non-breeding Light water reactors, a CO2-equivalent value of 12 g CO2-eq/kWh, which is the lowest global warming forcing of all baseload power sources, with comparable low carbon power baseload sources, such as hydropower and biomass, producing substantially more global warming forcing 24 and 230 g CO2-eq/kWh respectively.

In 2014, Brookings Institution published *The Net Benefits of Low and No-Carbon Electricity Technologies* which states, after performing an energy and emissions cost analysis, that "The net benefits of new nuclear, hydro, and natural gas combined cycle plants far outweigh the net benefits of new wind or solar plants", with the most cost effective low carbon power technology being determined to be nuclear power.

During his presidential campaign, Barack Obama stated, "Nuclear power represents more than 70% of our noncarbon generated electricity. It is unlikely that we can meet our aggressive climate goals if we eliminate nuclear power as an option."

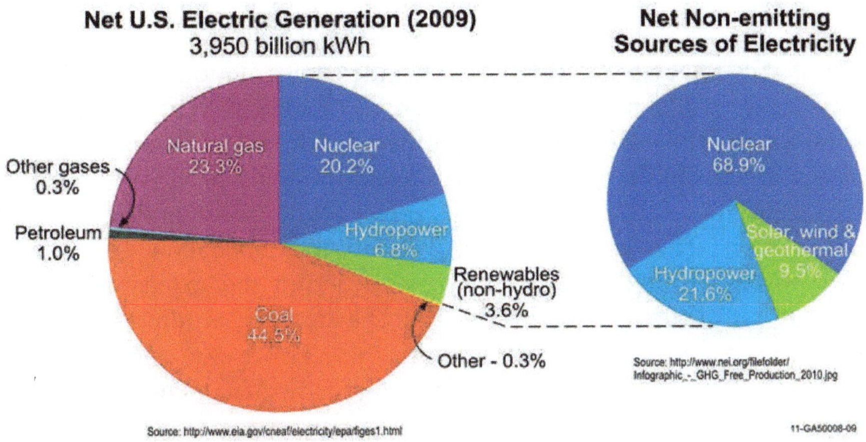

This graph illustrates nuclear power is the United States's largest contributor of non-greenhouse-gas-emitting electric power generation, comprising nearly three-quarters of the non-emitting sources.

Analysis in 2015 by Professor and Chair of Environmental Sustainability Barry W. Brook and his colleagues on the topic of replacing fossil fuels entirely, from the electric grid of the world, has determined that at the historically modest and proven-rate at which nuclear energy was added to and replaced fossil fuels in France and Sweden during each nation's building programs in the 1980s, within 10 years nuclear energy could displace or remove fossil fuels from the electric grid completely, "allow[ing] the world to meet the most stringent greenhouse-gas mitigation targets.". In a similar analysis, Brook had earlier determined that 50% of all global energy, that is not solely electricity, but transportation synfuels etc. could be generated within approximately 30 years, if the global nuclear fission build rate was identical to each of these nation's already proven decadal rates(in units of installed nameplate capacity, GW per year, per unit of global GDP(GW/year/$).

This is in contrast to the completely conceptual paper-studies for a *100% renewable energy* world, which would require an orders of magnitude more costly global investment per year, an investment rate that has no historical precedent, having never been attempted due to its prohibitive cost, and with far greater land area that would be required to be devoted to the wind, wave and solar projects, along with the inherent assumption that humanity will use less, and not more, energy in the future. As Brook notes the "principal limitations on nuclear fission are not technical, economic or fuel-related, but are instead linked to complex issues of societal acceptance, fiscal and political inertia, and inadequate critical evaluation of the real-world constraints facing [the other] low-carbon alternatives."

Nuclear power may be uncompetitive compared with fossil fuel energy sources in countries without a carbon tax program, and in comparison to a fossil fuel plant of the same power output, nuclear power plants take a longer amount of time to construct.

Two new, first of their kind, EPR reactors under construction in Finland and France have been delayed and are running over-budget. However learning from experience, two further EPR reactors under construction in China are on, and ahead, of schedule respectively. As of 2013, according to the IAEA and the European Nuclear Society, worldwide there were 68 civil nuclear power reactors under construction in 15 countries. China has 29 of these nuclear power reactors under construction, as of 2013, with plans to build many more, while in the US the licenses of almost half its reactors have been extended to 60 years, and plans to build another dozen are under serious consideration. There are also a considerable number of new reactors being built in South Korea, India, and Russia. At least 100 older and smaller reactors will "most probably be closed over the next 10–15 years". This is probable only if one does not factor in the ongoing Light Water Reactor Sustainability Program, created to permit the extension of the life span of the USA's 104 nuclear reactors to 60 years. The licenses of almost half of the USA's reactors have been extended to 60 years as of 2008. Two new "passive safety" AP1000 reactors are, as of 2013, being constructed at Vogtle Electric Generating Plant.

Public opinion about nuclear power varies widely between countries. A poll by Gallup International (2011) assessed public opinion in 47 countries. The poll was conducted following a tsunami and earthquake which caused an accident at the Fukushima nuclear power plant in Japan. 49% stated that they held favourable views about nuclear energy, while 43% held an unfavourable view.Another global survey by Ipsos (2011) assessed public opinion in 24 countries. Respondents to this survey showed a clear preference for renewable energy sources over coal and nuclear energy (refer to graph opposite). Ipsos (2012) found that solar and wind were viewed by the public as being more environmentally friendly and more viable long-term energy sources relative to nuclear power and natural gas. However, solar and wind were viewed as being less reliable relative to nuclear power and natural gas. In 2012 a poll done in the UK found that 63% of those surveyed support nuclear power, and with opposition to nuclear power at 11%. In Germany, strong anti-nuclear sentiment led to eight of the seventeen operating reactors being permanently shut down following the March 2011 Fukushima nuclear disaster.

Nuclear fusion research, in the form of the International Thermonuclear Experimental Reactor is underway. Fusion powered electricity generation was initially believed to be readily achievable, as fission power had been. However, the extreme requirements for continuous reactions and plasma containment led to projections being extended by several decades. In 2010, more than 60 years after the first attempts, commercial power production was still believed to be unlikely before 2050. Although rather than an either, or, issue economical fusion-fission hybrid reactors could be built before any attempt at this more demanding commercial "pure-fusion reactor"/DEMO reactor takes place.

Coal to Gas Fuel Switching

Most mitigation proposals imply—rather than directly state—an eventual reduction in global fossil fuel production. Also proposed are direct quotas on global fossil fuel production.

Natural gas emits far fewer greenhouse gases (i.e. CO_2 and methane—CH_4) than coal when burned at power plants, but evidence has been emerging that this benefit could be completely negated by methane leakage at gas drilling fields and other points in the supply chain.

A study performed by the Environmental Protection Agency (EPA) and the Gas Research Institute (GRI) in 1997 sought to discover whether the reduction in carbon dioxide emissions from increased natural gas (predominantly methane) use would be offset by a possible increased level of methane emissions from sources such as leaks and emissions. The study concluded that the reduction in emissions from increased natural gas use outweighs the detrimental effects of increased methane emissions. More recent peer-reviewed studies have challenged the findings of this study, with researchers from the National Oceanic and Atmospheric Administration (NOAA) reconfirming findings of high rates of methane (CH4) leakage from natural gas fields.

A 2011 study by noted climate research scientist, Tom Wigley, found that while carbon dioxide (CO_2) emissions from fossil fuel combustion may be reduced by using natural gas rather than coal to produce energy, it also found that additional methane (CH4) from leakage adds to the radiative forcing of the climate system, offsetting the reduction in CO_2 forcing that accompanies the transition from coal to gas. The study looked at methane leakage from coal mining; changes in radiative forcing due to changes in the emissions of sulfur dioxide and carbonaceous aerosols; and differences in the efficiency of electricity production between coal- and gas-fired power generation. On balance, these factors more than offset the reduction in warming due to reduced CO_2 emissions. When gas replaces coal there is additional warming out to 2,050 with an assumed leakage rate of 0%, and out to 2,140 if the leakage rate is as high as 10%. The overall effects on global-mean temperature over the 21st century, however, are small. Petron et al. (2013) and Alvarez et al. (2012) note that estimated that leakage from gas infrastructure is likely to be underestimated. These studies indicate that the exploitation of natural gas as a "cleaner" fuel is questionable. A 2014 meta-study of 20 years of natural gas technical literature shows that methane emissions are consistently underestimated but on a 100-year scale, the climate benefits of coal to gas fuel switching are likely larger than the negative effects of natural gas leakage.

Heat Pump

Outside unit of an air-source heat pump.

A heat pump is a device that provides heat energy from a source of heat to a destination called a "heat sink". Heat pumps are designed to move thermal energy opposite to the direction of spontaneous heat flow by absorbing heat from a cold space and releasing it to a warmer one. A heat pump uses some amount of external power to accomplish the work of transferring energy from the heat source to the heat sink.

While air conditioners and freezers are familiar examples of heat pumps, the term "heat pump" is more general and applies to many HVAC (heating, ventilating, and air conditioning) devices used for space heating or space cooling. When a heat pump is used for heating, it employs the same basic refrigeration-type cycle used by an air conditioner or a refrigerator, but in the opposite direction—releasing heat into the conditioned space rather than the surrounding environment. In this use, heat pumps generally draw heat from the cooler external air or from the ground. In heating mode, heat pumps are three to four times more efficient in their use of electric power than simple electrical resistance heaters.

It has been concluded that heat pumps are the single technology that could reduce the greenhouse gas emissions of households better than every other technology that is available on the market. With a market share of 30% and (potentially) clean electricity, heat pumps could reduce global CO2 emissions by 8% annually. Using ground source heat pumps could reduce around 60% of the primary energy demand and 90% of CO2 emissions in Europe in 2050 and make handling high shares of renewable energy easier. Using surplus renewable energy in heat pumps is regarded as the most effective household means to reduce global warming and fossil fuel depletion.

With significant amounts of fossil fuel used in electricity production, demands on the electrical grid also generate greenhouse gases. Without a high share of low-carbon electricity, a domestic heat pump will produce more carbon emissions than using natural gas.

Fossil fuel Phase-out: Carbon Neutral and Negative Fuels

3,500-4,000 environmental activists blocking a coal mine in Germany to limit climate change (Ende Gelände 2016).

Fossil fuel may be phased-out with carbon neutral and carbon negative pipeline and transportation fuels created with power to gas and gas to liquids technologies. Carbon dioxide from fossil fuel flue gas can be used to produce plastic lumber allowing carbon negative reforestation.

Demand Side Management

Energy efficiency and Conservation

A spiral-type integrated compact fluorescent lamp, use has grown among North American consumers since its introduction in the mid-1990s.

Efficient energy use, sometimes simply called "energy efficiency", is the goal of efforts to reduce the amount of energy required to provide products and services. For example, insulating a home allows a building to use less heating and cooling energy to achieve and maintain a comfortable temperature. Installing fluorescent lights or natural skylights reduces the amount of energy required to attain the same level of illumination compared to using traditional incandescent light bulbs. Compact fluorescent lights use two-thirds less energy and may last 6 to 10 times longer than incandescent lights.

Energy efficiency has proved to be a cost-effective strategy for building economies without necessarily growing energy consumption. For example, the state of California began implementing energy-efficiency measures in the mid-1970s, including building code and appliance standards with strict efficiency requirements. During the following years, California's energy consumption has remained approximately flat on a per capita basis while national U.S. consumption doubled. As part of its strategy, California implemented a "loading order" for new energy resources that puts energy efficiency first, renewable electricity supplies second, and new fossil-fired power plants last.

Energy conservation is broader than energy efficiency in that it encompasses using less energy to achieve a lesser energy service, for example through behavioural change, as well as encompassing energy efficiency. Examples of conservation without efficiency improvements would be heating a room less in winter, driving less, or working in a less brightly lit room. As with other definitions, the boundary between efficient energy use and energy conservation can be fuzzy, but both are important in environmental and economic terms. This is especially the case when actions are directed at the saving of fossil fuels.

Reducing energy use is seen as a key solution to the problem of reducing greenhouse gas emissions. According to the International Energy Agency, improved energy efficiency in buildings, industrial processes and transportation could reduce the world's energy needs in 2050 by one third, and help control global emissions of greenhouse gases.

Demand Side Switching Sources

Fuel switching on the demand side refers to changing the type of fuel used to satisfy a need for an energy service. To meet deep decarbonization goals, like the 80% reduction by 2050 goal being discussed in California and the European Union, many primary energy changes are needed. Energy efficiency alone may not be sufficient to meet these goals, switching fuels used on the demand side will help lower carbon emissions. Progressively coal, oil and eventually natural gas for space and water heating in buildings will need to be reduced. For an equivalent amount of heat, burning natural gas produces about 45 per cent less carbon dioxide than burning coal. There are various ways in which this could happen, and different strategies will likely make sense in different locations. While the system efficiency of a gas furnace may be higher than the combination of natural gas power plant and electric heat, the combination of the same natural gas power plant and an electric heat pump has lower emissions per unit of heat delivered in all but the coldest climates. This is possible because of the very efficient coefficient of performance of heat pumps.

At the beginning of this century 70% of all electricity was generated by fossil fuels, and as carbon free sources eventually make up half of the generation mix, replacing gas or oil furnaces and water heaters with electric ones will have a climate benefit. In areas like Norway, Brazil and Quebec that have abundant hydroelectricity, electric heat and hot water is common.

The economics of switching the demand side from fossil fuels to electricity for heating, will depend on the price of fuels vs electricity and the relative prices of the equipment. The EIA Annual Energy Outlook 2014 suggests that domestic gas prices will rise faster than electricity prices which will encourage electrification in the coming decades.

Electrifying heating loads may also provide a flexible resource that can participate in demand response. Since thermostatically controlled loads have inherent energy storage, electrification of heating could provide a valuable resource to integrate variable renewable resources into the grid.

Alternatives to electrification, include decarbonizing pipeline gas through power to gas, biogas, or other carbon neutral fuels. A 2015 study by Energy+Environmental Economics shows that a hybrid approach of decarbonizing pipeline gas, electrification, and energy efficiency can meet carbon reduction goals at a similar cost as only electrification and energy efficiency in Southern California.

Demand Side Grid Management

Expanding intermittent electrical sources such as wind power, creates a growing problem balancing grid fluctuations. Some of the plans include building pumped storage or continental super grids costing billions of dollars. However instead of building for more power,there are a variety of ways to affect the size and timing of electricity demand on the consumer side. Designing for reduced demands on a smaller power grid is more efficient and economic than having extra generation and transmission for intermittentcy, power failures and peak demands. Having these abilities is one of the chief aims of a smart grid.

Time of use metering is a common way to motivate electricity users to reduce their peak load consumption. For instance, running dishwashers and laundry at night after the peak has passed, reduces electricity costs.

Dynamic demand plans have devices passively shut off when stress is sensed on the electrical grid. This method may work very well with thermostats, when power on the grid sags a small amount, a low power temperature setting is automatically selected reducing the load on the grid. For instance millions of refrigerators reduce their consumption when clouds pass over solar installations. Consumers would need to have a smart meter in order for the utility to calculate credits. Demand response devices could receive all sorts of messages from the grid. The message could be a request to use a low power mode similar to dynamic demand, to shut off entirely during a sudden failure on the grid, or notifications about the current and expected prices for power. This would allow electric cars to recharge at the least expensive rates independent of the time of day. The vehicle-to-grid suggestion would use a car's battery or fuel cell to supply the grid temporarily.

Lifestyle and Behavior

The IPCC Fifth Assessment Report emphasises that behaviour, lifestyle and cultural change have a high mitigation potential in some sectors, particularly when complementing technological and structural change. In general, higher consumption lifestyles have a greater environmental impact. Overall, food accounts for the largest share of consumption-based GHG emissions with nearly 20% of the global carbon footprint, followed by housing, mobility, services, manufactured products, and construction. Food and services are more significant in poor countries, while mobility and manufactured goods are more significant in rich countries.

Dietary Change

A 2014 study into the real-life diets of British people estimates their greenhouse gas contributions (CO_2eq) to be: 7.19 kg/day for high meat-eaters through to 3.81 kg/day for vegetarians and 2.89 kg/day for vegans. The widespread adoption of a vegetarian diet could cut food-related greenhouse gas emissions by 63% by 2050. China introduced new dietary guidelines in 2016 which aim to cut meat consumption by 50% and thereby reduce greenhouse gas emissions by 1 billion tonnes by 2030. A 2016 study concluded that taxes on meat and milk could simultaneously result in reduced greenhouse gas emissions and healthier diets. The study analyzed surcharges of 40% on beef and 20% on milk and suggests that an optimum plan would reduce emissions by 1 billion tonnes per year.

Sinks and Negative Emissions

A carbon sink is a natural or artificial reservoir that accumulates and stores some carbon-containing chemical compound for an indefinite period, such as a growing forest. A negative carbon dioxide emission on the other hand is a permanent removal of carbon dioxide out of the atmosphere, such as directly capturing carbon dioxide in the atmosphere and storing it in geologic formations underground.

The Antarctic Climate and Ecosystems Cooperative Research Centre (ACE-CRC) notes that one third of humankind's annual emissions of CO_2 are absorbed by the oceans. However, this also leads to ocean acidification, with potentially significant impacts on marine life. Acidification lowers the level of carbonate ions available for calcifying organisms to form their shells. These organisms include plankton species that contribute to the foundation of the Southern Ocean food web. However acidification may impact on a broad range of other physiological and ecological processes, such as fish respiration, larval development and changes in the solubility of both nutrients and toxins.

Reforestation and Afforestation

Transferring land rights to indigenous inhabitants is argued to efficiently conserve forests. Regrowth of forests on abandoned farmland restores more forest than that lost to deforestation.

Almost 20 percent (8 $GtCO_2$/year) of total greenhouse-gas emissions were from deforestation in 2007. It is estimated that avoided deforestation reduces CO_2 emissions at a rate of 1 tonne of CO_2 per $1–5 in opportunity costs from lost agriculture. Reforestation and afforestation, where there was previously no forest, could save at least another 1 $GtCO_2$/year, at an estimated cost of $5–15/$tCO_2$.

Transferring rights over land from public domain to its indigenous inhabitants is argued to be a cost effective strategy to conserve forests. This includes the protection of such rights entitled in existing laws, such as India's Forest Rights Act. The transferring of such rights in China, perhaps the largest land reform in modern times, has been argued to have increased forest cover. In Brazil, forested areas given tenure to indigenous groups have even lower rates of clearing than national parks. A 2016 report concludes that modest investments in indigenous land rights will generate economic, social, and environmental returns for the communities involved and for climate protection. The report quantifies the economic value of securing such rights, with a focus on the Amazon region.

With increased intensive agriculture and urbanization, there is an increase in the amount of abandoned farmland. By some estimates, for every half a hectare of original old-growth forest cut down, more than 20 hectares of new secondary forests are growing, even though they do not have the same biodiversity as the original forests and original forests store 60% more carbon than these new secondary forests. According to a study in *Science*, promoting regrowth on abandoned farmland could offset years of carbon emissions.

Avoided Desertification

Managed grazing methods are argued to be able to restore grasslands, thereby significantly decreasing atmospheric CO_2levels.

Restoring grasslands store CO_2 from the air into plant material. Grazing livestock, usually not left to wander, would eat the grass and would minimize any grass growth.

However, grass left alone would eventually grow to cover its own growing buds, preventing them from photosynthesizing and the dying plant would stay in place. A method proposed to restore grasslands uses fences with many small paddocks and moving herds from one paddock to another after a day a two in order to mimick natural grazers and allowing the grass to grow optimally.Additionally, when part of leaf matter is consumed by a herding animal, a corresponding amount of root matter is sloughed off too as it would not be able to sustain the previous amount of root matter and while most of the lost root matter would rot and enter the atmosphere, part of the carbon is sequestered into the soil. It is estimated that increasing the carbon content of the soils in the world's 3.5 billion hectares of agricultural grassland by 1% would offset nearly 12 years of CO_2emissions. Allan Savory, as part of holistic management, claims that while large herds are often blamed for desertification, prehistoric lands supported large or larger herds and areas where herds were removed in the United States are still desertifying.

Carbon Capture and Storage

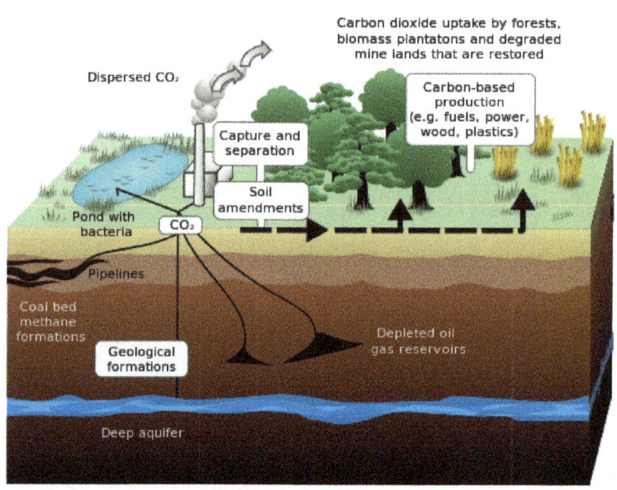

Schematic showing both terrestrial and geological sequestration of carbon dioxide emissions from a coal-fired plant.

Carbon capture and storage (CCS) is a method to mitigate climate change by capturing carbon dioxide (CO_2) from large point sources such as power plants and subsequently storing it away safely instead of releasing it into the atmosphere. The Intergovernmental Panel on Climate Change says CCS could contribute between 10% and 55% of the cumulative worldwide carbon-mitigation effort over the next 90 years. The International Energy Agency says CCS is "the most important single new technology for CO_2 savings" in power generation and industry. Though it requires up to 40% more energy to run a CCS coal power plant than a regular coal plant, CCS could potentially capture about 90% of all the carbon emitted by the plant. Norway, which first began storing CO_2, has cut its emissions by almost a million tons a year, or about 3% of the country's 1990 levels. As of late 2011, the total CO_2 storage capacity of all 14 projects in operation or under construction is over 33 million tonnes a year. This is broadly equivalent to preventing the emissions from more than six million cars from entering the atmosphere each year.

Negative Carbon Dioxide Emissions

Creating negative carbon dioxide emissions literally removes carbon from the atmosphere. Examples are direct air capture, biochar, bio-energy with carbon capture and storage and enhanced weathering technologies. These processes are sometimes considered as variations of sinks or mitigation,and sometimes as geoengineering.

In combination with other mitigation measures, sinks in combination with negative carbon emissions are considered crucial for meeting the 350 ppm target, and even the less conservative 450 ppm target.

Geoengineering

Geoengineering is seen by some as an alternative to mitigation and adaptation, but by others as an entirely separate response to climate change. In a literature assessment, Barker *et al.* (2007) described geoengineering as a type of mitigation policy. IPCC (2007) concluded that geoengineering options, such as ocean fertilization to remove CO_2 from the atmosphere, remained largely unproven.It was judged that reliable cost estimates for geoengineering had not yet been published.

Chapter 28 of the National Academy of Sciences report *Policy Implications of Greenhouse Warming: Mitigation, Adaptation, and the Science Base* (1992) defined geoengineering as "options that would involve large-scale engineering of our environment in order to combat or counteract the effects of changes in atmospheric chemistry." They evaluated a range of options to try to give preliminary answers to two questions: can these options work and could they be carried out with a reasonable cost. They also sought to encourage discussion of a third question — what adverse side effects might there be. The following types of option were examined: reforestation, increasing ocean absorption of carbon dioxide (carbon sequestration) and screening out some sunlight. NAS also argued "Engineered countermeasures need to be evaluated but should not be implemented without broad understanding of the direct effects and the potential side effects, the ethical issues, and the risks.». In July 2011 a report by the United States Government Accountability Office on geoengineering found that "[c]limate engineering technologies do not now offer a viable response to global climate change."

Carbon Dioxide Removal

Carbon dioxide removal has been proposed as a method of reducing the amount of radiative forcing. A variety of means of artificially capturing and storing carbon, as well as of enhancing natural sequestration processes, are being explored. The main natural process is photosynthesis by plants and single-celled organisms (see biosequestration). Artificial processes vary, and concerns have been expressed about the long-term effects of some of these processes. It is notable that the availability of cheap energy and appropriate sites for geological storage of carbon may make carbon dioxide air capture viable commercially. It is, however, generally expected that carbon dioxide air capture may be uneconomic when compared to carbon capture and storage from major sources — in particular, fossil fuel powered power stations, refineries, etc. In such cases, costs of energy produced will grow significantly. However, captured CO_2 can be used to force more crude oil out of oil fields, as Statoil and Shell have made plans to do. CO_2 can also be used in commercial greenhouses, giving an opportunity to kick-start the technology. Some attempts have been made to use algae to capture smokestack emissions, notably the GreenFuel Technologies Corporation, who have now shut down operations.

Solar Radiation Management

See also: Stratospheric sulfate aerosols (geoengineering)

The main purpose of solar radiation management seek to reflect sunlight and thus reduce global warming. The ability of stratospheric sulfate aerosols to create a global dimming effect has made them a possible candidate for use in climate engineering projects.

Non-CO_2 Greenhouse Gases

CO_2 is not the only GHG relevant to mitigation, and governments have acted to regulate the emissions of other GHGs emitted by human activities (anthropogenic GHGs). The emissions caps agreed to by most developed countries under the Kyoto Protocol regulate the emissions of almost all the anthropogenic GHGs. These gases are CO_2, methane (CH_4), nitrous oxide (N_2O), the hydrofluorocarbons (HFC), perfluorocarbons (PFC), and sulfur hexafluoride (SF_6).

Stabilizing the atmospheric concentrations of the different anthropogenic GHGs requires an understanding of their different physical properties. Stabilization depends both on how quickly GHGs are added to the atmosphere and how fast they are removed. The rate of removal is measured by the atmospheric lifetime of the GHG in question (see the main GHG article for a list). Here, the lifetime is defined as the time required for a given perturbation of the GHG in the atmosphere to be reduced to 37% of its initial amount. Methane has a relatively short atmospheric lifetime of about 12 years, while N_2O's lifetime is about 110 years. For methane, a reduction of about 30% below current emission levels would lead to a stabilization in its atmospheric concentration, while for N_2O, an emissions reduction of more than 50% would be required.

Methane is a significantly more potent greenhouse gas than carbon dioxide in the amount of heat it can trap, especially in the short term. Burning one molecule of methane generates one molecule of carbon dioxide, indicating there may be no net benefit in using gas as a fuel source. Reducing the amount of waste methane produced in the first place and moving away from use of gas as a fuel source will have a greater beneficial impact, as might other approaches to productive use of otherwise-wasted methane. In terms of prevention, vaccines are being developed in Australia to reduce the significant global warming contributions from methane released by livestock via flatulence and eructation.

Another physical property of the anthropogenic GHGs relevant to mitigation is the different abilities of the gases to trap heat (in the form of infrared radiation). Some gases are more effective at trapping heat than others, e.g., SF_6 is 22,200 times more effective a GHG than CO_2 on a per-kilogram basis. A measure for this physical property is the global warming potential (GWP), and is used in the Kyoto Protocol.

Although not designed for this purpose, the Montreal Protocol has probably benefited climate change mitigation efforts. The Montreal Protocol is an international treaty that has successfully reduced emissions of ozone-depleting substances (for example, CFCs), which are also greenhouse gases.

By Sector

Transport

The Tesla Roadster emits no tailpipe emissions, uses lithium ion batteries to achieve 220 mi (350 km) per charge, while also capable of going 0–60 in under 4 seconds.

Bicycles have almost no carbon footprint compared to cars, and canal transport may represent a positive option for certain types of freight in the 21st century

Modern energy-efficient technologies, such as plug-in hybrid electric vehicles, and development of new technologies, such as carbon-neutral synthetic gasoline & Jet fuel, may reduce the consumption of petroleum, land use changes and emissions of carbon dioxide. A shift from air transport and truck transport to electric rail transport would reduce emissions significantly. For electric vehicles, the reduction of carbon emissions will improve further if the way the required electricity is generated is low-carbon power in origin.

Urban Planning

Effective urban planning to reduce sprawl would decrease Vehicle Miles Travelled (VMT), lowering emissions from transportation. Increased use of public transport can also reduce greenhouse gas emissions per passenger kilometer. Between 1982 and 1997, the amount of land consumed for urban development in the United States increased by 47 percent while the nation's population grew by only 17 percent. Inefficient land use development practices have increased infrastructure costs as well as the amount of energy needed for transportation, community services, and buildings.

At the same time, a growing number of citizens and government officials have begun advocating a smarter approach to land use planning. These smart growth practices include compact community development, multiple transportation choices, mixed land uses, and practices to conserve green space. These programs offer environmental, economic, and quality-of-life benefits; and they also serve to reduce energy usage and greenhouse gas emissions.

Approaches such as New Urbanism and Transit-oriented development seek to reduce distances travelled, especially by private vehicles, encourage public transit and make walking and cycling more attractive options. This is achieved through "medium-density", mixed-use planning and the concentration of housing within walking distance of town centers and transport nodes.

Smarter growth land use policies have both a direct and indirect effect on energy consuming behavior. For example, transportation energy usage, the number one user of petroleum fuels, could be significantly reduced through more compact and mixed use land development patterns, which in turn could be served by a greater variety of non-automotive based transportation choices.

Building Design

Emissions from housing are substantial, and government-supported energy efficiency programmes can make a difference.

For institutions of higher learning in the United States, greenhouse gas emissions depend primarily on total area of buildings and secondarily on climate. If climate is not taken into account, annual greenhouse gas emissions due to energy consumed on campuses plus purchased electricity can be estimated with the formula, $E=aS^b$, where a =0.001621 metric tonnes of CO_2 equivalent/square foot or 0.0241 metric tonnes of CO_2 equivalent/square meter and b = 1.1354. New buildings can be constructed using passive solar building design, low-energy building, or zero-energy building techniques, using renewable heat sources. Existing buildings can be made more efficient through the use of insulation, high-efficiency appliances (particularly hot water heaters and furnaces), double- or triple-glazed gas-filled windows, external window shades, and building orientation and siting. Renewable heat sources such as shallow geothermal and passive solar energy reduce the amount of greenhouse gasses emitted. In addition to designing buildings which are more energy-efficient to heat, it is possible to design buildings that are more energy-efficient to cool by using lighter-coloured, more reflective materials in the development of urban areas (e.g. by painting roofs white) and planting trees. This saves energy because it cools buildings and reduces the urban heat island effect thus reducing the use of air conditioning.

Agriculture

According to the EPA, agricultural soil management practices can lead to production and emission of nitrous oxide (N2O), a major greenhouse gas and air pollutant. Activities that can contribute to N 2O emissions include fertilizer usage, irrigation and tillage. The management of soils accounts for over half of the emissions from the Agriculture sector. Cattle livestocks account for one third of emissions, through methane emissions. Manure management and rice cultivation also produce gaseous emissions.

Methods that significantly enhance carbon sequestration in soil include no-till farming, residue mulching, cover cropping, and crop rotation, all of which are more widely used in organic farming than in conventional farming. Because only 5% of US farmland currently uses no-till and residue mulching, there is a large potential for carbon sequestration.

A 2015 study found that farming can deplete soil carbon and render soil incapable of supporting life. Instead the study showed that conservation farming can protect carbon in soils, and repair damage over time.

The farming practise of cover crops has been recognized as climate-smart agriculture by the White House.

Societal Controls

Another method being examined is to make carbon a new currency by introducing tradeable “personal carbon credits”. The idea being it will encourage and motivate individuals to reduce their ‘carbon footprint’ by the way they live. Each citizen will receive a free annual quota of carbon that they can use to travel, buy food, and go about their business. It has been suggested that by using this concept it could actually solve two problems; pollution and poverty, old age pensioners will actually be better off because they fly less often, so they can cash in their quota at the end of the year to pay heating bills and so forth.

Population

Various organizations promote population control as a means for mitigating global warming. Proposed measures include improving access to family planning and reproductive health care and information, reducing natalistic politics, public education about the consequences of continued population growth, and improving access of women to education and economic opportunities.

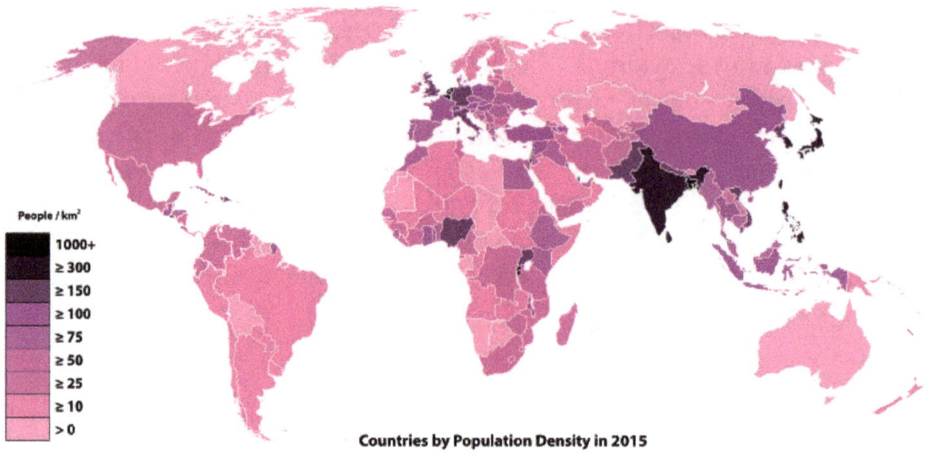

c

Population density by country.

Population control efforts are impeded by there being somewhat of a taboo in some countries against considering any such efforts. Also, various religions discourage or prohibit some or all forms of birth control.

Population size has a different per capita effect on global warming in different countries, since the per capita production of anthropogenic greenhouse gases varies greatly by country.

Costs and Benefits

Costs

The Stern Review proposes stabilising the concentration of greenhouse-gas emissions in the atmosphere at a maximum of 550ppm CO_2e by 2050. The Review estimates that this would mean cutting total greenhouse-gas emissions to three quarters of 2007 levels. The Review further estimates that the cost of these cuts would be in the range −1.0 to +3.5% of World GDP, (i.e. GWP), with an average estimate of approximately 1%. Stern has since revised his estimate to 2% of GWP. For comparison, the Gross World Product (GWP) at PPP was estimated at $74.5 trillion in 2010, thus 2% is approximately $1.5 trillion. The Review emphasises that these costs are contingent on steady reductions in the cost of low-carbon technologies. Mitigation costs will also vary according to how and when emissions are cut: early, well-planned action will minimise the costs.

One way of estimating the cost of reducing emissions is by considering the likely costs of potential technological and output changes. Policy makers can compare the marginal abatement costs of different methods to assess the cost and amount of possible abatement over time. The marginal abatement costs of the various measures will differ by country, by sector, and over time.

Benefits

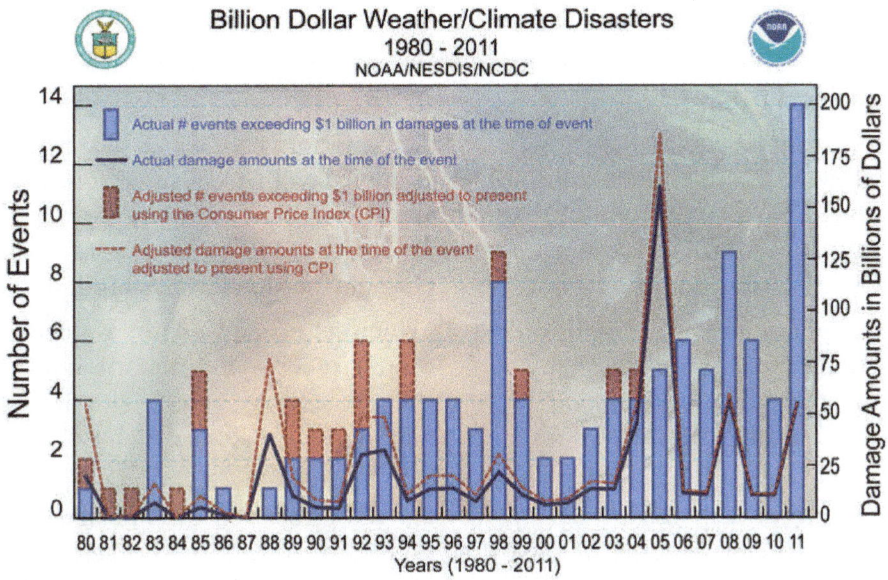

Total extreme weather cost and number of events costing more than $1 billion in the United States from 1980 to 2011.

Yohe *et al.* (2007) assessed the literature on sustainability and climate change.With high confidence, they suggested that up to the year 2050, an effort to cap greenhouse gas (GHG) emissions at 550 ppm would benefit developing countries significantly. This was judged to be especially the case when combined with enhanced adaptation. By 2100, however, it was still judged likely that there would be significant effects of global warming. This was judged to be the case even with aggressive mitigation and significantly enhanced adaptive capacity.

Sharing

One of the aspects of mitigation is how to share the costs and benefits of mitigation policies. There is no scientific consensus over how to share these costs and benefits (Toth *et al.*, 2001). In terms of the politics of mitigation, the UNFCCC's ultimate objective is to stabilize concentrations of GHG in the atmosphere at a level that would prevent "dangerous" climate change (Rogner *et al.*, 2007).

GHG emissions are an important correlate of wealth, at least at present (Banuri *et al.*, 1996, pp. 91–92). Wealth, as measured by per capita income (i.e., income per head of population), varies widely between different countries. Activities of the poor that involve emissions of GHGs are often associated with basic needs, such as heating to stay tolerably warm. In richer countries, emissions tend to be associated with things like cars, central heating, etc. The impacts of cutting emissions could therefore have different impacts on human welfare according to wealth.

Distributing Emissions Abatement Costs

There have been different proposals on how to allocate responsibility for cutting emissions (Banuri *et al.*, 1996, pp. 103–105):

- **Egalitarianism**: this system interprets the problem as one where each person has equal rights to a global resource, i.e., polluting the atmosphere.
- **Basic needs**: this system would have emissions allocated according to basic needs, as defined according to a minimum level of consumption. Consumption above basic needs would require countries to buy more emission rights. From this viewpoint, developing countries would need to be at least as well off under an emissions control regime as they would be outside the regime.
- **Proportionality and polluter-pays principle**: Proportionality reflects the ancient Aristotelian principle that people should receive in proportion to what they put in, and pay in proportion to the damages they cause. This has a potential relationship with the "polluter-pays principle", which can be interpreted in a number of ways:
- *Historical responsibilities*: this asserts that allocation of emission rights should be based on patterns of past emissions. Two-thirds of the stock of GHGs in the atmosphere at present is due to the past actions of developed countries (Goldemberg *et al.*, 1996, p. 29).
- *Comparable burdens and ability to pay*: with this approach, countries would reduce emissions based on comparable burdens and their ability to take on the costs of reduction. Ways to assess burdens include monetary costs per head of population, as well as other, more complex measures, like the UNDP's Human Development Index.
- *Willingness to pay*: with this approach, countries take on emission reductions based on their ability to pay along with how much they benefit from reducing their emissions.

Specific Proposals

- **Ad hoc**: Lashof (1992) and Cline (1992) (referred to by Banuri *et al.*, 1996, p. 106), for example, suggested that allocations based partly on GNP could be a way of sharing the burdens of emission reductions. This is because GNP and economic activity are partially tied to carbon emissions.
- **Equal per capita entitlements**: this is the most widely cited method of distributing abatement costs, and is derived from egalitarianism (Banuri *et al.*, 1996, pp. 106–107). This approach can be divided into two categories. In the first category, emissions are allocated according to national population. In the second category, emissions are allocated in a way that attempts to account for historical (cumulative) emissions.
- **Status quo**: with this approach, historical emissions are ignored, and current emission levels are taken as a status quo right to emit (Banuri *et al.*, 1996, p. 107). An analogy for this approach can be made with fisheries, which is a

common, limited resource. The analogy would be with the atmosphere, which can be viewed as an exhaustible natural resource (Goldemberg *et al.*, 1996, p. 27). In international law, one state recognized the long-established use of another state's use of the fisheries resource. It was also recognized by the state that part of the other state's economy was dependent on that resource.

Governmental and Intergovernmental Action

Many countries, both developing and developed, are aiming to use cleaner technologies (World Bank, 2010, p. 192). Use of these technologies aids mitigation and could result in substantial reductions in CO_2 emissions. Policies include targets for emissions reductions, increased use of renewable energy, and increased energy efficiency. It is often argued that the results of climate change are more damaging in poor nations, where infrastructures are weak and few social services exist. The Commitment to Development Index is one attempt to analyze rich country policies taken to reduce their disproportionate use of the global commons. Countries do well if their greenhouse gas emissions are falling, if their gas taxes are high, if they do not subsidize the fishing industry, if they have a low fossil fuel rate per capita, and if they control imports of illegally cut tropical timber.

Kyoto Protocol

The main current international agreement on combating climate change is the Kyoto Protocol, which came into force on 16 February 2005. The Kyoto Protocol is an amendment to the United Nations Framework Convention on Climate Change (UNFCCC). Countries that have ratified this protocol have committed to reduce their emissions of carbon dioxide and five other greenhouse gases, or engage in emissions trading if they maintain or increase emissions of these gases.

Temperature Targets

The graph on the right shows three "pathways" to meet the UNFCCC's 2 °C target, labelled "global technology", "decentralised solutions", and "consumption change". Each pathway shows how various measures (e.g., improved energy efficiency, increased use of renewable energy) could contribute to emissions reductions. Image credit: PBL Netherlands Environmental Assessment Agency.

Actions to mitigate climate change are sometimes based on the goal of achieving a particular temperature target. One of the targets that has been suggested is to limit the future increase in global mean temperature (global warming) to below 2 °C, relative to the pre-industrial level. The 2 °C target was adopted in 2010 by Parties to the United Nations Framework Convention on Climate Change. Most countries of the world are Parties to the UNFCCC. The target had been adopted in 1996 by the European Union Council.

Feasibility of 2 °C

Temperatures have increased by 0.8 °C compared to the pre-industrial level, and another 0.5–0.7 °C is already committed.The 2 °C rise is typically associated in climate models with a carbon dioxide equivalent concentration of 400–500 ppm by volume; the current (January 2015) level of carbon dioxide alone is 400 ppm by volume, and rising at 1–3 ppm annually. Hence, to avoid a very likely breach of the 2 °C target, CO_2levels would have to be stabilised very soon; this is generally regarded as unlikely, based on current programs in place to date. The importance of change is illustrated by the fact that world economic energy efficiency is improving at only half the rate of world economic growth.

Views in the Literature

There is disagreement among experts over whether or not the 2 °C target can be met. For example, according to Anderson and Bows (2011), "there is little to no chance" of meeting the target. On the other hand, according to Alcamo *et al.* (2013):

- Policies adopted by parties to the UNFCCC are too weak to meet a 2 or 1.5 °C target. However, these targets might still be achievable if more stringent mitigation policies are adopted immediately.
- Cost-effective 2 °C scenarios project annual global greenhouse gas emissions to peak before the year 2020, with deep cuts in emissions thereafter, leading to a reduction in 2050 of 41% compared to 1990 levels.

Discussion on Other Targets

Scientific analysis can provide information on the impacts of climate change and associated policies, such as reducing GHG emissions. However, deciding what policies are best requires value judgements. For example, limiting global warming to 1 °C relative to pre-industrial levels may help to reduce climate change damages more than a 2 °C limit. However, a 1 °C limit may be more costly to achieve than a 2 °C limit.

According to some analysts, the 2 °C "guardrail" is inadequate for the needed degree and timeliness of mitigation. On the other hand, some economic studies suggest more modest mitigation policies.For example, the emissions reductions proposed by Nordhaus (2010) might lead to global warming (in the year 2100) of around 3 °C, relative to pre-industrial levels.

Official Long-Term Target of 1.5 °C

In 2015, two official UNFCCC scientific expert bodies came to the conclusion that, "in some regions and vulnerable ecosystems, high risks are projected even for warming above 1.5°C". This expert position was, together with the strong diplomatic voice of the poorest countries and the island nations in the Pacific, the driving force leading to the decision of the Paris Conference 2015, to lay down this 1.5 °C long-term target on top of the existing 2 °C goal.

Encouraging use Changes

Emissions Tax

An emissions tax on greenhouse gas emissions requires individual emitters to pay a fee, charge or tax for every tonne of greenhouse gas released into the atmosphere. Most environmentally related taxes with implications for greenhouse gas emissions in OECD countries are levied on energy products and motor vehicles, rather than on CO_2 emissions directly.

Emission taxes can be both cost-effective and environmentally effective. Difficulties with emission taxes include their potential unpopularity, and the fact that they cannot guarantee a particular level of emissions reduction. Emissions or energy taxes also often fall disproportionately on lower income classes. In developing countries, institutions may be insufficiently developed for the collection of emissions fees from a wide variety of sources.

Subsidies

According to Mark Z. Jacobson, a program of subsidization balanced against expected flood costs could pay for conversion to 100% renewable power by 2030. Jacobson, and his colleague Mark Delucchi, suggest that the cost to generate and transmit power in 2020 will be less than 4 cents per kilowatt hour (in 2007 dollars) for wind, about 4 cents for wave and hydroelectric, from 4 to 7 cents for geothermal, and 8 cents per kWh for solar, fossil, and nuclear power.

Investment

Another indirect method of encouraging uses of renewable energy, and pursue sustainability and environmental protection, is that of prompting investment in this area through legal means, something that is already being done at national level as well as in the field of international investment.

Carbon emissions trading

With the creation of a market for trading carbon dioxide emissions within the Kyoto Protocol, it is likely that London financial markets will be the centre for this potentially highly lucrative business; the New York and Chicago stock markets may have a lower trade volume than expected as long as the US maintains its rejection of the Kyoto.

However, emissions trading may delay the phase-out of fossil fuels.

In the north-east United States, a successful cap and trade program has shown potential for this solution.

The European Union Emission Trading Scheme (EU ETS) is the largest multi-national, greenhouse gas emissions trading scheme in the world. It commenced operation on 1 January 2005, and all 28 member states of the European Union participate in the scheme which has created a new market in carbon dioxide allowances estimated at 35 billion Euros (US$43 billion) per year. The Chicago Climate Exchange was the first (voluntary) emissions market, and is soon to be followed by Asia's first market (Asia Carbon Exchange). A total of 107 million metric tonnes of carbon dioxide equivalent have been exchanged through projects in 2004, a 38% increase relative to 2003 (78 Mt CO_2e).

Twenty three multinational corporations have come together in the G8 Climate Change Roundtable, a business group formed at the January 2005 World Economic Forum. The group includes Ford, Toyota, British Airways and BP. On 9 June 2005 the Group published a statement stating that there was a need to act on climate change and claiming that market-based solutions can help. It called on governments to establish "clear, transparent, and consistent price signals" through "creation of a long-term policy framework" that would include all major producers of greenhouse gases.

The Regional Greenhouse Gas Initiative is a proposed carbon trading scheme being created by nine North-eastern and Mid-Atlantic American states; Connecticut, Delaware, Maine, Massachusetts, New Hampshire, New Jersey, New York, Rhode Island and Vermont. The scheme was due to be developed by April 2005 but has not yet been completed.

Implementation

Implementation puts into effect climate change mitigation strategies and targets. These can be targets set by international bodies or voluntary action by individuals or institutions. This is the most important, expensive and least appealing aspect of environmental governance.

Funding

Implementation requires funding sources but is often beset by disputes over who should provide funds and under what conditions. A lack of funding can be a barrier to successful strategies as there are no formal arrangements to finance climate change development and implementation. Funding is often provided by nations, groups of nations and increasingly NGO and private sources. These funds are often channelled through the Global Environmental Facility (GEF). This is an environmental funding mechanism in the World Bank which is designed to deal with global environmental issues. The GEF was originally designed to tackle four main areas: biological diversity, climate change, international waters and ozone layer depletion, to which land degradation and persistent organic pollutant were added. The GEF funds projects that are agreed to achieve global environmental benefits that are endorsed by governments and screened by one of the GEF's implementing agencies.

Problems

There are numerous issues which result in a current perceived lack of implementation. It has been suggested that the main barriers to implementation are, Uncertainty, Fragmentation, Institutional void, Short time horizon of policies and politicians and Missing motives and willingness to start adapting. The relationships between many climatic processes can cause large levels of uncertainty as they are not fully understood and can be a barrier to implementation. When information on climate change is held between the large numbers of actors involved it can be highly dispersed, context specific or difficult to access causing fragmentation to be a barrier. Institutional void is the lack of commonly accepted rules and norms for policy processes to take place, calling into question the legitimacy and efficacy of policy processes. The Short time horizon of policies and politicians often means that climate change policies are not implemented in favour of socially favoured societal issues. Statements are often posed to keep the illusion of political action to prevent or postpone decisions being made. Missing motives and willingness to start adapting is a large barrier as it prevents any implementation.

The issues that arise with a system which involves international government cooperation, such as Cap and Trade, could potentially be improved with a polycentric approach where the rules are enforced by many small sections of authority as apposed to one overall enforcement agency.

Occurrence

Despite a perceived lack of occurrence, evidence of implementation is emerging internationally. Some examples of this are the initiation of NAPA's and of joint implementation. Many developing nations have made National Adaptation Programs of Action (NAPAs) which are frameworks to prioritize adaption needs. The implementation of many of these is supported by GEF agencies. Many developed countries are implementing 'first generation' institutional adaption plans particularly at the state and local government scale. There has also been a push towards joint implementation between countries by the UNFCC as this has been suggested as a cost-effective way for objectives to be achieved.

Territorial Policies

United States

Efforts to reduce greenhouse gas emissions by the United States include energy policies which encourage efficiency through programs like Energy Star, Commercial Building Integration, and the Industrial Technologies Program. On 12 November 1998, Vice President Al Gore symbolically signed the Kyoto Protocol, but he indicated participation by the developing nations was necessary prior its being submitted for ratification by the United States Senate.

In 2007, Transportation Secretary Mary Peters, with White House approval, urged governors and dozens of members of the House of Representatives to block California's first-in-the-nation limits on greenhouse gases from cars and trucks, according to e-mails obtained by Congress. The U.S. Climate Change Science Program is a group of about twenty federal agencies and US Cabinet Departments, all working together to address global warming.

The Bush administration pressured American scientists to suppress discussion of global warming, according to the testimony of the Union of Concerned Scientists to the Oversight and Government Reform Committee of the U.S. House of Representatives. "High-quality science" was "struggling to get out," as the Bush administration pressured scientists to tailor their writings on global warming to fit the Bush administration's skepticism, in some cases at the behest of an ex-oil industry lobbyist. "Nearly half of all respondents perceived or personally experienced pressure to eliminate the words 'climate change,' 'global warming' or other similar terms from a variety of communications." Similarly, according to the testimony of senior officers of the Government Accountability Project, the White House attempted to bury the report "National Assessment of the Potential Consequences of Climate Variability and Change," produced by U.S. scientists pursuant to U.S. law. Some U.S. scientists resigned their jobs rather than give in to White House pressure to underreport global warming.

In the absence of substantial federal action, state governments have adopted emissions-control laws such as the Regional Greenhouse Gas Initiative in the Northeast and the Global Warming Solutions Act of 2006 in California.

Developing Countries

In order to reconcile economic development with mitigating carbon emissions, developing countries need particular support, both financial and technical. One of the means of achieving this is the Kyoto Protocol's Clean Development Mechanism (CDM). The World Bank's Prototype Carbon Fund is a public private partnership that operates within the CDM.

An important point of contention, however, is how overseas development assistance not directly related to climate change mitigation is affected by funds provided to climate change mitigation. One of the outcomes of the UNFCC Copenhagen Climate Conference was the Copenhagen Accord, in which developed countries promised to provide US $30 million between 2010 and 2012 of new and additional resources. Yet it remains unclear what exactly the definition of additional is and the European Commission has requested its member states to define what they understand to be additional, and researchers at the Overseas Development Institute have found four main understandings:

- Climate finance classified as aid, but additional to (over and above) the '0.7%' ODA target;
- Increase on previous year's Official Development Assistance (ODA) spent on climate change mitigation;
- Rising ODA levels that include climate change finance but where it is limited to a specified percentage; and
- Increase in climate finance not connected to ODA.

The main point being that there is a conflict between the OECD states budget deficit cuts, the need to help developing countries adapt to develop sustainably and the need to ensure that funding does not come from cutting aid to other important Millennium Development Goals.

However, none of these initiatives suggest a quantitative cap on the emissions from developing countries. This is considered as a particularly difficult policy proposal as the economic growth of developing countries are proportionally reflected in the growth of greenhouse emissions. Critics of mitigation often argue that, the developing countries' drive to attain a comparable living standard to the developed countries would doom the attempt at mitigation of global warming. Critics also argue that holding down emissions would shift the human cost of global warming from a general one to one that was borne most heavily by the poorest populations on the planet.

In an attempt to provide more opportunities for developing countries to adapt clean technologies, UNEP and WTO urged the international community to reduce trade barriers and to conclude the Doha trade round "which includes opening trade in environmental goods and services".

Non-governmental Approaches

While many of the proposed methods of mitigating global warming require governmental funding, legislation and regulatory action, individuals and businesses can also play a part in the mitigation effort.

Choices in Personal Actions and Business Operations

Environmental groups encourage individual action against global warming, often aimed at the consumer. Common recommendations include lowering home heating and cooling usage, burning less gasoline, supporting renewable energy sources, buying local products to reduce transportation, turning off unused devices, and various others.

A geophysicist at Utrecht University has urged similar institutions to hold the vanguard in voluntary mitigation, suggesting the use of communications technologies such as videoconferencing to reduce their dependence on long-haul flights.

Air travel and Shipment

In 2008, climate scientist Kevin Anderson raised concern about the growing effect of rapidly increasing global air transport on the climate in a paper, and a presentation, suggesting that reversing this trend is necessary to reduce emissions.

Part of the difficulty is that when aviation emissions are made at high altitude, the climate impacts are much greater than otherwise. Others have been raising the related concerns of the increasing hypermobility of individuals, whether traveling for business or pleasure, involving frequent and often long distance air travel, as well as air shipment of goods.

Business Opportunities and Risks

On 9 May 2005 Jeff Immelt, the chief executive of General Electric (GE), announced plans to reduce GE's global warming related emissions by one percent by 2012. "GE said that given its projected growth, those emissions would have risen by 40 percent without such action."

On 21 June 2005 a group of leading airlines, airports and aerospace manufacturers pledged to work together to reduce the negative environmental impact of aviation, including limiting the impact of air travel on climate change by improving fuel efficiency and reducing carbon dioxide emissions of new aircraft by fifty percent per seat kilometre by 2020 from 2000 levels. The group aims to develop a common reporting system for carbon dioxide emissions per aircraft by the end of 2005, and pressed for the early inclusion of aviation in the European Union's carbon emission trading scheme.

Investor Response

Climate change is also a concern for large institutional investors who have a long term time horizon and potentially large exposure to the negative impacts of global warming because of the large geographic footprint of their multi-national holdings. SRI (Socially responsible investing) Funds allow investors to invest in funds that meet high ESG (environmental, social, governance) standards as such funds invest in companies that are aligned with these goals. Proxy firms can be used to draft guidelines for investment managers that take these concerns into account.

Legal Action

In some countries, those affected by climate change may be able to sue major producers. Attempts at litigation have been initiated by entire peoples such as Palau and the Inuit, as well as non-governmental organizations such as the Sierra Club. Although proving that particular weather events are due specifically to global warming may never be possible, methodologies have been developed to show the increased risk of such events caused by global warming.

For a legal action for negligence (or similar) to succeed, "Plaintiffs ... must show that, more probably than not, their individual injuries were caused by the risk factor in question, as opposed to any other cause. This has sometimes been translated to a requirement of a relative risk of at least two." Another route (though with little legal bite) is the World Heritage Convention, if it can be shown that climate change is affecting World Heritage Sites like Mount Everest.

Besides countries suing one another, there are also cases where people in a country have taken legal steps against their own government. Legal action for instance has been taken to try to force the U.S. Environmental Protection Agency to regulate greenhouse gas emissions under the Clean Air Act,and against the Export-Import Bank and OPIC for failing to assess environmental impacts (including global warming impacts) under NEPA.

In the Netherlands and Belgium, organisations as Urgenda and the vzw Klimaatzaak in Belgium have also sued their governments as they believe their governments aren't meeting the emission reductions they agreed to. Urgenda has all ready won their case against the Dutch government.

According to a 2004 study commissioned by Friends of the Earth, ExxonMobil and its predecessors caused 4.7 to 5.3 percent of the world's man-made carbon dioxide emissions between 1882 and 2002. The group suggested that such studies could form the basis for eventual legal action.

In 2015, Exxon, received a subpoena. According to the Washington Post and confirmed by the company, the attorney general of New York, Eric Schneiderman, opened an investigation into the possibility that the company had mislead the public and investors about the risks of climate change.

ENVIRONMENTAL MIGRANT

Drought refugees from Oklahoma camping by the roadside, California, 1936

Environmental migrants are people who are forced to leave their home region due to sudden or long-term changes to their local environment which compromise their well being or secure livelihood, such changes are held to include increased droughts, desertification, sea level rise, and disruption of seasonal weather patterns such as monsoons. Environmental migrants may flee to or migrate to another country or they may migrate internally within their own country. However, the term "environmental migrant" is used somewhat interchangeably with a range of similar terms, such as **environmental refugee** or **climate refugee**, although the distinction between these terms is contested. Despite problems in formulating a uniform and clear-cut definition of 'environmental migration', such a concept has increased as an issue of concern in the 2000s as policy-makers, environmental and social scientists attempt to conceptualise the potential societal effects of climate change and general environmental degradation.

Definition and Concept

The term "environmental refugee" was first proposed by Lester Brown in 1976, since then there has been a proliferation in the use of the term at which "environmental migrant" and a cluster of similar categories, including "forced environmental migrant", "environmentally motivated migrant", "climate refugee", "climate change refugee", "environmentally displaced person (EDP)», "disaster refugee", "environmental displacee", "eco-refugee", "ecologically displaced person" and "environmental-refugee-to-be (ERTB)" have been utilized. The differences between these terms are less important than what they have in common: they all suggest that there is a determinable relationship between environmental drivers and human migration which is analytically useful, policy-relevant and possibly grounds for the expansion of refugee law.

Under the United Nations Convention Relating to the Status of Refugees of 1951, a refugee is more narrowly defined (in Article 1A) as a person who "owing to a well-founded fear of being persecuted for reasons of race, religion, nationality, membership of a particular social group, or political opinion, is outside the country of his nationality, and is unable to or, owing to such fear, is unwilling to avail himself of the protection of that country". While the concept of a refugee was expanded by the Convention's 1967 Protocol and by regional conventions in Africa and Latin America to include persons who had fled war or other violence in their home country, in its present state the convention does not provide long-term legal protection to refugees due to environmental change.

The International Organization for Migration proposes the following definition for environmental migrants:

"Environmental migrants are persons or groups of persons who, for compelling reasons of sudden or progressive changes in the environment that adversely affect their lives or living conditions, are obliged to leave their habitual homes, or choose to do so, either temporarily or permanently, and who move either within their country or abroad."

The term *climate refugees* or *climate migrants* refers to the subset of environmental migrants forced to move "due to sudden or gradual alterations in the natural environment related to at least one of three impacts of climate change: sea-level rise, extreme weather events, and drought and water scarcity" (Global Governance Project 2012). The term *climate exiles* has been used to refer to those climate migrants who may be in danger of becoming stateless.

However, there is yet to be a universally accepted definition of "environmental migration" or "climate refugee". Thus, the International Organization for Migration formulated a working definition which encompasses the complexity of the topic.

This working definition recognizes that

- Environmental migrants are not only those displaced by the environmental event but also those whose migration is triggered by deteriorating environmental conditions
- Environmentally induced movement can take place within as well as across international borders;
- It can be both long and short term; and
- Population movements triggered by environmental forces can be forced as well as a matter of choice

Types

The International Organisation for Migration proposes three types of environmental migrants:

- *Environmental emergency migrants*: people who flee temporarily due to an environmental disaster or sudden environmental event. (Examples: someone forced to leave due to hurricane, tsunami, earthquake, etc.)
- *Environmental forced migrants*: people who have to leave due to deteriorating environmental conditions. (Example: someone forced to leave due to a slow deterioration of their environment such as deforestation, coastal deterioration, etc.)
- *Environmental motivated migrants* also known as *environmentally induced economic migrants*: people who choose to leave to avoid possible future problems. (Example: someone who leaves due to declining crop productivity caused by desertification)

Enumeration

There have been a number of attempts over the decades to enumerate environmental migrants and refugees. Jodi Jacobson (1988) is cited as the first researcher to enumerate the issue, stating that there were already up to 10 million 'Environmental Refugees'. Drawing on 'worst-case scenarios' about sea-level rise, she argued that all forms of 'Environmental Refugees' would be six times as numerous as political refugees. (1988: 38). By 1989, Mustafa Tolba, Executive Director of UNEP, was claiming that 'as many as 50 million people could become environmental refugees' if the world did not act to support sustainable development (Tolba 1989: 25). In 1990, the Intergovernmental Panel on Climate Change (IPCC 1990: 20) declared that the greatest single consequence of climate change could be migration, 'with millions of people displaced by shoreline erosion, coastal flooding and severe drought' (Warner & Laczko: 2008: 235). In the mid-1990s, British environmentalist, Norman Myers, became the most prominent proponent of this 'maximalist' school (Suhrke 1993). Noting, that "environmental refugees will soon become the largest group of involuntary refugees". Additionally, he stated that there were 25 million environmental refugees in the mid-1990s, further claiming that this figure could double by 2010, with an upper limit of 200 million by 2050 (Myers 1997). Myers argued that the causes of environmental displacement would include desertification, lack of water, salination of irrigated lands and the depletion of bio-diversity. He also hypothesised that displacement would amount to 30m in China, 30m in India, 15m in Bangladesh, 14m in Egypt, 10m in other delta areas and coastal zones, 1m in island states, and with otherwise agriculturally displaced people totalling 50m (Myers & Kent 1995) by 2050. More recently, Myers has suggested that the figure by 2050 might be as high as 250 million (Christian Aid 2007: 6).

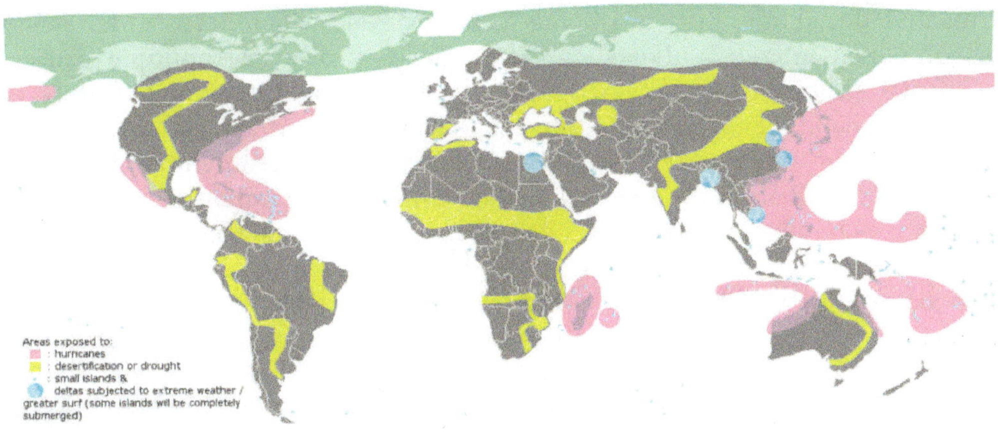

A map showing where natural disasters caused/aggravated by global warming may occur, and thus where environmental refugees would be created

These claims have gained significant currency, with the most common projection being that the world will have 150-200 million climate change refugees by 2050. Variations of this claim have been made in influential reports on climate change by the IPCC (Brown 2008: 11) and the Stern Review on the Economics of Climate Change (Stern et al. 2006: 3), as well as by NGOs such as Friends of the Earth, Greenpeace Germany (Jakobeit and Methmann 2007) and Christian Aid; and inter-governmental organisations such as the Council of Europe, UNESCO, IOM (Brown 2008) and UNHCR.

Norman Myers is the most cited researcher in this field, who found that 25 million environmental migrants existed in 1995 in his work (Myers & Kent 1995), which drew upon over 1000 sources.However, Vikram Kolmannskog has stated that Myers' work can be 'criticized for being inconsistent, impossible to check and failing to take proper account of opportunities to adapt' (2008: 9).Furthermore, Myers himself has acknowledged that his figures are based upon 'heroic extrapolation' (Brown 2008: 12). More generally, Black has argued that there is 'surprisingly little scientific evidence' that indicates that the world is 'filling-up with environmental refugees' (1998: 23). Indeed, Francois Gemenne has stated that: ‹When it comes to predictions, figures are usually based on the number of people living in regions at risk, and not on the number of people actually expected to migrate. Estimates do not account for adaptation strategies [or] different levels of vulnerability' (Gemenne 2009: 159).

Asia and the Pacific

According to the Internal Displacement Monitoring Centre, more than 42 million people were displaced in Asia and the Pacific during 2010 and 2011, more than twice the population of Sri Lanka. This figure includes those displaced by storms, floods, and heat and cold waves. Still others were displaced by drought and sea-level rise. Most of those compelled to leave their homes eventually returned when conditions improved, but an undetermined number became migrants, usually within their country, but also across national borders.

Climate-induced migration is a highly complex issue which needs to be understood as part of global migration dynamics. Migration typically has multiple causes, and environmental factors are intertwined with other social and economic factors, which themselves can be influenced by environmental changes. Environmental migration should not be treated solely as a discrete category, set apart from other migration flows. A 2012 Asian Development Bank study argues that climate-induced migration should be addressed as part of a country's development agenda, given the major implications of migration on economic and social development. The report recommends interventions both to address the situation of those who have migrated, as well as those who remain in areas subject to environmental risk. It says: "To reduce migration compelled by worsening environmental conditions, and to strengthen resilience of at-risk communities, governments should adopt polices and commit financing to social protection, livelihoods development, basic urban infrastructure development, and disaster risk management."

Additionally, it is maintained that the poor populate areas that are most at risk for environmental destruction and climate change, including coastlines, flood-lines and steep slopes. As a result, climate change threatens areas already suffering from extreme poverty. "The issue of equity is crucial. Climate affects us all, but does not affect us all equally," UN Secretary-General Ban Ki-moon told delegates at a climate conference in Indonesia. Africa is also one of the world regions where environmental displacement is critical largely due to droughts and other climate related eventualities.

In Minqin county, Gansu Province, "10,000 people have left the area and have become *shengtai yimin*, 'ecological migrants'".

In 2013 a claim of a Kiribati man of being a "climate change refugee" under the Convention relating to the Status of Refugees (1951) was determined by the New Zealand High Court to be untenable.The Refugee Convention did not apply as there is no persecution or serious harm related to any of the five stipulated convention grounds. The Court rejected the argument that the international community itself (or countries which can be said to have been historically high emitters of carbon dioxide or other greenhouse gases) were the "persecutor" for the purposes of the Refugee Convention. This analysis of the need for the person to identify persecution of the type descried in the Refugee Convention does not exclude the possibility that a people for countries experiencing severe impacts of climate change can come with the Refugee Convention. However it is not the climate change event itself, rather the social and political response to climate change, which is likely to create the pathway for a successful claim. The New Zealand Immigration and Protection Tribunal and the High Court, "there is a complex inter-relationship between natural disasters, environmental degradation and human vulnerability. Sometimes a tenable pathway to international protection under the Refugee Convention can result. Environmental issues sometimes lead to armed conflict. There may be ensuing violence towards or direct repression of an entire section of a population. Humanitarian relief can become politicised, particularly in situations where some group inside a disadvantaged country is the target of direct discrimination." The New Zealand Court of Appeal also rejected the claim in a 2014 decision. On further appeal the New Zealand Supreme Court confirmed the earlier adverse rulings against the application for refugee status, with the Supreme Court also rejecting the proposition "that environmental degradation resulting from climate change or other natural disasters could never create a pathway into the Refugee Convention or protected person jurisdiction."

In 2014 attention was drawn to an appeal to the New Zealand Immigration and Protection Tribunal against the deportation of a Tuvaluan family on the basis that they were "climate change refugees", who would suffer hardship resulting from the environmental degradation of Tuvalu. However the subsequent grant of residence permits to the family was made on grounds unrelated to the refugee claim. The family was successful in their appeal because, under the relevant immigration legislation, there were "exceptional circumstances of a humanitarian nature" that justified the grant of resident permits as the family was integrated into New Zealand society with a sizeable extended family which had effectively relocated to New Zealand.

North America

Isle de Jean Charles, Louisiana, home of a tribe of Biloxi Chitimacha-Choctaw Indians, is being depopulated with federal grant money, due to saltwater intrusion and sea level rise. The states of New York and New Jersey implemented buyout programs for some homeowners displaced by Hurricane Sandy, which was worsened by global warming's effect on sea surface temperatures.

Political and Legal Perspectives

The International Organization for Migration (IOM) expects the scale of global migration to rise as a result of accelerated climate change. It therefore recommends policymakers around the world to take a proactive stance on the matter. The IOM is composed of 146 member states and 13 observer states and "works closely with governments in promoting migration management that ensures humane and orderly migration that is beneficial to migrants and societies." Additionally, When interviewing Oliver- Smith, an anthropologist and member of the UN group, National Geographic Magazine noted that "there are at least 20 million environmental refugees worldwide, the [UN] group says – more than those displaced by war and political repression combined." Therefore, it is imperative that we begin to recognize this recent division of refugee.

The Environmental Justice Foundation (EJF) has argued that the people who will be forced to move due to climate change currently have no adequate recognition in international law. The EJF contends that a new multilateral legal instrument is required to specifically address the needs of 'climate refugees' in order to confer protection to those fleeing environmental degradation and climate change. They have also asserted that additional funding is needed to enable developing countries to adapt to climate change. Sujatha Byravan and Sudhir Chella Rajan have argued for the use of the term ‹climate exiles' and for international agreements to provide them political and legal rights, including citizenship in other countries, bearing in mind those countries' responsibilities and capabilities.

In some cases, climate change may lead to conflict arising between countries that as a result of flooding or other conditions produce a large number of refugees, and bordering countries that build fences to keep out these refugees. The Bangladesh–India border is largely separated via a fence, and case studies suggest the possibility of violent conflict arising due to people fleeing from areas suffering from destruction of arable land. Current migration has already resulted in low-scale conflicts.

Popular Culture

German artist Hermann Josef Hack's *World Climate Refugee Camp* in Hannover displaying 600 small climate refugee tents.

The notion of ‹environmental migrant', and particularly ‹climate refugee', has gained traction in popular culture. A documentary entitled *Climate Refugees* has been released. "Climate Refugees" is an Official Selection for the 2010 Sundance Film Festival.More recently, Short Documentary Academy Award Nominee, Sun Come Up (2011), tells the story of Carteret islanders who are forced to leave their ancestral land in response to climate change and migrate to war-torn Bougainville

Since 2007, German artist Hermann Josef Hack has shown his World Climate Refugee Camp in the centers of various European cities. The model camp, made of roughly 1000 miniature tents, is a public art intervention that depicts the social impacts of climate change.

ABRUPT CLIMATE CHANGE

An **abrupt climate change** occurs when the climate system is forced to transition to a new climate state at a rate that is determined by the climate system energy-balance, and which is more rapid than the rate of change of the external forcing. Past events include the end of the Carboniferous Rainforest Collapse, Younger Dryas, Dansgaard-Oeschger events, Heinrich events and possibly also the Paleocene-Eocene thermal maximum. The term is also used within the context of global warming to describe sudden climate change that is detectable over the time-scale of a human lifetime. One proposed reason for the observed abrupt climate change is that feedback loops within the climate system both enhance small perturbations and cause a variety of stable states.

Timescales of events described as ‹abrupt’ may vary dramatically. Changes recorded in the climate of Greenland at the end of the Younger Dryas, as measured by ice-cores, imply a sudden warming of +10 °C (+18 °F) within a timescale of a few years. Other abrupt changes are the +4 °C (+7.2 °F) on Greenland 11,270 years ago or the abrupt +6 °C (11 °F) warming 22,000 years ago on Antarctica. By contrast, the Paleocene-Eocene thermal maximum may have initiated anywhere between a few decades and several thousand years. Finally, Earth Systems models project that under ongoing greenhouse gas emissions as early as 2047, the Earth’s near surface temperature could depart from the range of variability in the last 150 years, affecting over 3 billion people and most places of great species diversity on Earth.

According to the Committee on Abrupt Climate Change of the National Research Council:

There are essentially two definitions of abrupt climate change:

- In terms of physics, it is a *transition of the climate system into a different mode on a time scale that is faster than the responsible forcing.*
- In terms of impacts, *“an abrupt change is one that takes place so rapidly and unexpectedly that human or natural systems have difficulty adapting to it”.*

These definitions are complementary: the former gives some insight into how abrupt climate change comes about ; the latter explains why there is so much research devoted to it.

Current Situation

The IPCC states that global warming “could lead to some effects that are abrupt or irreversible”.

In an article in *Science*, Richard Alley et al. said "it is conceivable that human forcing of climate change is increasing the probability of large, abrupt events. Were such an event to recur, the economic and ecological impacts could be large and potentially serious."

A 2013 report from the U.S. National Research Council called for attention to the abrupt impacts of climate change, stating that even steady, gradual change in the physical climate system can have abrupt impacts elsewhere—in human infrastructure and ecosystems for example—if critical thresholds are crossed. The report emphasizes the need for an early warning system that could help society better anticipate sudden changes and emerging impacts.

Regional Changes

Lenton et al. investigated tipping elements in the climate system. These were regional effects of global warming, some of which had abrupt onset and may therefore be regarded as abrupt climate change. They found that "Our synthesis of present knowledge suggests that a variety of tipping elements could reach their critical point within this century under anthropogenic climate change."

Ocean Effects

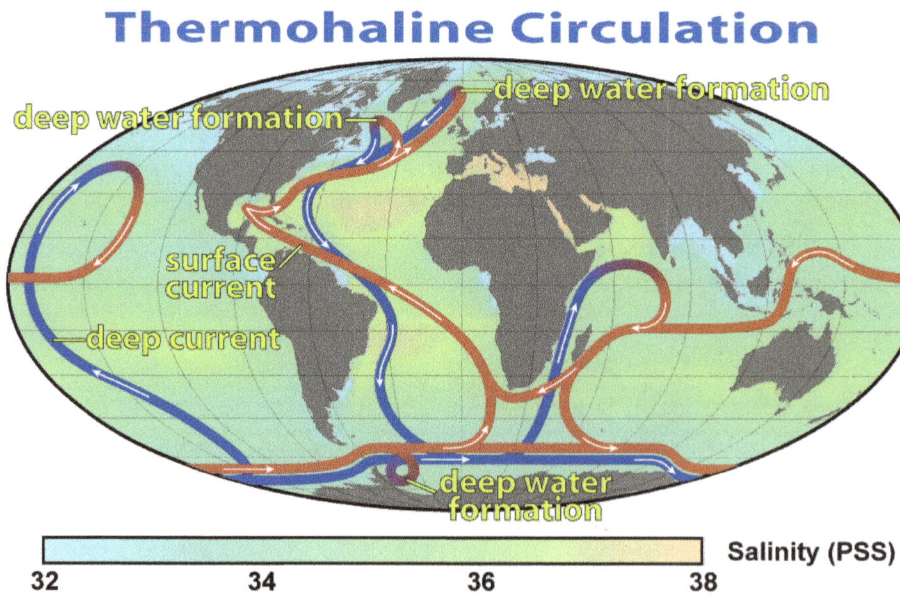

A summary of the path of the thermohaline circulation. Blue paths represent deep-water currents, while red paths represent surface currents.

See also: Shutdown of thermohaline circulation

See also: Cold blob (North Atlantic)

Global oceans have established patterns of currents. Several potential disruptions to this system of currents have been identified as a result of global warming:

- Increasing frequency of El Niño events
- Potential disruption to the thermohaline circulation, such as that which may have occurred during the Younger Dryas event.
- Changes to the North Atlantic oscillation

Effects on Weather

Hansen et al. 2015 found, that the shutdown or substantial slowdown of the Atlantic Meridional Overturning Circulation (AMOC), besides possibly contributing to extreme end-Eemian events, will cause a more general increase of severe weather. Additional surface cooling from ice melt increases surface and lower tropospheric temperature gradients, and causes in model simulations a large increase of mid-latitude eddy energy throughout the midlatitude troposphere. This in turn leads to an increase of baroclinicity produced by stronger temperature gradients, which provides energy for more severe weather events.

Many of the most memorable and devastating storms in eastern North America and western Europe, popularly known as superstorms, have been winter cyclonic storms, though sometimes occurring in late fall or early spring, that generate near-hurricane-force winds and often large amounts of snowfall. Continued warming of low latitude oceans in coming decades will provide more water vapor to strengthen such storms. If this tropical warming is combined with a cooler North Atlantic Ocean from AMOC slowdown and an increase in midlatitude eddy energy, we can anticipate more severe baroclinic storms.

Our results at least imply that strong cooling in the North Atlantic from AMOC shutdown does create higher wind speed. The increment in seasonal mean wind speed of the northeasterlies relative to preindustrial conditions is as much as 10–20 %. Such a percentage increase of wind speed in a storm translates into an increase of storm power dissipation by a factor ~1.4–2, because wind power dissipation is proportional to the cube of wind speed. However, our simulated changes refer to seasonal mean winds averaged over large grid-boxes, not individual storms.

Climate Feedback Effects

One source of abrupt climate change effects is a feedback process, in which a warming event causes a change which leads to further warming. This can also apply to cooling. Example of such feedback processes are:

- Ice-albedo feedback, where the advance or retreat of ice cover alters the ‹whiteness' of the earth, and its ability to absorb the sun's energy
- The dying and burning of forests, as a result of global warming

Past Events

The Younger Dryas period of abrupt climate change is named after the Alpine flower, Dryas.

Several periods of abrupt climate change have been identified in the paleoclimatic record. Notable examples include:

- About 25 climate shifts, called Dansgaard-Oeschger cycles, which have been identified in the ice core record during the glacial period over the past 100,000 years.
- The Younger Dryas event, notably its sudden end. It is the most recent of the Dansgaard-Oeschger cycles and began 12,900 years ago and moved back into a warm-and-wet climate regime about 11,600 years ago. It has been suggested that: "The extreme rapidity of these changes in a variable that directly represents regional climate implies that the events at the end of the last glaciation may have been responses to some kind of threshold or trigger in the North Atlantic climate system." A model for this event based on disruption to the thermohaline circulation has been supported by other studies.
- The Paleocene-Eocene Thermal Maximum, timed at 55 million years ago, which may have been caused by the clathrate gun effect, although potential alternative mechanisms have been identified. This was associated with rapid ocean acidification
- The Permian-Triassic Extinction Event, also known as the great dying, in which up to 95% of all species became extinct, has been hypothesized to be related to

a rapid change in global climate. Life on land took 30 million years to recover.

- The Carboniferous Rainforest Collapse occurred 300 million years ago, at which time tropical rainforests were devastated by climate change. The cooler, drier climate had a severe effect on the biodiversity of amphibians, the primary form of vertebrate life on land.

There are also abrupt climate changes associated with the catastrophic draining of glacial lakes. One example of this is the 8.2 kiloyear event, which associated with the draining of Glacial Lake Agassiz.Another example is the Antarctic Cold Reversal, c. 14,500 years before present (BP), which is believed to have been caused by a meltwater pulse from the Antarctic ice sheet. These rapid meltwater release events have been hypothesized as a cause for Dansgaard-Oeschger cycles,

Abrupt Climate Shifts Since 1976

Had the 1997 El Niño lasted twice as long, the rain forests of the Amazon basin and Southeast Asia could have quickly added much additional carbon dioxide to the air from burning and rotting, with heat waves and extreme weather quickly felt around the world (The "Burn Locally, Crash Globally" scenario.)

Most abrupt climate shifts, however, are likely due to sudden circulation shifts, analogous to a flood cutting a new river channel. The best-known examples are the several dozen shutdowns of the North Atlantic Ocean's Meridional Overturning Circulation during the last ice age, affecting climate worldwide. But there have been a series of less dramatic abrupt climate shifts since 1976, along with some near misses.

- The circulation shift in the western Pacific in the winter of 1976-1977 proved to have much wider impacts.
- Since 1950, El Niños had been weak and short, but La Niñas were often big and long, This pattern reversed after 1977.
- Land temperatures had remained relatively trendless from 1950 to 1976, despite the CO_2 rising from 310 to 332 ppm as fossil fuel emissions tripled. Then in 1977 there was a marked shift in observed global mean surface temperature to a rising fever of about 2 °C/century.
- The expansion of the tropics from overheating is usually thought to be gradual, but the percentage of the land surface in the two most extreme classifications of drought suddenly doubled in 1982 and stayed there until 1997 when it jumped to triple (after six years, it stepped down to double). While their inceptions correlate with the particularly large El Niños of 1982 and 1997, the global drought steps far outlast the 13-month durations of those El Niños.
- There were near-misses for "Burn Locally, Crash Globally" in Amazonia in 1998, 2005, and 2007, each with higher flammability than its predecessor.
- There have also been two occasions when the Atlantic's Meridional Overturning Circulation lost a crucial safety factor. The Greenland Sea flushing at 75 °N shut down in 1978, recovering over the next decade. Then the second-largest flushing site, the Labrador Sea, shut down in 1997 for ten years. While shutdowns

overlapping in time have not been seen during the fifty years of observation, previous total shutdowns had severe worldwide climate consequences.

This makes abrupt climate shifts more like a heart attack than like a chronic disease whose course can be extrapolated. Like heart attacks, some abrupt climate shifts are minor, some are catastrophic—and one cannot predict which or when. The recent track record, however, is that there have been several sudden shifts and several near-misses in each decade since 1976.

Consequential Effects

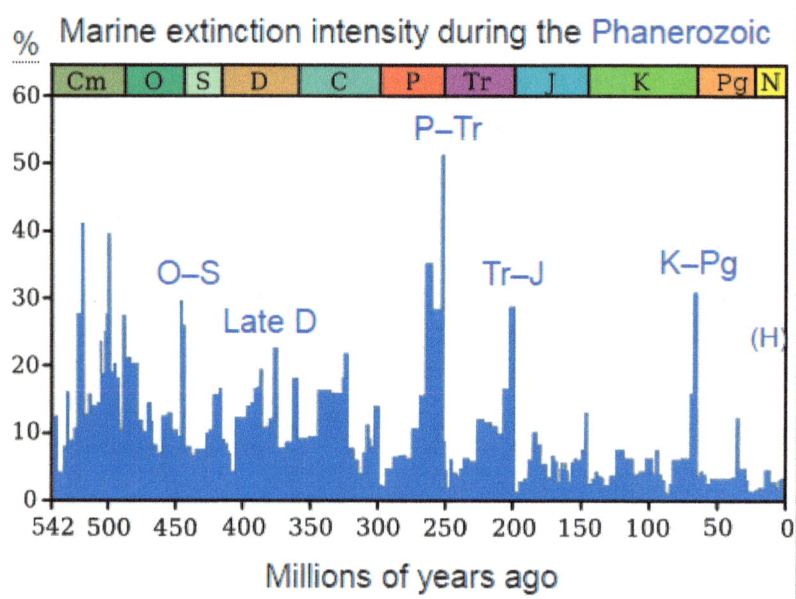

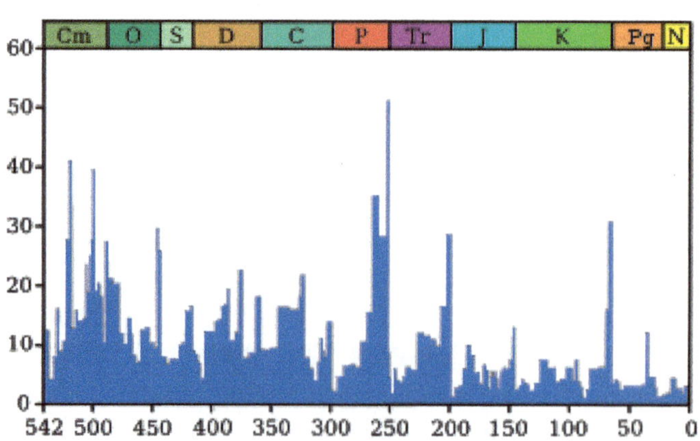

The Permian–Triassic extinction event, labelled "P-Tr" here, is the most significant extinction event in this plot for marine genera.

Abrupt climate change has likely been the cause of wide ranging and severe effects:

- Mass extinctions in the past, most notably the Permian-Triassic Extinction event (often referred to as the great dying) and the Carboniferous Rainforest Collapse, have been suggested as a consequence of abrupt climate change.
- Loss of biodiversity. Without interference from abrupt climate change and other extinction events the biodiversity of this planet would continue to grow.
- Rapid Ocean acidification, which can harm marine life (such as corals).

REFERENCES

1. Marland, G., T.A. Boden, and R. J. Andres. 2007. Global, Regional, and National CO2 Emissions. In Trends: A Compendium of Data on Global Change. Carbon Dioxide Information Analysis Center, Oak Ridge National Laboratory, United States Department of Energy, Oak Ridge, Tenn., U.S.A.
2. ^ Jump up to:[a] [b] Ipsos 2011, p. 3
3. Fisher, B.S.; et al., "Ch. 3: Issues related to mitigation in the long-term context", Contribution of Working Group III to the Fourth Assessment Report of the Intergovernmental Panel on Climate Change, 2007, 3.5 Interaction between mitigation and adaptation, in the light of climate change impacts and decision-making under long-term uncertainty, in IPCC AR4 WG3 2007
4. "Summary for policymakers", Climate Change 2007: Working Group III: Mitigation of Climate Change, Table SPM.3, C. Mitigation in the short and medium term (until 2030), in IPCC AR4 WG3 2007
5. Oppenheimer, M., et al., Section 19.7.1: Relationship between Adaptation Efforts, Mitigation Efforts, and Residual Impacts, in: Chapter 19: Emergent risks and key vulnerabilities (archived July 8 2014), pp.46–49, in IPCC AR5 WG2 A 2014
6. "Social, Economic, and Ethical Concepts and Methods, Executive Summary" (PDF), Climate Change 2014: Mitigation of Climate Change, in IPCC AR5 WG3 2014, p. 211
7. Technology flows and development", Climate Change 2007: Synthesis Report, in IPCC AR4 SYR 2007, p. 68
8. Levine, M.; et al., "Ch 6: Residential and commercial buildings", Climate Change 2007: Working Group III: Mitigation of Climate Change, Sec 6.4.2 Thermal envelope, in IPCC AR4 WG3 2007
9. UNFCCC (5 March 2013), Introduction to the Convention, UNFCCC
10. UNFCCC (2002), Full Text of the Convention, Article 2: Objectives, UNFCCC
11. Oppenheimer, M., et al., FAQ 19.1, in: Chapter 19: Emergent risks and key vulnerabilities (archived July 8 2014), p.52, in IPCC AR5 WG2 A 2014
12. UNFCCC. Conference of the Parties (COP) (15 March 2011), Report of the Conference of the Parties on its sixteenth session, held in Cancun from 29 November to 10 December 2010. Addendum. Part two: Action taken by the Conference of the Parties at its sixteenth session (PDF), Geneva, Switzerland:

United Nations, p.3, paragraph 4. Document available in UN languages and text format.

13. Sutter, John D.; Berlinger, Joshua (12 December 2015). "Final draft of climate deal formally accepted in Paris". CNN. Cable News Network, Turner Broadcasting System, Inc. Retrieved 12 December 2015.
14. Victor, D., et al., Executive summary, in: Chapter 1: Introductory Chapter, p.4 (archived 3 July 2014), in IPCC AR5 WG3 2014
15. The Radical Emissions Reduction Conference, 2013. Held at the Royal Society, Carlton House Terrace, London. 10–11 Dec.
16. The Guardian, 2013. 'Every little helps' is a dangerous mantra for climate change. By Adam Corner, 13 Dec.
17. New Statesman, 2013. How science is telling us all to revolt. By Naomi Klein, 29 Oct.
18. Anderson, K. & Bows, A., 2011. Beyond 'dangerous' climate change: emission scenarios for a new world. Philos. Trans. Royal Society A.
19. Anderson K. (2011, July). Climate Change: Going beyond dangerous – Brutal numbers, tenuous hope, or cognitive dissonance? Video of presentation made to UK Dept. for Internal Development.
20. 4 Degrees and Beyond Conference, official website: Programme, abstracts, presentations, and audio recordings.
21. Meehl, G.A.; et al., "Ch. 10: Global Climate Projections", Climate Change 2007: Working Group I: The Physical Science Basis, FAQ 10.3: If Emissions of Greenhouse Gases are Reduced, How Quickly do Their Concentrations in the Atmosphere Decrease?, in IPCC AR4 WG1 2007, pp. 824–825
22. Rogner, H.-H.; et al. (2007). "1.2 Ultimate objective of the UNFCCC". In B. Metz et al. Introduction. Climate Change 2007: Mitigation. Contribution of Working Group III to the Fourth Assessment Report of the Intergovernmental Panel on Climate Change. Print version: Cambridge University Press, Cambridge, United Kingdom and New York, NY, USA. This version: IPCC website. Retrieved 2011-06-07.
23. Forster, P.; et al., "Ch. 2: Changes in Atmospheric Constituents and in Radiative Forcing", Contribution of Working Group I to the Fourth Assessment Report of the Intergovernmental Panel on Climate Change, 2007, FAQ 2.1 How do Human Activities Contribute to Climate Change and How do They Compare with Natural Influences?, in IPCC AR4 WG1 2007, p. 135
24. IPCC, "Summary for Policymakers", Contribution of Working Group I to the Fourth Assessment Report of the Intergovernmental Panel on Climate Change, 2007, Human and Natural Drivers of Climate Change, in IPCC AR4 WG1 2007
25. U.S. Climate Change Science Program and the Subcommittee on Global Change Research (January 2009). Granger Morgan; H. Dowlatabadi; M. Henrion; D. Keith; R. Lempert; S. McBride; M. Small; T. Wilbanks, eds. "Best practice approaches for characterizing, communicating, and incorporating scientific uncertainty in

decisionmaking". National Oceanic and Atmospheric Administration, Washington D.C., USA. pp. 10–11. Retrieved 2010-06-07.

26. Sterman, J.D.; L.B. Sweeney (2007). "Understanding public complacency about climate change: adults' mental models of climate change violate conservation of matter" (PDF). Climatic Change. 80 (3–4): 221–222. doi:10.1007/s10584-006-9107-5. Retrieved 2011-05-10.

27. 2. Stabilization and Climate Change of the Next Few Decades and Next Several Centuries, p.21, in: Summary, in US NRC 2011

28. Anderson, Kevin; Bows, Alice (13 January 2011). "Beyond ‹dangerous' climate change: emission scenarios for a new world". Philosophical transactions. Series A, Mathematical, physical, and engineering sciences. 369 (1934): 20–44. doi:10.1098/rsta.2010.0290. PMID 21115511.

29. Anderson, Kevin; Bows, Alice (2012). "A new paradigm for climate change". Nature Climate Change. 2 (9): 639–640. doi:10.1038/nclimate1646.

30. Anderson K. (2012). Real clothes for the Emperor: Facing the challenges of climate change. The Cabot annual lecture, Univ. of Bristol. Video, Transcript

31. The Radical Emission Reduction Conference: 10–11 December 2013, sponsored by the Tyndall Centre. Video proceedings on-line.

32. Fisher, B.S.; et al., "Ch 3: Issues related to mitigation in the long-term context", Contribution of Working Group III to the Fourth Assessment Report of the Intergovernmental Panel on Climate Change, 2007, Sec 3.1 Emissions scenarios, in IPCC AR4 WG3 2007

33. Rogner, H.-H.; et al., "Ch 1: Introduction", Contribution of Working Group III to the Fourth Assessment Report of the Intergovernmental Panel on Climate Change, 2007, Sec 1.3.2.4 Total GHG emissions, in IPCC AR4 WG3 2007, p. 111

34. Fisher, B.S.; et al., "Ch 3: Issues related to mitigation in the long-term context", Contribution of Working Group III to the Fourth Assessment Report of the Intergovernmental Panel on Climate Change, 2007, Sec 3.3 Mitigation scenarios, in IPCC AR4 WG3 2007

35. Fisher, B.S.; et al., "Ch 3: Issues related to mitigation in the long-term context", Contribution of Working Group III to the Fourth Assessment Report of the Intergovernmental Panel on Climate Change, 2007, Table 3.5, in: Sec 3.3.5 Long-term stabilization scenarios, in IPCC AR4 WG3 2007

36. BP: Statistical Review of World Energy, Workbook (xlsx), London, 2012

37. World energy outlook 2012 (IEA)

38. World Consumption of Primary Energy by Energy Type and Selected Country Groups 31 December 2008 Microsoft Excel file format

39. Eenergiläget in Sweden 2011 figure 49 and 53

40. Figure 4.10, in: Chapter 4: Stabilization Scenarios, in Clarke & others 2007, p. 103

41. Sec 5.5 Technology flows and development", Climate Change 2007: Synthesis Report, in IPCC AR4 SYR 2007

42. Issues in Science & Technology Online; “Promoting Low-Carbon Electricity Production”

43. Mitigation options”, Climate Change 2007: Synthesis Report, in IPCC AR4 SYR 2007

44. Kahn Ribeiro, S.; et al., “Ch 5: Transport and its infrastructure”, Contribution of Working Group III to the Fourth Assessment Report of the Intergovernmental Panel on Climate Change, 2007, Non-motorized transport (NMT), in: Sec 5.3.1.5 Road transport: mode shifts, in IPCC AR4 WG3 2007

45. Impacts assessment of plug-in hybrid vehicles on electric utilities and regional u.s. power grids” (PDF). Pacific Northwest National Laboratory. 2010.

46. Pacala, Stephen; Socolow, Robert H. (2004). “Stabilization Wedges: Solving the Climate Problem for the Next 50 Years with Current Technologies” (PDF). Science. AAAS. 305 (5686): 968–972. Bibcode:2004Sci...305..968P. doi:10.1126/science.1100103. PMID 15310891. See also: “Stabwedge”. CMI (Carbon Mitigation Initiative) at Princeton University. Resources for Pacala & Socolow(2004)

47. Romm, Joe (19 June 2008). “Cleaning up on carbon”. Nature Reports Climate Change. Retrieved 2 January 2013.

48. Sathaye, J.; et al., “Ch 12: Sustainable Development and mitigation”, Contribution of Working Group III to the Fourth Assessment Report of the Intergovernmental Panel on Climate Change, 2007, Sec 12.2.1.1 Development paths as well as climate policies determine GHG emissions, in IPCC AR4 WG3 2007, pp. 701–703

49. Morita, T.; et al., “Ch 2. Greenhouse Gas Emission Mitigation Scenarios and Implications”, Climate Change 2001: Working Group III: Mitigation, Sec 2.5.2.2 Storylines of Post-SRES Mitigation Scenarios, in IPCC TAR WG3 2001, pp. 149–150

50. Edwin Cartlidge (18 November 2011). “Saving for a rainy day”. Science (Vol 334). pp. 922–924.

51. IEA Renewable Energy Working Party (2002). Renewable Energy... into the mainstream, p. 9.

52. HM Treasury (2006). Stern Review on the Economics of Climate Change.

53. International Energy Agency. IEA urges governments to adopt effective policies based on key design principles to accelerate the exploitation of the large potential for renewable energy 29 September 2008.

54. REN21 (2006). Changing climates: The Role of Renewable Energy in a Carbon-constrained World(PDF) p. 2.

55. New UN report points to power of renewable energy to mitigate carbon emissions UN News Centre, 8 December 2007.

56. Joel Makower, Ron Pernick and Clint Wilder (2008). Clean Energy Trends 2008, Clean Edge, p. 2.

57. United Nations Environment Programme and New Energy Finance Ltd. (2007). Global Trends in Sustainable Energy Investment 2007: Analysis of Trends and

Issues in the Financing of Renewable Energy and Energy Efficiency in OECD and Developing Countries (PDF) p. 3.

58. REN21 (2010). Renewables 2010 Global Status Report p. 15.

59. Conclusion". Worldwide electricity production from renewable energy sources. Paris: Observ'ER. 2012. Retrieved 28 March 2013.

60. http://www.ren21.net/Portals/0/documents/Resources/GSR/2014/GSR2014_full%20report_low%20res.pdf

61. Paul Gipe (4 April 2013). "100 Percent Renewable Vision Building". Renewable Energy World.

62. IPCC (2011). "Special Report on Renewable Energy Sources and Climate Change Mitigation" (PDF). Cambridge University Press, Cambridge, United Kingdom and New York, NY, USA. p. 17. Archived from the original (PDF) on 2014-01-11.

63. International Renewable Energy Agency (2012). "Renewable Power Generation Costs in 2012: An Overview" (PDF).

64. Donald W. Aitken. Transitioning to a Renewable Energy Future, International Solar Energy Society, January 2010, p. 3.

65. REN21 (2012). Renewables Global Status Report 2012 Archived December 15, 2012, at the Wayback Machine. p. 17.

66. REN21 (2011). Renewables 2011: Global Status Report" (PDF). pp. 11–13. Archived from the original (PDF) on 2011-09-05.

67. Top of the list, Renewable Energy World, 2 January 2006.

68. Keith Johnson, Wind Shear: GE Wins, Vestas Loses in Wind-Power Market Race, Wall Street Journal, March 25th 2009, accessed on January 7th 2010.

69. Mark A. Delucchi; Mark Z. Jacobson (2011). "Providing all global energy with wind, water, and solar power, Part II: Reliability, system and transmission costs, and policies" (PDF). Energy Policy. Elsevier Ltd. pp. 1170–1190.

70. Ben Sills (August 29, 2011). "Solar May Produce Most of World's Power by 2060, IEA Says". Bloomberg.

71. Smil, Vaclav (2012). "A Skeptic Looks at Alternative Energy". Spectrum magazine. IEEE. Retrieved 28 March 2013.

72. Japan's nuclear disaster boosts renewables". UPI.com. March 21, 2011.

73. John Vidall (15 March 2011). "Japan nuclear crisis prompts surging investor confidence in renewables". The Guardian. London.

74. Lindsay Morris (25 January 2012). "Obama: Sticking to "Promise of Clean Energy""". Renewable Energy World.

75. REN21 (2010). Renewables 2010 Global Status Report Archived April 16, 2012, at the Wayback Machine. p. 9 & 34.

76. REN21 (2010). Renewables 2010 Global Status Report p. 53.

77. The Nuclear Renaissance – World Nuclear Association".

78. Nuclear Renaissance Threatened as Japan's Reactor Struggles Bloomberg, published March 2011, accessed 2011-03-14
79. Analysis: Nuclear renaissance could fizzle after Japan quake Reuters, published 2011-03-14, accessed 2011-03-14
80. Japan nuclear woes cast shadow over U.S. energy policy Reuters, published 2011-03-13, accessed 2011-03-14
81. NEWS ANALYSIS: Japan crisis puts global nuclear expansion in doubt". Platts. 21 March 2011.
82. WNA (20 June 2013). "Nuclear power down in 2012". World Nuclear News.
83. Sims, R.E.H.; et al., "Ch. 4: Energy supply", Climate Change 2007: Working Group III: Mitigation of Climate Change, Executive summary, in IPCC AR4 WG3 2007
84. IEA (16 June 2010), Technology Roadmap: Nuclear Energy. 2010 Edition (PDF), Organization for Economic Co-operation and Development (OECD) / International Energy Agency (IEA) and OECD / Nuclear Energy Agency (NEA), pp.5–6. Also available in Chinese and Italian
85. Kharecha, Pushker A.; Hansen, James E. (7 May 2013). "Prevented Mortality and Greenhouse Gas Emissions from Historical and Projected Nuclear Power". Environ. Sci. Technol. 47 (9): 4889–4895. doi:10.1021/es3051197.
86. Brian Martin. Opposing nuclear power: past and present, Social Alternatives, Vol. 26, No. 2, Second Quarter 2007, pp. 43–47.
87. M.V. Ramana (July 2011). Nuclear power and the public". Bulletin of the Atomic Scientists. p. 44.
88. Mark Cooper (July 2011). The implications of Fukushima: The US perspective". Bulletin of the Atomic Scientists. p. 9.
89. Good reasons not to waste nuclear 'waste'. Mark Lynas 2011
90. Radioactive Waste Management – Nuclear Waste Disposal". World Nuclear Association. 2016. A typical 1000 MWe light water reactor will generate (directly and indirectly) 200-350 m3 low- and intermediate-level waste per year. It will also discharge about 20 m3 (27 tonnes) of used fuel per year, which corresponds to a 75 m3 disposal volume following encapsulation if it is treated as waste. Where that used fuel is reprocessed, only 3 m3 of vitrified waste (glass) is produced, which is equivalent to a 28 m3 disposal volume following placement in a disposal canister.
91. Choi, Hangbok; Baxter, Alan (1 May 2010). "A comparative study on recycling spent fuels in gas-cooled fast reactors". Annals of Nuclear Energy. 37 (5): 723–729. doi:10.1016/j.anucene.2010.01.014.
92. Publication: Key World Energy Statistics 2016" (PDF). p. 25.
93. World Power Reactors". Archived from the original on March 3, 2008. Retrieved 2016-05-06.
94. Warner, E. S.; Heath, G. A. (2012). "Life Cycle Greenhouse Gas Emissions of Nuclear Electricity Generation". Journal of Industrial Ecology. 16: S73.

doi:10.1111/j.1530-9290.2012.00472.x.

95. IPCC Working Group III – Mitigation of Climate Change, Annex II I: Technology – specific cost and performance parameters" (PDF). IPCC. 2014. p. 10. Retrieved 1 August 2014.

96. Free exchange: Sun, wind and drain". The Economist. 26 July 2014. Sun, wind and drain Wind and solar power are even more expensive than is commonly thought

97. THE NET BENEFITS OF LOW AND NO-CARBON ELECTRICITY TECHNOLOGIES. MAY 2014, Charles Frank PDF

98. Comparing the Costs of Intermittent and Dispatchable Electricity-Generating Technologies", by Paul Joskow, Massachusetts Institute of Technology, September 2011

99. Archived copy" (PDF). Archived from the original (PDF) on October 17, 2008. Retrieved November 21, 2012.

100. Qvist, Staffan A.; Brook, Barry W. (13 May 2015). "Potential for Worldwide Displacement of Fossil-Fuel Electricity by Nuclear Energy in Three Decades Based on Extrapolation of Regional Deployment Data". PLoS ONE. 10 (5): e0124074. doi:10.1371/journal.pone.0124074. PMC 4429979. PMID 25970621.

101. Brook, Barry W. (1 March 2012). "Could nuclear fission energy, etc., solve the greenhouse problem? The affirmative case". Energy Policy. 42: 4–8. doi:10.1016/j.enpol.2011.11.041.

102. Loftus, Peter J.; Cohen, Armond M.; Long, Jane C. S.; Jenkins, Jesse D. (1 January 2015). "A critical review of global decarbonization scenarios: what do they tell us about feasibility?". WIREs Clim Change. 6 (1): 93–112. doi:10.1002/wcc.324.

103. A critical review of global decarbonization scenarios: what do they tell us about feasibility? Open access PDF

104. A critical review of global decarbonization scenarios: what do they tell us about feasibility? Open access PDF. Figure 6

105. Trevor Findlay. The Future of Nuclear Energy to 2030 and its Implications for Safety, Security and Nonproliferation February 4, 2010.

106. Trevor Findlay (2010). The Future of Nuclear Energy to 2030 and its Implications for Safety, Security and Nonproliferation: Overview, The Centre for International Governance Innovation (CIGI), Waterloo, Ontario, Canada, pp. 10–11.

107. M.V. Ramana. Nuclear Power: Economic, Safety, Health, and Environmental Issues of Near-Term Technologies, Annual Review of Environment and Resources, 2009, 34, pp. 144–145.

108. International Energy Agency, World Energy Outlook, 2009, p. 160.

109. James Kanter. In Finland, Nuclear Renaissance Runs Into Trouble New York Times, May 28, 2009.

110. James Kanter. Is the Nuclear Renaissance Fizzling? Green, 29 May 2009.

111. Rob Broomby. Nuclear dawn delayed in Finland BBC News, 8 July 2009.

112. China Builds Nuclear Reactor for 40% Less Than Cost in France, Areva Says". Bloomberg.

113. PRIS – Home". Iaea.org. Retrieved 2016-05-06.

114. Author: Marion Brünglinghaus, ENS, European Nuclear Society. "Nuclear power plants, world-wide". Euronuclear.org. Retrieved 2016-05-06.

115. World Nuclear Association (December 10, 2010). Nuclear Power in China

116. Nuclear Power in the USA". World Nuclear Association. June 2008. Retrieved 2008-07-25.

117. Matthew L. Wald (December 7, 2010). Nuclear 'Renaissance' Is Short on Largess The New York Times.

118. Michael Dittmar. Taking stock of nuclear renaissance that never was Sydney Morning Herald, August 18, 2010.

119. Ipsos (9 March 2012), After Fukushima: Global Opinion on Energy Policy (PDF), p.7. Survey website: After Fukushima: Global Opinion on Energy Policy: Ipsos Public Affairs

120. UK popular support for nuclear power rises -poll". Reuters. 2 July 2016.

121. Annika Breidthardt (May 30, 2011). "German government wants nuclear exit by 2022 at latest". Reuters.

122. Beyond ITER". The ITER Project. Information Services, Princeton Plasma Physics Laboratory. Archived from the original on 7 November 2006. Retrieved 5 February 2011. – Projected fusion power timeline

123. Stewart, C. L.; Stacey, W. M. (1 July 2014). "The SABrR Concept for A Fission-Fusion Hybrid ^{238}U-to-^{239}Pu Fissile Production Reactor". NT. 187 (1). doi:10.13182/NT13-102.

124. Climate Control: a proposal for controlling global greenhouse gas emissions" (PDF). Sustento Institute. Retrieved 2007-12-10.

125. Hackney, Thomas (July 2009). "#7: Moratorium on New Projects for Fossil Fuel Production & Declining Cap on Existing Production" (PDF). BCSEA's Climate Action Portfolio. BC Sustainable Energy Association. Retrieved 2008-04-24.

126. http://www.usclimatenetwork.org/resource-database/report-coal-to-gas-the-influence-of-methane-leakage

127. Tom Wigley

128. http://www.epa.gov/ttnchie1/conference/ei20/session6/gpetron.pdf

129. Greater focus needed on methane leakage from natural gas infrastructure". Pnas.org. Retrieved 2016-05-06.

130. Brandt, A. R.; Heath, G. A.; Kort, E. A.; O'Sullivan, F.; Pétron, G.; Jordaan, S. M.; Tans, P.; Wilcox, J.; Gopstein, A. M.; Arent, D.; others (2014). "Methane leaks from North American natural gas systems"(PDF). Science. 343 (6172): 733–735. Bibcode:2014Sci...343..733B. doi:10.1126/science.1247045. Retrieved 2015-04-29.

131. Air-source heat pumps National Renewable Energy Laboratory June 2011

132. Staffell Iain; et al. (2012). "A review of domestic heat pumps". Energy and Environmental Science. 5 (11): 9291–9306. doi:10.1039/c2ee22653g.

133. Carvalho et al, Ground source heat pump carbon emissions and primary energy reduction potential for heating in buildings in Europe—results of a case study in Portugal. Renewable and Sustainable Energy Reviews 2015; 45,, 755–768 doi:10.1016/j.rser.2015.02.034.

134. Sternberg André, Bardow André (2015). "Power-to-What? – Environmental assessment of energy storage systems. In". Energy and Environmental Science. 8 (2): 389–400. doi:10.1039/c4ee03051f.

135. pg 7

136. Dodge, Edward (December 6, 2014). "Power-to-Gas Enables Massive Energy Storage". TheEnergyCollective.com. Retrieved 25 May 2015.

137. Ground, Lukas; Schulze, Paula; Holstein, Johan (June 20, 2013). Systems Analysis Power to Gas(PDF). Groningen: DNV, KEMA Nederland B.V. Retrieved 25 May 2015.

138. Shell Pearl GTL – Andy Brown," Royal Dutch Shell Company video, March 18, 2011.

139. Scott, Mark (October 7, 2014). "Energy for a Rainy Day, or a Windless One". New York Times. Retrieved 26 May 2015.

140. Randall, Tom (January 30, 2015). "Seven Reasons Cheap Oil Can't Stop Renewables Now". BloombergBusiness. Bloomberg L.P. Retrieved 26 May 2015.

141. Home – 5th Conference on Carbon Dioxide as Feedstock for Fuels, Chemistry and Polymers". Co2-chemistry.eu. 2008-12-01. Retrieved 2016-05-06.

142. Philips Tornado Asian Compact Fluorescent". Philips. Retrieved 2007-12-24.

143. Diesendorf, Mark (2007). Greenhouse Solutions with Sustainable Energy, UNSW Press, p. 86.

144. Loading Order White Paper" (PDF). Retrieved 2010-07-16.

145. Diesendorf, Mark (2007). Greenhouse Solutions with Sustainable Energy, UNSW Press, p. 87.

146. Sophie Hebden (2006-06-22). "Invest in clean technology says IEA report". Scidev.net. Retrieved 2010-07-16.

147. Climate Change". California Air Resources Board. Retrieved 2015-04-29.

148. Roadmap for moving to a low-carbon economy in 2050 – European Commission". Retrieved 2015-04-29.

149. Wei, Max; Nelson, James H; Greenblatt, Jeffery B; Mileva, Ana; Johnston, Josiah; Ting, Michael; Yang, Christopher; Jones, Chris; McMahon, James E; Kammen, Daniel M (2013-03-01). "Deep carbon reductions in California require electrification and integration across economic sectors". Environmental Research Letters. 8 (1): 014038. Bibcode:2013ERL.....8a4038W. doi:10.1088/1748-9326/8/1/014038. ISSN 1748-9326. Retrieved 2015-03-21.

150. Williams, James H. (2013). "The technology path to deep greenhouse gas emissions cuts by 2050: The pivotal role of electricity" (PDF). Science, 335, 6064, 53–59, 2012. Retrieved 2015-03-21.

151. Natural Gas and the Environment". Naturalgas.org. Retrieved 2011-02-06.

152. Annual Energy Outlook 2015 – Energy Information Administration". Eia.gov. 2015-04-14. Retrieved 2016-05-06.

153. E3 Decarbonizing Pipeline 01-27-2015.pdf (PDF), retrieved 2015-04-14

154. Edenhofer, Ottmar; Pichs-Madruga, Ramón; et al. (2014). "Summary for Policymakers". In IPCC. Climate change 2014: mitigation of climate change. Contribution of Working Group III to the Fifth Assessment Report of the Intergovernmental Panel on Climate Change (PDF). Cambridge, UK and New York, NY, USA: Cambridge University Press. ISBN 978-1-107-65481-5. Retrieved 2016-06-21.

155. Fleurbaey, Marc; Kartha, Sivan; et al. (2014). "Chapter 4: Sustainable Development and Equity". In IPCC. Climate change 2014: mitigation of climate change. Contribution of Working Group III to the Fifth Assessment Report of the Intergovernmental Panel on Climate Change (PDF). Cambridge, UK and New York, NY, USA: Cambridge University Press. ISBN 978-1-107-65481-5. Retrieved 2016-06-21.

156. Scarborough, Peter; Appleby, Paul N.; Mizdrak, Anja; Briggs, Adam D.M.; Travis, Ruth C.; Bradbury, Kathryn E.; Key, Timothy J. (July 2014). "Dietary greenhouse gas emissions of meat-eaters, fish-eaters, vegetarians and vegans in the UK" (PDF). Climatic Change. 125 (2): 179–192. doi:10.1007/s10584-014-1169-1. PMC 4372775. PMID 25834298. Retrieved 2016-06-21.

157. Harvey, Fiona (21 March 2016). "Eat less meat to avoid dangerous global warming, scientists say". The Guardian. Retrieved 2016-06-20.

158. Milman, Oliver (20 June 2016). "China's plan to cut meat consumption by 50% cheered by climate campaigners". The Guardian. Retrieved 2016-06-20.

159. Carrington, Damian (7 November 2016). "Tax meat and dairy to cut emissions and save lives, study urges". The Guardian. London, United Kingdom. ISSN 0261-3077. Retrieved 2016-11-07.

160. Springmann, Marco; Mason-D'Croz, Daniel; Robinson, Sherman; Wiebe, Keith; Godfray, H Charles J; Rayner, Mike; Scarborough, Peter (7 November 2016). "Mitigation potential and global health impacts from emissions pricing of food commodities". Nature Climate Change. doi:10.1038/nclimate3155. ISSN 1758-678X.

161. Archived copy". Archived from the original on August 11, 2013. Retrieved July 21, 2013.

162. Archived copy". Archived from the original on May 14, 2013. Retrieved July 21, 2013.

163. Archived copy". Archived from the original on August 11, 2013. Retrieved July 21, 2013.

164. Stern, N. (2006). Stern Review on the Economics of Climate Change: Part III: The Economics of Stabilisation. HM Treasury, London: http://hm-treasury.gov.uk/sternreview_index.htm

165. India should follow China to find a way out of the woods on saving forest people". The Guardian. 22 July 2016. Retrieved 2 November 2016.

166. China's forest tenure reforms". rightsandresources.org. Retrieved 7 August 2016.

167. Ding, Helen; Veit, Peter; Gray, Erin; Reytar, Katie; Altamirano, Juan-Carlos; Blackman, Allen; Hodgdon, Benjamin (October 2016). "Climate benefits, tenure costs: The economic case for securing indigenous land rights in the Amazon". World Resources Institute (WRI). Washington DC, USA. Retrieved 2016-11-02.

168. Ding, Helen; Veit, Peter G; Blackman, Allen; Gray, Erin; Reytar, Katie; Altamirano, Juan-Carlos; Hodgdon, Benjamin (2016). Climate benefits, tenure costs: The economic case for securing indigenous land rights in the Amazon (PDF). Washington DC, USA: World Resources Institute (WRI). ISBN 978-1-56973-894-8. Retrieved 2016-11-02.

169. New Jungles Prompt a Debate on Rain Forests". New York Times. 29 January 2009. Retrieved 18 July 2016.

170. Young, E. (2008). IPCC Wrong On Logging Threat to Climate. New Scientist, August 5, 2008. Retrieved on August 18, 2008, from http://environment.newscientist.com/article/dn14466-ipcc-wrong-on-logging-threat-toclimate.html

171. In Latin America, Forests May Rise to Challenge of Carbon Dioxide". New York Times. 16 May 2016. Retrieved 18 July 2016.

172. How cows could repair the world". nationalgeographic.com. March 6, 2013. Retrieved May 5, 2013.

173. Restoring soil carbon can reverse global warming, desertification and biodiversity". mongabay.com. February 21, 2008. Retrieved May 5, 2013.

174. How eating grass-fed beef could help fight climate change". time.com. January 25, 2010. Retrieved May 11, 2013.

175. Global Status of CCS Report:2011". Global CCS Institute. Retrieved 2011-12-14.

176. OECD Environmental Outlook to 2050, Climate Change Chapter, pre-release version" (PDF). OECD. 2011. Retrieved 2012-04-23.

177. IEA Technology Roadmap Carbon Capture and Storage 2009" (PDF). OECD/IEA. 2009. Retrieved 2012-04-23.

178. Geoengineering the climate: science, governance and uncertainty". The Royal Society. 2009. Retrieved 2012-04-23.

179. Hare, B.; Meinshausen, M. (2006). "How Much Warming are We Committed to and How Much can be Avoided?". Climatic Change. 75: 111–149. doi:10.1007/s10584-005-9027-9.

180. Azar, C., Lindgren, K., Larson, E.D. and Möllersten, K.: (2006)"Carbon capture and storage from fossil fuels and biomass – Costs and potential role in stabilising the atmosphere", Climatic Change, 74, 47–79.

181. OECD Environmental Outlook to 2050, Climate Change Chapter, pre-release version" (PDF). OECD. 2011. Retrieved 2012-01-16.

182. Barker, T.; et al. (2007). 11.2.2 Ocean fertilization and other geo-engineering options. In (book chapter): Mitigation from a cross-sectoral perspective. In: Climate Change 2007: Mitigation. Contribution of Working Group III to the Fourth Assessment Report of the Intergovernmental Panel on Climate Change (B. Metz et al.(eds.)). Print version: Cambridge University Press, Cambridge, U.K., and New York, N.Y., U.S.A.. This version: IPCC website. ISBN 978-0-521-88011-4. Retrieved 2010-04-05.

183. IPCC (2007). C. Mitigation in the short and medium term (until 2030). In (book section): Summary for Policymakers. In: Climate Change 2007: Mitigation. Contribution of Working Group III to the Fourth Assessment Report of the Intergovernmental Panel on Climate Change (B. Metz et al. (eds.)). Print version: Cambridge University Press, Cambridge, U.K., and New York, N.Y., U.S.A.. This version: IPCC website. ISBN 978-0-521-88011-4. Retrieved 2010-05-15.

184. Policy Implications of Greenhouse Warming: Mitigation, Adaptation, and the Science Base (1992), Committee on Science, Engineering, and Public Policy (COSEPUP)

185. GAO (2011). Technical status, future directions, and potential responses. July 2011. GAO-11-71

186. The Royal Society, (2009) Geoengineering the climate: science, governance and uncertainty". Retrieved 2009-09-12.

187. Statoil, Shell in plan to raise oil output by injecting CO_2 – report, AFX News via Forbes, March 8, 2006, checked 2009-01-15

188. MIT Energy Research Council : Research Spotlight". Web.mit.edu. Retrieved 2010-08-26.

189. GreenFuel Technologies Closing Down : Greentech Media". Greentechmedia.com. 2009-05-13. Retrieved 2010-08-26.

190. Launder B.; J.M.T. Thompson (2008). "Global and Arctic climate engineering: numerical model studies". Phil. Trans. R. Soc. A. 366 (1882): 4039–4056. Bibcode:2008RSPTA.366.4039C. doi:10.1098/rsta.2008.0132. PMID 18757275.

191. Shindell, Drew (2012-01-13). "Simultaneously Mitigating Near-Term Climate Change and Improving Human Health and Food Security | Science". Sciencemag.org. Retrieved 2016-05-06.

192. Grubb, M. (July–September 2003). "The Economics of the Kyoto Protocol" (PDF). World Economics. 4 (3): 146–147. Retrieved 2010-03-25.

193. Methane vs. Carbon Dioxide: A Greenhouse Gas Showdown". One Green Planet. Retrieved 2015-11-15.

194. Burp vaccine cuts greenhouse gas emissions Rachel Nowak for NewScientist September 2004

195. Albrittion, D.L.; et al. (2001). "Halocarbons and related compounds". In J.T. Houghton; et al. Technical summary. Climate Change 2001: The physical

science basis. Contribution of Working Group I to the Third Assessment Report of the Intergovernmental Panel on Climate Change. Print version: Cambridge University Press, Cambridge, United Kingdom and New York, NY, USA. This version: GRID-Arendal website. p. 43. Retrieved 2011-06-07.

196. Glossary: A.P.M. Baede. of the main report: S. Solomon et al., eds. (2007). "Definition of "Global Warming Potential"". Annex I: Glossary. Climate Change 2007: The Physical Science Basis. Contribution of Working Group I to the Fourth Assessment Report of the Intergovernmental Panel on Climate Change. Print version: Cambridge University Press, Cambridge, United Kingdom and New York, NY, USA. This version: IPCC website. Retrieved 2011-06-07.

197. Velders, G.J.M.; et al. (20 March 2007). "The importance of the Montreal Protocol in protecting climate". PNAS. 104 (12): 4814–4819. Bibcode:2007PNAS..104.4814V. doi:10.1073/pnas.0610328104. PMC 1817831. PMID 17360370. Retrieved 2011-06-07.

198. The Future of the Canals" (PDF). London Canal Museum. Retrieved 8 September 2013.

199. Lowe, Marcia D. (April 1994). "Back on Track: The Global Rail Revival". Retrieved 2007-02-15.

200. Schwartzman, Peter. "TRUCKS VS. TRAINS—WHO WINS?". Retrieved 2007-02-15.

201. Fulton, William; Pendall, Rolf; Nguyen, Mai; Harrison, Alicia (2001). "Who Sprawls Most? How Growth Patterns Differ Across the U.S" (PDF). Survey Series. Washington D.C.: The Brookings Institution Center on Urban and Metropolitan Policy.

202. Energy Saving Trust: Home and the environment". Energy Saving Trust. Retrieved 2010-08-26.

203. Osborne, Hilary (2005-08-02). "Energy efficiency ‹saves £350m a year'". Guardian Unlimited. London.

204. Fetcher, Ned (December 2009). "Effects of climate and institution size on greenhouse gas emissions from colleges and universities in the United States". Sustainability: the Journal of Record. 2 (6): 362–367. doi:10.1089/SUS.2009.9820. Retrieved 2 March 2011.

205. Greenhouse Gas Emissions of Colleges and Universities". Retrieved 2 March 2011.

206. Rosenfeld, Arthur H.; Romm, Joseph J.; Akbari, Hashem; Lloyd, Alan C. (February–March 1997). "Technology Review". Painting the Town White – and Green. Massachusetts Institute of Technology.

207. Committee on Science, Engineering; Public Policy (1992). Policy Implications of Greenhouse Warming: Mitigation, Adaptation, and the Science Base. Washington, D.C.: National Academy Press. ISBN 0-309-04386-7.

208. Agriculture: Sources of Greenhouse Gas Emissions". EPA. 2015.

209. Susan S. Lang (13 July 2005). "Organic farming produces same corn and soybean

yields as conventional farms, but consumes less energy and no pesticides, study finds". Retrieved 8 July 2008.

210. Pimentel, David; Hepperly, Paul; Hanson, James; Douds, David; Seidel, Rita (2005). "Environmental, Energetic, and Economic Comparisons of Organic and Conventional Farming Systems". BioScience. 55 (7): 573–82. doi:10.1641/0006-3568(2005)055[0573:EEAECO]2.0.CO;2.

211. Lal, Rattan; Griffin, Michael; Apt, Jay; Lave, Lester; Morgan, M. Granger (2004). "Ecology: Managing Soil Carbon". Science. 304 (5669): 393. doi:10.1126/science.1093079. PMID 15087532.

212. A. N. (Thanos) Papanicolaou, Kenneth M. Wacha, Benjamin K. Abban, Christopher G. Wilson, Jerry L. Hatfield, Charles O. Stanier, Timothy R. Filley (2015). "Conservation Farming Shown to Protect Carbon in Soil". Journal of Geophysical Research: Biogeosciences. 120 (11): 2375–2401. doi:10.1002/2015JG003078.

213. Cover Crops, a Farming Revolution With Deep Roots in the Past". The New York Times. 2016.

214. Facing a changing world: women, population and climate, United Nations Population Fund

215. To the point of farce: a martian view of the hardinian taboo—the silence that surrounds population controlMaurice King, Charles Elliott BMJ

216. Who is Heating Up the Planet? A Closer Look at Population and Global Warming from Sierra Club

217. Jowit, Juliette; Wintour, Patrick (26 June 2008). "Cost of tackling global climate change has doubled, warns Stern". The Guardian. London.

218. CIA World Factbook". US CIA. 18 Oct 2011. Retrieved 21 Oct 2011.

219. Yohe, G.W.; et al. (2007). Executive summary. In (book chapter): Perspectives on climate change and sustainability. In: Climate Change 2007: Impacts, Adaptation and Vulnerability. Contribution of Working Group II to the Fourth Assessment Report of the Intergovernmental Panel on Climate Change (M.L. Parry et al., (eds.)). Print version: Cambridge University Press, Cambridge, U.K., and New York, N.Y., U.S.A.. Web version: IPCC website. ISBN 978-0-521-88010-7. Retrieved 2010-05-15.

220. Toth, F.L.; et al. (2001). 10.4.7 Emerging Conclusions with Respect to Policy-relevant Scientific Questions. In (book chapter): Decision-making Frameworks. In: Climate Change 2001: Mitigation. Contribution of Working Group III to the Third Assessment Report of the Intergovernmental Panel on Climate Change (B. Metz et al. Eds.). Print version: Cambridge University Press, Cambridge, U.K., and New York, N.Y., U.S.A.. This version: GRID-Arendal website]. ISBN 978-0-521-01502-8. Retrieved 2010-01-10.

221. Rogner, H.-H.; et al. (2007). Executive Summary. In (book chapter): Introduction. In: Climate Change 2007: Mitigation. Contribution of Working Group III to the Fourth Assessment Report of the Intergovernmental Panel on Climate Change (B. Metz et al. (eds)). Print version: Cambridge University Press, Cambridge,

United Kingdom and New York, NY, USA. Web version: IPCC website. ISBN 978-0-521-88011-4. Retrieved 2010-05-05.

222. Banuri, T.; et al. (1996). Equity and Social Considerations. In: Climate Change 1995: Economic and Social Dimensions of Climate Change. Contribution of Working Group III to the Second Assessment Report of the Intergovernmental Panel on Climate Change (J.P. Bruce et al. Eds.) (PDF). This version: Printed by Cambridge University Press, Cambridge, U.K., and New York, N.Y., U.S.A.. PDF version: IPCC website. doi:10.2277/0521568544. ISBN 978-0-521-56854-8.

223. Goldemberg, J.; et al. (1996). Introduction: scope of the assessment. In: Climate Change 1995: Economic and Social Dimensions of Climate Change. Contribution of Working Group III to the Second Assessment Report of the Intergovernmental Panel on Climate Change (J.P. Bruce et al. Eds.) (PDF). This version: Printed by Cambridge University Press, Cambridge, U.K., and New York, N.Y., U.S.A.. Web version: IPCC website. doi:10.2277/0521568544. ISBN 978-0-521-56854-8.

224. Article Adaptation. If you can't stand the heat", The Economist, special report on "Climate change", 28 November 2015, page 10-12.

225. World Bank (2010). World Development Report 2010: Development and Climate Change. The International Bank for Reconstruction and Development / The World Bank, 1818 H Street NW, Washington DC 20433. doi:10.1596/978-0-8213-7987-5. ISBN 978-0-8213-7987-5. Retrieved 2010-04-06.

226. PBL Netherlands Environment Agency (15 June 2012), "Figure 6.14, in: Chapter 6: The energy and climate challenge", in van Vuuren, D.; M. Kok, Roads from Rio+20 (PDF), ISBN 978-90-78645-98-6, p.177, Report no: 500062001. Report website.

227. Jaeger, C.C.; J. Jaeger (2011), "Three views of two degrees" (PDF), Regional Environmental Change, Springer-Verlag, 11 (1), pp. 15–26, doi:10.1007/s10113-010-0190-9, ISSN 1436-3798

228. Rijsberman, F.J.; R.J. Stewart, eds. (1990), Targets and Indicators of Climate Change, Stockholm, Sweden: Stockholm Environment Institute, ISBN 91-88116-21-2. Summary available from the Climate Emergency Institute.

229. UNFCCC (15 March 2011), FCCC/CP/2010/7/Add.1: Report of the Conference of the Parties on its sixteenth session, held in Cancun from 29 November to 10 December 2010. Addendum. Part two: Action taken by the Conference of the Parties at its sixteenth session (PDF), Geneva, Switzerland: UN Office. Available as a PDF in English, Spanish, French, Arabic, and Russian.

230. UNFCCC (3 May 2012), Essential Background, UNFCCC

231. Oliver Geden (2013), Modifying the 2°C Target. Climate Policy Objectives in the Contested Terrain of Scientific Policy Advice, Political Preferences, and Rising Emissions, SWP Research Paper 5/13

232. Oliver Geden (2010), What Comes After the Two-Degree Target?, SWP Comments 19

233. EU climate change target "unfeasible"". EurActiv.com. 2006-02-01. Retrieved 2007-02-21.

234. Adam, David (14 April 2009). "World will not meet 2C warming target, climate change experts agree". London: Guardian News and Media Limited. Retrieved 2009-04-14.

235. United States Department of Energy World Trends

236. Oppenheimer, M., et al., Section 19.7.2: Limits to Mitigation, in: Chapter 19: Emergent risks and key vulnerabilities (archived July 8 2014), pp.49–50, in IPCC AR5 WG2 A 2014

237. Oliver Geden/Silke Beck: Renegotiating the global climate stabilization target. In: Nature Climate Change, 4, 2014, pp. 747–748

238. Anderson, K. & Bows, A., 2011. Beyond ‹dangerous' climate change: emission scenarios for a new world. Philos. Trans. Royal Society A.

239. Alcamo & others 2013, p. xi

240. Alcamo & others 2013, pp. xiii–xiv

241. "Implication for Carbon Emissions Target," in: Hansen & others 2013, p. 15

242. "4. Discussion and conclusions," in: Luderer & others 2013, p. 6

243. Report on the structured expert dialogue on the 2013–2015 review" (PDF). UNFCCC, Subsidiary Body for Scientific and Technological Advice & Subsidiary Body for Implementation. 2015-04-04. Retrieved 2016-06-21.

244. 1.5°C temperature limit – key facts". Climate Analytics. Retrieved 2016-06-21.

245. Gupta, S.; et al. (2007). "13.2.1.2 Taxes and charges". In B. Metz; et al. Policies, instruments, and co-operative arrangements. Climate Change 2007: Mitigation. Contribution of Working Group III to the Fourth Assessment Report of the Intergovernmental Panel on Climate Change. Print version: Cambridge University Press, Cambridge, U.K., and New York, N.Y., U.S.A.. This version: IPCC website. Retrieved 2010-03-18.

246. Jacobson, M.Z. (2009). "Review of solutions to global warming, air pollution, and energy security" (PDF). Energy and Environmental Science. 2 (2): 148–73. doi:10.1039/b809990c.

247. Jacobson, M.Z.; Delucchi, M.A. (2009). "A Plan to Power 100 Percent of the Planet with Renewables" (originally published as "A Path to Sustainable Energy by 2030")". Scientific American. 301 (5): 58–65. doi:10.1038/scientificamerican1109-58. PMID 19873905.

248. Farah, Paolo Davide (2015). "Sustainable Energy Investments and National Security: Arbitration and Negotiation Issues". Journal of World Energy Law and Business. 8 (6). Retrieved 26 November 2015.

249. How high-pressure politics threatens action on climate The Observer June 2005

250. StoryOfStuff.com (2009) The Story of Cap and Trade"

251. Success of Northeast Cap-and-Trade System Shows Market-Based Climate Policy Is Well Within Reach".

252. Emission Trading Scheme (EU ETS) from ec.europa.eu

253. The $20,000,000,000,000 question Robins, Nick for Opendemocracy

254. State and Trends of the Carbon Market International Emissions Trading Association 2005
255. Statement of G8 Climate Change Roundtable Archived May 8, 2013, at the Wayback Machine. Convened by the World Economic Forum June 2005
256. Evans. J (forthcoming 2012) Environmental Governance, Routledge, Oxon
257. Biesbroek. G.R, Termeer. C.J.A.M, Kabat. P, Klostermann.J.E.M (unpublished) Institutional governance barriers for the development and implementation of climate adaptation strategies, Working paper for the International Human Dimensions Programme (IHDP) conference "Earth System Governance: People, Places, and the Planet", December 2–4, Amsterdam, the Netherlands
258. Mee. L. D, Dublin. H. T, Eberhard. A. A (2008) Evaluating the Global Environment Facility: A goodwill gesture or a serious attempt to deliver global benefits?, Global Environmental Change 18, 800–810
259. http://www10.iadb.org/intal/intalcdi/pe/2009/04268.pdf
260. Preston. B. L, Westaway. R. M, Yuen. E. Y (2004) Climate adaptation planning in practice: an evaluation of adaptation plans from three developed nations, European Management Journal, 22(3) 304–314
261. UNFCCC (2011) Report on the twentieth meeting of the Least Developed Countries Expert Group, Subsidiary Body for Implementation, United Nations Framework Convention on Climate Change
262. UNFCCC (2011) Annual report of the Joint Implementation Supervisory Committee to the Conference of the Parties serving as the meeting of the Parties to the Kyoto Protocol, United Nations Framework Convention on Climate Change
263. Industrial Technologies Program: BestPractices". Eere.energy.gov. Retrieved 2010-08-26.
264. Clinton Hails Global Warming Pact". All Politics. CNN. 1997-12-11. Retrieved 2006-11-05.
265. How the White House Worked to Scuttle California's Climate Law", San Francisco Chronicle, September 25, 2007
266. Reuters, January 30, 2007, free archived version at http://www.commondreams.org/headlines07/0130-10.htm, last visited Jan. 30, ‹07
267. Written testimony of Dr. Grifo before the Committee on Oversight and Government Reform of the U.S. House of Representatives on January 30, 2007, archived at "Archived copy" (PDF). Archived from the original (PDF) on 2009-08-05. Retrieved 2009-12-15.
268. written testimony of Rick Piltz before the Committee on Oversight and Government Reform of the U.S. House of Representatives on January 30, 2007, archived at http://oversight.house.gov/Documents/20070130113813-92288.pdf last visited Jan. 30, 07
269. Barringer, Felicity (2012-10-13). "In California, a Grand Experiment to Rein in Climate Change". The New York Times.
270. Prototype Carbon Fund from the World Bank Carbon Finance Unit

271. Jessica Brown, Neil Bird and Liane Schalatek (2010) Climate finance additionality: emerging definitions and their implications Overseas Development Institute

272. Free trade can help combat global warming, finds UN report UN News Centre, 26 June 2009

273. Andrew Biggin (16 August 2007). "Scientific bodies must take own action on emissions". Nature. 448 (7155): 749. Bibcode:2007Natur.448..749B. doi:10.1038/448749a. PMID 17700677.

274. Anderson, K; Bows, A (2008). "Reframing the climate change challenge in light of post-2000 emission trends". Philosophical Transactions of the Royal Society A: Mathematical, Physical and Engineering Sciences. 366 (1882): 3863–3882. Bibcode:2008RSPTA.366.3863A. doi:10.1098/rsta.2008.0138.

275. Anderson, K (June 17, 2008). "Reframing climate change: from long-term targets to emission pathways". (esp. slide 24 onward)

276. Gössling S, Ceron JP, Dubois G, Hall CM, Gössling IS, Upham P, Earthscan London (2009). Hypermobile travellers. and Implications for Carbon Dioxide Emissions Reduction. In: Climate Change and Aviation: Issues, Challenges and Solutions, London.

277. Video: Paradise lost? - Need to Know". PBS. Palau suing the industrialized countries over global warming

278. Environmental Integrity Project, Sierra Club Announce Plans to Sue EPA Unless It Revises Nitrogen Oxide Emissions Standard, Curbs Nitrous Oxide Pollution Linked to Global Warming – NewsOn6.com – Tulsa, OK – News, Weather, Video and Sports – KOTV.com -".

279. Edward Lorenz (1982): "Climate is what you expect, weather is what you get"

280. Stott, et al. (2004), "Human contribution to the European heatwave of 2003", Nature, Vol. 432, 2 December 2004

281. Grossman, Columbia J. of Env. Law, 2003

282. Climate change ‹ruining' Everest". Heatisonline.org. 2004-11-17. Retrieved 2010-08-26.

283. Climate change ‹ruining' Belize BBC November 2004

284. Climate Justice Ongoing Cases

285. Hague, Arthur Neslen The (24 June 2015). "Dutch government ordered to cut carbon emissions in landmark ruling". The Guardian.

286. Press release (29 January 2004). Archived press release: Exxonmobil's contribution to global warming revealed. Friends of the Earth Trust. Retrieved May 25, 2015.

287. New York is investigating Exxon Mobil for allegedly misleading the public about climate change". The Washington Post. 2015-11-05. Retrieved 2015-12-29.

CHAPTER 5

Climate Models

Chapter Contents

A climate model is a tool used by scientists to examine the climate of the earth. These models can be small and used to examine the climate of a small area, but more often they are large, global climate models that examine the interrelatedness of the earth's climate across all oceans and land masses. In order to make one of these models, scientists rely on supercomputers that solve various complex equations for every given point in a three dimensional grid of atmosphere. The sophistication of these models has improved a great deal over the past years and scientists are often able to accurately predict the global effects of local events such as volcanic eruptions or spilled pollutants.

INTRODUCTION

Climate models use quantitative methods to simulate the interactions of the important drivers of climate, including atmosphere, oceans, land surface and ice. They are used for a variety of purposes from study of the dynamics of the climate system to projections of future climate.

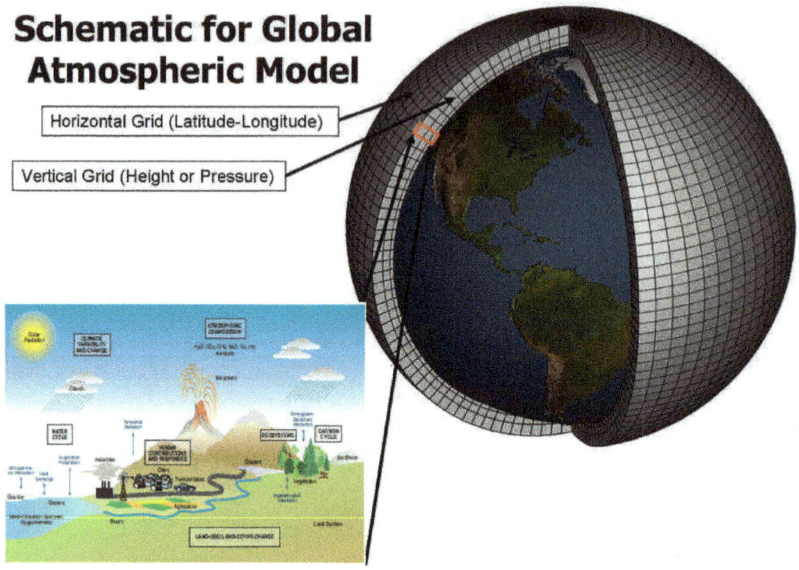

Climate models are systems of differential equations based on the basic laws of physics, fluid motion, and chemistry. To "run" a model, scientists divide the planet into a 3-dimensional grid, apply the basic equations, and evaluate the results. Atmospheric models calculate winds, heat transfer, radiation, relative humidity, and surface hydrology within each grid and evaluate interactions with neighboring points.

All climate models take account of incoming energy from the sun as short wave electromagnetic radiation, chiefly visible and short-wave (near) infrared, as well as outgoing long wave (far) infrared electromagnetic. Any imbalance results in a change in temperature.

Models vary in complexity:

- A simple radiant heat transfer model treats the earth as a single point and averages outgoing energy
- This can be expanded vertically (radiative-convective models) and/or horizontally
- Finally, (coupled) atmosphere–ocean–sea ice **global climate models** solve the full equations for mass and energy transfer and radiant exchange.
- Box models can treat flows across and within ocean basins.
- Other types of modelling can be interlinked, such as land use, allowing researchers to predict the interaction between climate and ecosystems.

BOX MODELS

Box models are simplified versions of complex systems, reducing them to boxes (or reservoirs) linked by fluxes. The boxes are assumed to be mixed homogeneously. Within a given box, the concentration of any chemical species is therefore uniform. However, the abundance of a species within a given box may vary as a function of time due to the input to (or loss from) the box or due to the production, consumption or decay of this species within the box.

Simple box models, i.e. box model with a small number of boxes whose properties (e.g. their volume) do not change with time, are often useful to derive analytical formulas describing the dynamics and steady-state abundance of a species. More complex box models are usually solved using numerical techniques.

Box models are used extensively to model environmental systems or ecosystems and in studies of ocean circulation and the carbon cycle.

ZERO-DIMENSIONAL MODELS

A very simple model of the radiative equilibrium of the Earth is

$$(1-a)S\pi r^2 = 4\pi r^2 \epsilon\sigma T^4$$

where

- the left hand side represents the incoming energy from the Sun
- the right hand side represents the outgoing energy from the Earth, calculated from the Stefan-Boltzmann law assuming a model-fictive temperature, *T*, sometimes called the 'equilibrium temperature of the Earth', that is to be found,

and

- *S* is the solar constant – the incoming solar radiation per unit area—about 1367 $W \cdot m^{-2}$
- a is the Earth›s average albedo, measured to be 0.3.
- *r* is Earth's radius—approximately 6.371×10^6m
- ϖ is the mathematical constant (3.141...)
- σ is the Stefan-Boltzmann constant—approximately $5.67 \times 10^{-8}\ J \cdot K^{-4} \cdot m^{-2} \cdot s^{-1}$
- ϵ is the effective emissivity of earth, about 0.612

The constant ϖr^2 can be factored out, giving

$$(1-a)S = 4\epsilon\sigma T^4$$

Solving for the temperature,

$$T = \sqrt[4]{\frac{(1-a)S}{4\epsilon\sigma}}$$

This yields an apparent effective average earth temperature of 288 K (15 °C; 59 °F). This is because the above equation represents the effective *radiative* temperature of the Earth (including the clouds and atmosphere). The use of effective emissivity and albedo account for the greenhouse effect.

This very simple model is quite instructive, and the only model that could fit on a page. For example, it easily determines the effect on average earth temperature of changes in solar constant or change of albedo or effective earth emissivity.

The average emissivity of the earth is readily estimated from available data. The emissivities of terrestrial surfaces are all in the range of 0.96 to 0.99 (except for some small desert areas which may be as low as 0.7). Clouds, however, which cover about half of the earth's surface, have an average emissivity of about 0.5 (which must be reduced by the fourth power of the ratio of cloud absolute temperature to average earth absolute temperature) and an average cloud temperature of about 258 K (−15 °C; 5 °F).Taking all this properly into account results in an effective earth emissivity of about 0.64 (earth average temperature 285 K (12 °C; 53 °F)).

This simple model readily determines the effect of changes in solar output or change of earth albedo or effective earth emissivity on average earth temperature. It says nothing, however about what might cause these things to change. Zero-dimensional models do not address the temperature distribution on the earth or the factors that move energy about the earth.

RADIATIVE-CONVECTIVE MODELS

The zero-dimensional model above, using the solar constant and given average earth temperature, determines the effective earth emissivity of long wave radiation emitted to space. This can be refined in the vertical to a one-dimensional radiative-convective model, which considers two processes of energy transport:

- upwelling and downwelling radiative transfer through atmospheric layers that both absorb and emit infrared radiation
- upward transport of heat by convection (especially important in the lower troposphere).

The radiative-convective models have advantages over the simple model: they can determine the effects of varying greenhouse gas concentrations on effective emissivity and therefore the surface temperature. But added parameters are needed to determine local emissivity and albedo and address the factors that move energy about the earth.

Effect of ice-albedo feedback on global sensitivity in a one-dimensional radiative-convective climate model.

HIGHER-DIMENSION MODELS

The zero-dimensional model may be expanded to consider the energy transported horizontally in the atmosphere. This kind of model may well be zonally averaged. This model has the advantage of allowing a rational dependence of local albedo and emissivity on temperature – the poles can be allowed to be icy and the equator warm – but the lack of true dynamics means that horizontal transports have to be specified.

EMICs (Earth-system models of intermediate complexity)

Depending on the nature of questions asked and the pertinent time scales, there are, on the one extreme, conceptual, more inductive models, and, on the other extreme, general circulation models operating at the highest spatial and temporal resolution currently feasible. Models of intermediate complexity bridge the gap. One example is the Climber-3 model. Its atmosphere is a 2.5-dimensional statistical-dynamical model with 7.5° × 22.5° resolution and time step of half a day; the ocean is MOM-3 (Modular Ocean Model) with a 3.75° × 3.75° grid and 24 vertical levels.

GCMS (GLOBAL CLIMATE MODELS OR GENERAL CIRCULATION MODELS)

General Circulation Models (GCMs) discretise the equations for fluid motion and energy transfer and integrate these over time. Unlike simpler models, GCMs divide the atmosphere and/or oceans into grids of discrete "cells", which represent computational units. Unlike simpler models which make mixing assumptions, processes internal to a cell—such as convection—that occur on scales too small to be resolved directly are parameterised at the cell level, while other functions govern the interface between cells.

Atmospheric GCMs (AGCMs) model the atmosphere and impose sea surface temperatures as boundary conditions. Coupled atmosphere-ocean GCMs (AOGCMs, e.g. HadCM3, EdGCM, GFDL CM2.X, ARPEGE-Climat) combine the two models. The first general circulation climate model that combined both oceanic and atmospheric processes was developed in the late 1960s at the NOAA Geophysical Fluid Dynamics Laboratory AOGCMs represent the pinnacle of complexity in climate models and internalise as many processes as possible. However, they are still under development and uncertainties remain. They may be coupled to models of other processes, such as the carbon cycle, so as to better model feedback effects. Such integrated multi-system models are sometimes referred to as either "earth system models or "global climate models.

RESEARCH AND DEVELOPMENT

There are three major types of institution where climate models are developed, implemented and used:

- National meteorological services. Most national weather services have a climatology section.
- Universities. Relevant departments include atmospheric sciences, meteorology, climatology, and geography.
- National and international research laboratories. Examples include the National Center for Atmospheric Research (NCAR, in Boulder, Colorado, USA), the Geophysical Fluid Dynamics Laboratory (GFDL, in Princeton, New Jersey, USA), the Hadley Centre for Climate Prediction and Research (in Exeter, UK), the Max Planck Institute for Meteorology in Hamburg, Germany, or the Laboratoire des Sciences du Climat et de l›Environnement (LSCE), France, to name but a few.

The World Climate Research Programme (WCRP), hosted by the World Meteorological Organization (WMO), coordinates research activities on climate modelling worldwide. A 2012 U.S. National Research Council report discussed how the large and diverse U.S.

climate modeling enterprise could evolve to become more unified. Efficiencies could be gained by developing a common software infrastructure shared by all U.S. climate researchers, and holding an annual climate modeling forum, the report found.

GENERAL CIRCULATION MODEL

A **general circulation model (GCM)** is a type of climate model. It employs a mathematical model of the general circulation of a planetary atmosphere or ocean. It uses the Navier–Stokes equations on a rotating sphere with thermodynamic terms for various energy sources (radiation, latent heat). These equations are the basis for computer programs used to simulate the Earth's atmosphere or oceans. Atmospheric and oceanic GCMs (AGCM and OGCM) are key components along with sea ice and land-surface components.

GCMs and global climate models are used for weather forecasting, understanding the climate and forecasting climate change.

Versions designed for decade to century time scale climate applications were originally created by Syukuro Manabe and Kirk Bryan at the Geophysical Fluid Dynamics Laboratory in Princeton, New Jersey. These models are based on the integration of a variety of fluid dynamical, chemical and sometimes biological equations.

Terminology

The acronym *GCM* originally stood for *General Circulation Model*. Recently, a second meaning came into use, namely *Global Climate Model*. While these do not refer to the same thing, General Circulation Models are typically the tools used for modelling climate, and hence the two terms are sometimes used interchangeably. However, the term "global climate model" is ambiguous and may refer to an integrated framework that incorporates multiple components including a general circulation model, or may refer to the general class of climate models that use a variety of means to represent the climate mathematically.

In 1956, Norman Phillips developed a mathematical model that could realistically depict monthly and seasonal patterns in the troposphere. It became the first successful climate model. Following Phillips›s work, several groups began working to create GCMs. The first to combine both oceanic and atmospheric processes was developed in the late 1960s at the NOAA Geophysical Fluid Dynamics Laboratory. By the early 1980s, the United States› National Center for Atmospheric Research had developed the Community Atmosphere Model; this model has been continuously refined. In 1996, efforts began to model soil and vegetation types. Later the Hadley Centre for Climate Prediction and Research›s HadCM3 model coupled ocean-atmosphere elements. The role of gravity waves was added in the mid-1980s. Gravity waves are required to simulate regional and global scale circulations accurately.

Atmospheric and Oceanic Models

Atmospheric (AGCMs) and oceanic GCMs (OGCMs) can be coupled to form an atmosphere-ocean coupled general circulation model (CGCM or AOGCM). With the addition of submodels such as a sea ice model or a model for evapotranspiration over land, AOGCMs become the basis for a full climate model.

Trends

A recent trend in GCMs is to apply them as components of Earth system models, e.g. by coupling ice sheet models for the dynamics of the Greenland and Antarctic ice sheets, and one or more chemical transport models (CTMs) for species important to climate. Thus a carbon CTM may allow a GCM to better predict anthropogenic changes in carbon dioxide concentrations. In addition, this approach allows accounting for inter-system feedback: e.g. chemistry-climate models allow the possible effects of climate change on ozone hole to be studied.

Climate prediction uncertainties depend on uncertainties in chemical, physical and social models (see IPCC scenarios below). Significant uncertainties and unknowns remain, especially regarding the future course of human population, industry and technology.

Structure

Three-dimensional (more properly four-dimensional) GCMs apply discrete equations for fluid motion and integrate these forward in time. They contain parameterisations for processes such as convection that occur on scales too small to be resolved directly.

A simple general circulation model (SGCM) consists of a dynamic core that relates properties such as temperature to others such as pressure and velocity. Examples are programs that solve the primitive equations, given energy input and energy dissipation in the form of scale-dependent friction, so that atmospheric waves with the highest wavenumbers are most attenuated. Such models may be used to study atmospheric processes, but are not suitable for climate projections.

Atmospheric GCMs (AGCMs) model the atmosphere (and typically contain a land-surface model as well) using imposed sea surface temperatures (SSTs). They may include atmospheric chemistry.

AGCMs consist of a dynamical core which integrates the equations of fluid motion, typically for:

- surface pressure
- horizontal components of velocity in layers
- temperature and water vapor in layers
- radiation, split into solar/short wave and terrestrial/infra-red/long wave
- parameters for:
- convection
- land surface processes
- albedo
- hydrology
- cloud cover

A GCM contains prognostic equations that are a function of time (typically winds, temperature, moisture, and surface pressure) together with diagnostic equations that are evaluated from them for a specific time period. As an example, pressure at any height can be diagnosed by applying the hydrostatic equation to the predicted surface pressure and the predicted values of temperature between the surface and the height of interest. Pressure is used to compute the pressure gradient force in the time-dependent equation for the winds.

OGCMs model the ocean (with fluxes from the atmosphere imposed) and may contain a sea ice model. For example, the standard resolution of HadOM3 is 1.25 degrees in latitude and longitude, with 20 vertical levels, leading to approximately 1,500,000 variables.

AOGCMs (e.g. HadCM3, GFDL CM2.X) combine the two submodels. They remove the need to specify fluxes across the interface of the ocean surface. These models are the basis for model predictions of future climate, such as are discussed by the IPCC. AOGCMs internalise as many processes as possible. They have been used to provide predictions at a regional scale. While the simpler models are generally susceptible to analysis and their results are easier to understand, AOGCMs may be nearly as hard to analyse as the climate itself.

Grid

The fluid equations for AGCMs are made discrete using either the finite difference method or the spectral method. For finite differences, a grid is imposed on the atmosphere. The simplest grid uses constant angular grid spacing (i.e., a latitude / longitude grid). However, non-rectangular grids (e.g., icosahedral) and grids of variable resolution are more often used. The LMDz model can be arranged to give high resolution over any given section of the planet. HadGEM1 (and other ocean models) use an ocean grid with higher resolution in the tropics to help resolve processes believed to be important for the El Niño Southern Oscillation (ENSO). Spectral models generally use a gaussian grid, because of the mathematics of transformation between spectral and grid-point space. Typical AGCM resolutions are between 1 and 5 degrees in latitude or longitude: HadCM3, for example, uses 3.75 in longitude and 2.5 degrees in latitude, giving a grid of 96 by 73 points (96 x 72 for some variables); and has 19 vertical levels. This results in approximately 500,000 "basic variables, since each grid point has four variables (u,v, T, Q), though a full count would give more (clouds; soil levels). HadGEM1 uses a grid of 1.875 degrees in longitude and 1.25 in latitude in the atmosphere; HiGEM, a high-resolution variant, uses 1.25 x 0.83 degrees respectively. These resolutions are lower than is typically used for weather forecasting. Ocean resolutions tend to be higher, for example HadCM3 has 6 ocean grid points per atmospheric grid point in the horizontal.

For a standard finite difference model, uniform gridlines converge towards the poles. This would lead to computational instabilities (see CFL condition) and so the model variables must be filtered along lines of latitude close to the poles. Ocean models suffer from this problem too, unless a rotated grid is used in which the North Pole is shifted onto a nearby landmass. Spectral models do not suffer from this problem. Some experiments use geodesic grids and icosahedral grids, which (being more uniform) do not have pole-problems. Another approach to solving the grid spacing problem is to deform a Cartesian cube such that it covers the surface of a sphere.

Flux Buffering

Some early versions of AOGCMs required an *ad hoc* process of "flux correction" to achieve a stable climate. This resulted from separately prepared ocean and atmospheric models that each used an implicit flux from the other component different than that component could produce. Such a model failed to match observations. However, if the fluxes were 'corrected', the factors that led to these unrealistic fluxes might be unrecognised, which could affect model sensitivity. As a result, the vast majority of models used in the current round of IPCC reports do not use them. The model improvements that now make flux corrections unnecessary include improved ocean physics, improved resolution in both atmosphere and ocean, and more physically consistent coupling between atmosphere and ocean submodels. Improved models now maintain stable, multi-century simulations of surface climate that are considered to be of sufficient quality to allow their use for climate projections.

Convection

Moist convection releases latent heat and is important to the Earth's energy budget. Convection occurs on too small a scale to be resolved by climate models, and hence it must be handled via parameters. This has been done since the 1950s. Akio Arakawa did much of the early work, and variants of his scheme are still used, although a variety of different schemes are now in use. Clouds are also typically handled with a parameter, for a similar lack of scale. Limited understanding of clouds has limited the success of this strategy, but not due to some inherent shortcoming of the method.

Software

Most models include software to diagnose a wide range of variables for comparison with observations or study of atmospheric processes. An example is the 1.5-metre temperature, which is the standard height for near-surface observations of air temperature. This temperature is not directly predicted from the model but is deduced from surface and lowest-model-layer temperatures. Other software is used for creating plots and animations.

Projections

Projected annual mean surface air temperature from 1970-2100, based on SRES emissions scenario A1B, using the NOAA GFDL CM2.1 climate model (credit: NOAAGeophysical Fluid Dynamics Laboratory).

Coupled AOGCMs use transient climate simulations to project/predict climate changes under various scenarios. These can be idealised scenarios (most commonly, CO_2emissions increasing at 1%/yr) or based on recent history (usually the "IS92a" or more recently the SRES scenarios). Which scenarios are most realistic remains uncertain.

The 2001 IPCC Third Assessment Report F igure 9.3 shows the global mean response of 19 different coupled models to an idealised experiment in which emissions increased at 1% per year. Figure 9.5shows the response of a smaller number of models to more recent trends. For the 7 climate models shown there, the temperature change to 2100 varies from 2 to 4.5 °C with a median of about 3 °C.

Future scenarios do not include unknown events – for example, volcanic eruptions or changes in solar forcing. These effects are believed to be small in comparison to greenhouse gas (GHG) forcing in the long term, but large volcanic eruptions, for example, can exert a substantial temporary cooling effect.

Human GHG emissions are a model input, although it is possible to include an economic/technological submodel to provide these as well. Atmospheric GHG levels are usually supplied as an input, though it is possible to include a carbon cycle model that reflects vegetation and oceanic processes to calculate such levels.

Emissions Scenarios

Projected change in annual mean surface air temperature from the late 20th century to the middle 21st century, based on SRES emissions scenario A1B (credit: NOAA Geophysical Fluid Dynamics Laboratory).

For the six SRES marker scenarios, IPCC (2007:7–8) gave a "best estimate of global mean temperature increase (2090–2099 relative to the period 1980–1999) of 1.8 °C to 4.0 °C.Over the same time period, the "likely range (greater than 66% probability, based on expert judgement) for these scenarios was for a global mean temperature increase of 1.1 to 6.4 °C.

In 2008 a study made climate projections using several emission scenarios. In a scenario where global emissions start to decrease by 2010 and then declined at a sustained rate of 3% per year, the likely global average temperature increase was predicted to be 1.7 °C above pre-industrial levels by 2050, rising to around 2 °C by 2100. In a projection designed to simulate a future where no efforts are made to reduce global emissions, the likely rise in global average temperature was predicted to be 5.5 °C by 2100. A rise as high as 7 °C was thought possible, although less likely.

Another no-reduction scenario resulted in a median warming over land (2090–99 relative to the period 1980–99) of 5.1 °C. Under the same emissions scenario but with a different model, the predicted median warming was 4.1 °C.

Model Accuracy

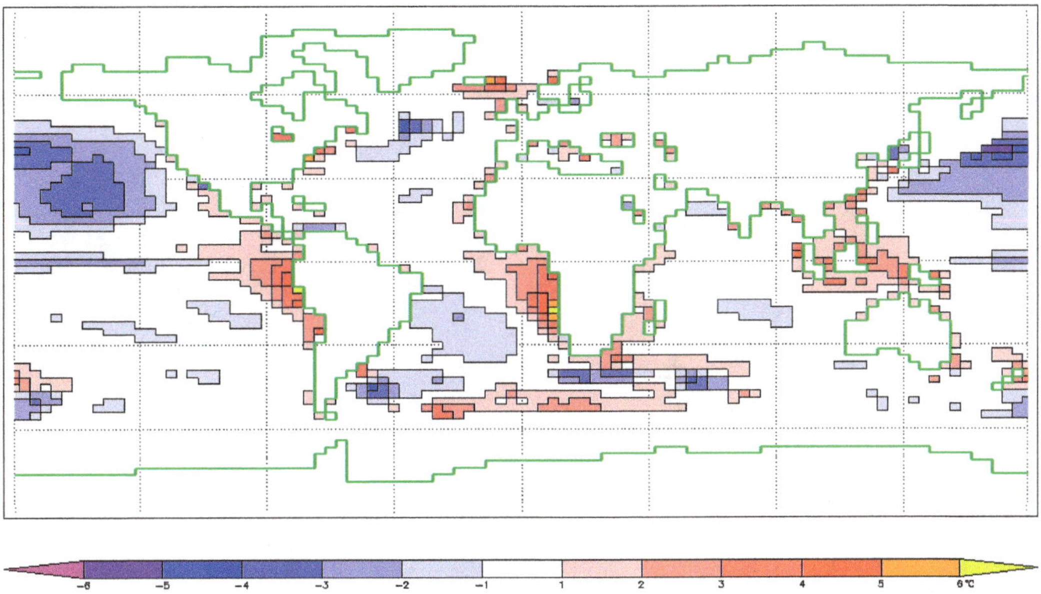

SST errors in HadCM3.

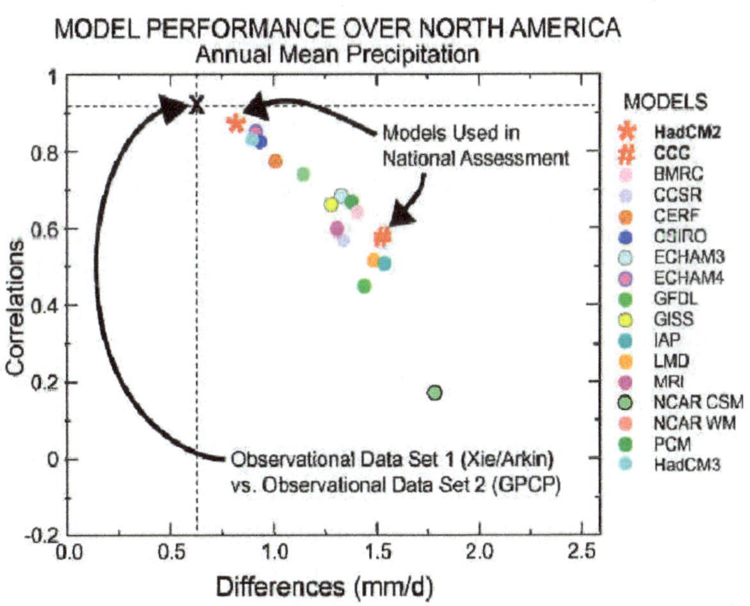

North American precipitation from various models.

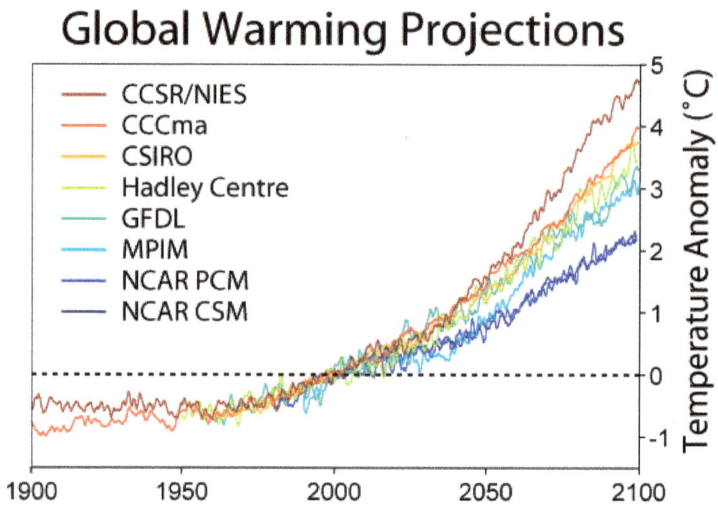

Temperature predictions from some climate models assuming the SRES A2 emissions scenario.

AOGCMs internalise as many processes as are sufficiently understood. However, they are still under development and significant uncertainties remain. They may be coupled to models of other processes, such as the carbon cycle, so as to better model feedbacks. Most recent simulations show "plausible" agreement with the measured temperature anomalies over the past 150 years, when driven by observed changes in greenhouse gases and aerosols. Agreement improves by including both natural and anthropogenic forcings.

Imperfect models may nevertheless produce useful results. GCMs are capable of reproducing the general features of the observed global temperature over the past century.

A debate over how to reconcile climate model predictions that upper air (tropospheric) warming should be greater than observed surface warming, some of which appeared to show otherwise, was resolved in favour of the models, following data revisions.

Cloud effects are a significant area of uncertainty in climate models. Clouds have competing effects on climate. They cool the surface by reflecting sunlight into space; they warm it by increasing the amount of infrared radiation transmitted from the atmosphere to the surface. In the 2001 IPCC report possible changes in cloud cover were highlighted as a major uncertainty in predicting climate.

Climate researchers around the world use climate models to understand the climate system. Thousands of papers have been published about model-based studies. Part of this research is to improve the models.

In 2000, a comparison between measurements and dozens of GCM simulations of ENSO-driven tropical precipitation, water vapor, temperature, and outgoing longwave radiation found similarity between measurements and simulation of most factors. However the simulated change in precipitation was about one-fourth less than what was observed. Errors in simulated precipitation imply errors in other processes, such as errors in the evaporation rate that provides moisture to create precipitation. The other possibility is that the satellite-based measurements are in error. Either indicates progress is required in order to monitor and predict such changes. A more complete discussion of climate models is provided in the IPCC›s Third Assessment Report.

- The model mean exhibits good agreement with observations.
- The individual models often exhibit worse agreement with observations.
- Many of the non-flux adjusted models suffered from unrealistic climate drift up to about 1 °C/century in global mean surface temperature.
- The errors in model-mean surface air temperature rarely exceed 1 °C over the oceans and 5 °C over the continents; precipitation and sea level pressure errors are relatively greater but the magnitudes and patterns of these quantities are recognisably similar to observations.
- Surface air temperature is particularly well simulated, with nearly all models closely matching the observed magnitude of variance and exhibiting a correlation > 0.95 with the observations.
- Simulated variance of sea level pressure and precipitation is within ±25% of observed.
- All models have shortcomings in their simulations of the present day climate of the stratosphere, which might limit the accuracy of predictions of future climate change.
- There is a tendency for the models to show a global mean cold bias at all levels.
- There is a large scatter in the tropical temperatures.
- The polar night jets in most models are inclined poleward with height, in noticeable contrast to an equatorward inclination of the observed jet.
- There is a differing degree of separation in the models between the winter sub-tropical jet and the polar night jet.
- For nearly all models the r.m.s. error in zonal- and annual-mean surface air temperature is small compared with its natural variability.
- There are problems in simulating natural seasonal variability. (2000)
- In flux-adjusted models, seasonal variations are simulated to within 2 K of observed values over the oceans. The corresponding average over non-flux-adjusted models shows errors up to about 6 K in extensive ocean areas.
- Near-surface land temperature errors are substantial in the average over flux-adjusted models, which systematically underestimates (by about 5 K) temperature in areas of elevated terrain. The corresponding average over non-flux-adjusted models forms a similar error pattern (with somewhat increased amplitude) over land.
- In Southern Ocean mid-latitudes, the non-flux-adjusted models overestimate the magnitude of January-minus-July temperature differences by ~5 K due to an overestimate of summer (January) near-surface temperature. This error is common to five of the eight non-flux-adjusted models.
- Over Northern Hemisphere mid-latitude land areas, zonal mean differences between July and January temperatures simulated by the non-flux-adjusted

models show a greater spread (positive and negative) about observed values than results from the flux-adjusted models.

- The ability of coupled GCMs to simulate a reasonable seasonal cycle is a necessary condition for confidence in their prediction of long-term climatic changes (such as global warming), but it is not a sufficient condition unless the seasonal cycle and long-term changes involve similar climatic processes.
- Coupled climate models do not simulate with reasonable accuracy clouds and some related hydrological processes (in particular those involving upper tropospheric humidity). Problems in the simulation of clouds and upper tropospheric humidity, remain worrisome because the associated processes account for most of the uncertainty in climate model simulations of anthropogenic change.

The precise magnitude of future changes in climate is still uncertain; for the end of the 21st century (2071 to 2100), for SRES scenario A2, the change of global average SAT change from AOGCMs compared with 1961 to 1990 is +3.0 °C (5.4 °F) and the range is +1.3 to +4.5 °C (+2.3 to 8.1 °F).

The IPCC›s Fifth Assessment Report asserted "...very high confidence that models reproduce the general features of the global-scale annual mean surface temperature increase over the historical period." However, the report also observed that the rate of warming over the period 1998-2012 was lower than that predicted by 111 out of 114 Coupled Model Intercomparison Project climate models.

Relation to Weather Forecasting

The global climate models used for climate projections are similar in structure to (and often share computer code with) numerical models for weather prediction, but are nonetheless logically distinct.

Most weather forecasting is done on the basis of interpreting numerical model results. Since forecasts are short—typically a few days or a week—such models do not usually contain an ocean model but rely on imposed SSTs. They also require accurate initial conditions to begin the forecast—typically these are taken from the output of a previous forecast, blended with observations. Predictions must require only a few hours; but because they only cover a one-week the models can be run at higher resolution than in climate mode. Currently the ECMWF runs at 40 km (25 mi) resolution as opposed to the 100-to-200 km (62-to-124 mi) scale used by typical climate model runs. Often local models are run using global model results for boundary conditions, to achieve higher local resolution: for example, the Met Office runs a mesoscale model with an 11 km (6.8 mi) resolution covering the UK, and various agencies in the US employ models such as the NGM and NAM models. Like most global numerical weather prediction models such as the GFS, global climate models are often spectral models instead of grid models. Spectral models are often used for global models because some computations in modeling can be performed faster, thus reducing run times.

Computations

Climate models use quantitative methods to simulate the interactions of the atmosphere, oceans, land surface and ice.

All climate models take account of incoming energy as short wave electromagnetic radiation, chiefly visible and short-wave (near) infrared, as well as outgoing energy as long wave (far) infrared electromagnetic radiation from the earth. Any imbalance results in a change in temperature.

The most talked-about models of recent years relate temperature to emissions of greenhouse gases. These models project an upward trend in the surface temperature record, as well as a more rapid increase in temperature at higher altitudes.

Three (or more properly, four since time is also considered) dimensional GCM›s discretise the equations for fluid motion and energy transfer and integrate these over time. They also contain parametrisations for processes such as convection that occur on scales too small to be resolved directly.

Atmospheric GCMs (AGCMs) model the atmosphere and impose sea surface temperatures as boundary conditions. Coupled atmosphere-ocean GCMs (AOGCMs, e.g. HadCM3, EdGCM, GFDL CM2.X, ARPEGE-Climat) combine the two models.

Models range in complexity:

- A simple radiant heat transfer model treats the earth as a single point and averages outgoing energy
- This can be expanded vertically (radiative-convective models), or horizontally
- Finally, (coupled) atmosphere–ocean–sea ice global climate models discretise and solve the full equations for mass and energy transfer and radiant exchange.
- Box models treat flows across and within ocean basins.

Other submodels can be interlinked, such as land use, allowing researchers to predict the interaction between climate and ecosystems.

Climate Models

Earth-system models of intermediate complexity (EMICs)

The Climber-3 model uses a 2.5-dimensional statistical-dynamical model with 7.5° × 22.5° resolution and time step of 1/2 a day. An oceanic submodel is MOM-3 (Modular Ocean Model) with a 3.75° × 3.75° grid and 24 vertical levels.

Radiative-Convective Models (RCM)

One-dimensional, radiative-convective models were used to verify basic climate assumptions in the '80s and '90s.

Tropical Cyclone Forecast Model

A **tropical cyclone forecast model** is a computer program that uses meteorological data to forecast aspects of the future state of tropical cyclones. There are three types of models: statistical, dynamical, or combined statistical-dynamic. Dynamical models utilize powerful supercomputers with sophisticated mathematical modeling software and meteorological data to calculate future weather conditions. Statistical models forecast the evolution of a tropical cyclone in a simpler manner, by extrapolating from historical datasets, and thus can be run quickly on platforms such as personal computers. Statistical-dynamical models use aspects of both types of forecasting. Four primary types of forecasts exist for tropical cyclones: track, intensity, storm surge, and rainfall. Dynamical models were not developed until the 1970s and the 1980s, with earlier efforts focused on the storm surge problem.

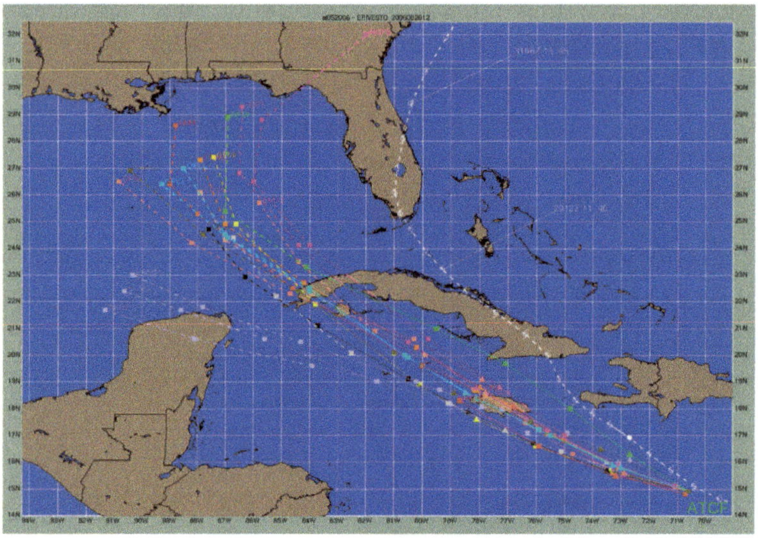

Significant track errors still occur on occasion, as seen in this Ernesto (2006) early forecast. The NHC official forecast is light blue, while the storm's actual track is the white line over Florida.

Track models did not show forecast skill when compared to statistical models until the 1980s. Statistical-dynamical models were used from the 1970s into the 1990s. Early models use data from previous model runs while late models produce output after the official hurricane forecast has been sent. The use of consensus, ensemble, and superensemble forecasts lowers errors more than any individual forecast model. Both consensus and superensemble forecasts can use the guidance of global and regional models runs to improve the performance more than any of their respective components. Techniques used at the Joint Typhoon Warning Center indicate that superensemble forecasts are a very powerful tool for track forecasting.

Statistical Guidance

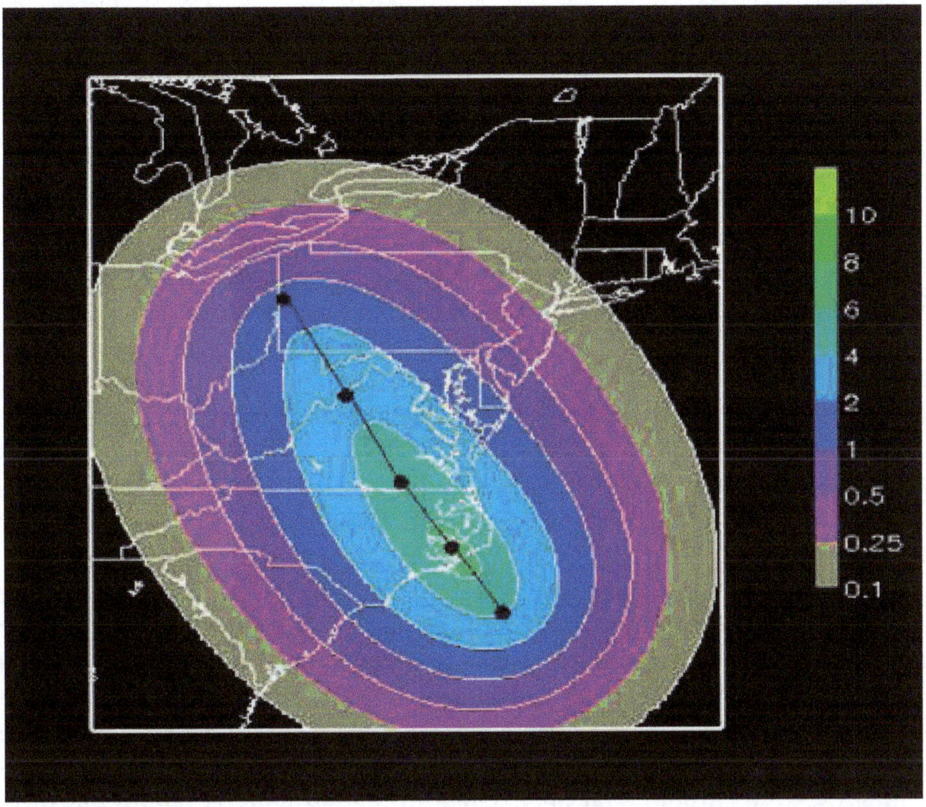

r-CLIPER for Hurricane Isabel (2003).

The first statistical guidance used by the National Hurricane Center was the Hurricane Analog Technique (HURRAN), which was available in 1969. It used the newly developed North Atlantic tropical cyclone database to find storms with similar tracks. It then shifted their tracks through the storm's current path, and used location, direction and speed of motion, and the date to find suitable analogs. The method did well with storms south of the 25th parallel which had not yet turned northward, but poorly with systems near or after recurvature. Since 1972, the Climatology and Persistence (CLIPER) statistical model has been used to help generate tropical cyclone track forecasts. In the era of skillful dynamical forecasts, CLIPER is now being used as the baseline to show model and forecaster skill. The Statistical Hurricane Intensity Forecast (SHIFOR) has been used since 1979 for tropical cyclone intensity forecasting. It uses climatology and persistence to predict future intensity, including the current Julian day, current cyclone intensity, the cyclone›s intensity 12 hours ago, the storm›s initial latitude and longitude, as well as its zonal (east-west) and meridional (north-south) components of motion.

A series of statistical-dynamical models, which used regression equations based upon CLIPER output and the latest output from primitive equation models run at the National Meteorological Center, then National Centers for Environmental Prediction, were developed between the 1970s and 1990s and were named NHC73, NHC83, NHC90, NHC91, and NHC98. Within the field of tropical cyclone track forecasting, despite the ever-improving dynamical model guidance which occurred with increased computational power, it was not until the decade of the 1980s when numerical weather prediction showed skill, and until the 1990s when it consistently outperformed statistical or simple dynamical models. In 1994, a version of SHIFOR was created for the northwest Pacific Ocean for typhoon forecasting, known as the Statistical Typhoon Intensity Forecast (STIFOR), which used the 1971–1990 data for that region to develop intensity forecasts out to 72 hours into the future. In regards to intensity forecasting, the Statistical Hurricane Intensity Prediction Scheme (SHIPS) utilizes relationships between environmental conditions from the Global Forecast System (GFS) such as vertical wind shear and sea surface temperatures, climatology, and persistence (storm behavior) via multiple regression techniques to come up with an intensity forecast for systems in the northern Atlantic and northeastern Pacific oceans. A similar model was developed for the northwest Pacific Ocean and Southern Hemisphere known as the Statistical Intensity Prediction System (STIPS), which accounts for land interactions through the input environmental conditions from the Navy Operational Global Prediction System (NOGAPS) model. The version of SHIPS with an inland decay component is known as Decay SHIPS (DSHIPS). The Logistic. Growth Equation Model (LGEM) uses the same input as SHIPS but within a simplified dynamical prediction system. Within tropical cyclone rainfall forecasting, the Rainfall Climatology and Persistence (r-CLIPER) model was developed using microwave rainfall data from polar orbiting satellites over the ocean and first-order rainfall measurements from the land, to come up with a realistic rainfall distribution for tropical cyclones based on the National Hurricane Center›s track forecast. It has been operational since 2004. A statistical-parametric wind radii model has been developed for use at the National Hurricane Center and Joint Typhoon Warning Center which uses climatology and persistence to predict wind structure out to five days into the future.

Dynamical Guidance

1972, the first model to forecast storm surge along the continental shelf of the United States was developed, known as the Special Program to List the Amplitude of Surges from Hurricanes (SPLASH). In 1978, the first hurricane-tracking model based on atmospheric dynamics – the movable fine-mesh (MFM) model – began operating. The Quasi-Lagrangian Limited Area (QLM) model is a multi-level primitive equation model using a Cartesian grid and the Global Forecasting System (GFS) for boundary conditions. In the early 1980s, the assimilation of satellite-derived winds from water vapor, infrared, and visible satellite imagery was found to improve tropical cyclones track forecasting. The Geophysical Fluid Dynamics Laboratory (GFDL) hurricane model was used for research purposes between 1973 and the mid-1980s. Once it was determined that it could show skill in hurricane prediction, a multi-year transition transformed the research model into an operational model which could be used by the National Weather Service for both track and intensity forecasting in 1995. By 1985, the Sea Lake and Overland Surges from Hurricanes (SLOSH) Model had been developed for use in areas of the Gulf of Mexico and near the United States' East coast, which was more robust than the SPLASH model.

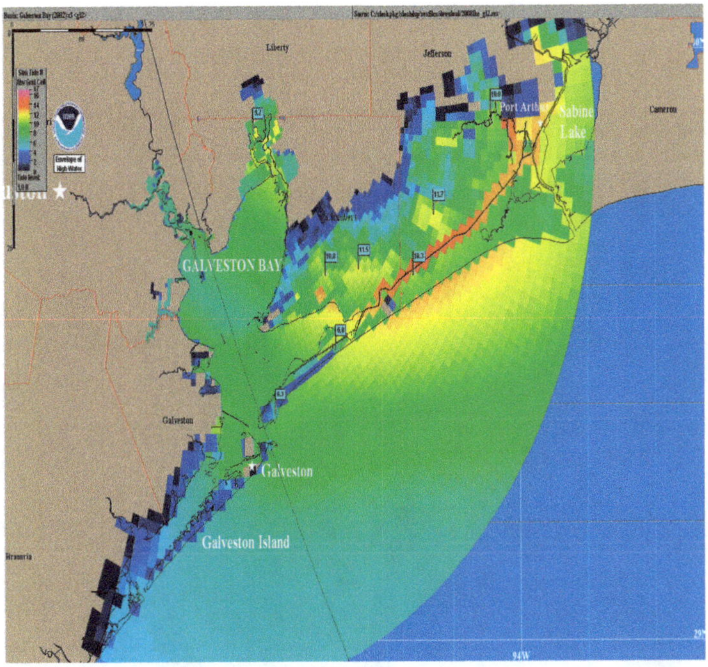

Example of a SLOSH run.

The Beta Advection Model (BAM) has been used operationally since 1987 using steering winds averaged through the 850 hPa to 200 hPa layer and the Beta effect which causes a storm to drift northwest due to differences in the coriolis effect across the tropical cyclone. The larger the cyclone, the larger the impact of the beta effect is likely to be. Starting in 1990, three versions of the BAM were run operationally: the BAM shallow (BAMS) average winds in an 850 hPa to 700 hPa layer, the BAM Medium (BAMM) which uses average winds in an 850 hPa to 400 hPa layer, and the BAM Deep (BAMD) which is the same as the pre-1990 BAM. For a weak hurricane without well-developed central thunderstorm activity, BAMS works well, because weak storms tend to be steered by low-level winds.As the storm grows stronger and associated thunderstorm activity near its center gets deeper, BAMM and BAMD become more accurate, as these types of storms are steered more by the winds in the upper-level. If the forecast from the three versions is similar, then the forecaster can conclude that there is minimal uncertainty, but if the versions vary by a great deal, then the forecaster has less confidence in the track predicted due to the greater uncertainty. Large differences between model predictions can also indicate wind shear in the atmosphere, which could affect the intensity forecast as well.

Tested in 1989 and 1990, The Vic Ooyama Barotropic (VICBAR) model used a cubic-B spline representation of variables for the objective analysis of observations and solutions to the shallow-water prediction equations on nested domains, with the boundary conditions defined as the global forecast model. It was implemented operationally as the Limited Area Sine Transform Barotropic (LBAR) model in 1992, using the GFS for boundary conditions. By 1990, Australia had developed its own storm surge model which was able to be run in a few minutes on a personal computer. The Japan Meteorological

Agency (JMA) developed its own Typhoon Model (TYM) in 1994, and in 1998, the agency began using its own dynamic storm surge model.

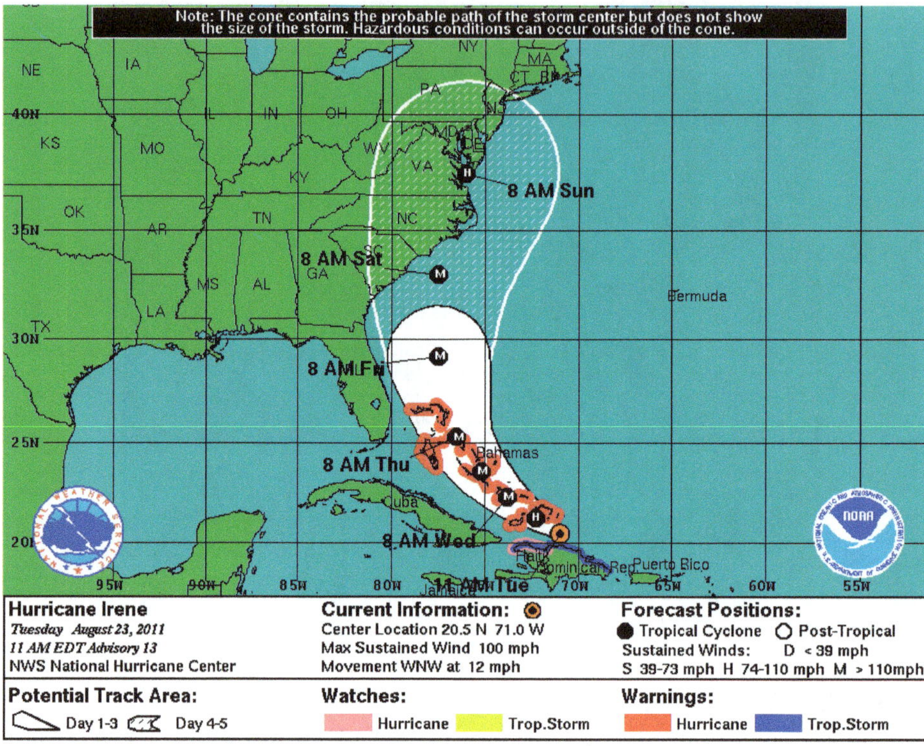

A NOAA prediction for Hurricane Irene.

The Hurricane Weather Research and Forecasting (HWRF) model is a specialized version of the Weather Research and Forecasting (WRF) model and is used to forecast the track and intensity of tropical cyclones. The model was developed by the National Oceanic and Atmospheric Administration (NOAA), the U.S. Naval Research Laboratory, the University of Rhode Island, and Florida State University. It became operational in 2007. Despite improvements in track forecasting, predictions of the intensity of a tropical cyclone based on numerical weather prediction continue to be a challenge, since statiscal methods continue to show higher skill over dynamical guidance. Other than the specialized guidance, global guidance such as the GFS, Unified Model (UKMET), NOGAPS, Japanese Global Spectral Model (GSM), European Centre for Medium-Range Weather Forecasts model, France's Action de Recherche Petite Echelle Grande Echelle (ARPEGE) and Aire Limit′ee Adaptation Dynamique Initialisation (ALADIN) models, India›s National Centre for Medium Range Weather Forecasting (NCMWRF) model, Korea›s Global Data Assimilation and Prediction System (GDAPS) and Regional Data Assimilation and Prediction System (RDAPS) models, Hong Kong/China›s Operational Regional Spectral Model (ORSM) model, and Canadian Global Environmental Multiscale Model (GEM) model are used for track and intensity purposes.

Timeliness

Some models do not produce output quickly enough to be used for the forecast cycle immediately after the model starts running (including HWRF, GFDL, and FSSE). Most of the above track models (except CLIPER) require data from global weather models, such as the GFS, which produce output about four hours after the synoptic times of 0000, 0600, 1200, and 1800 Universal Coordinated Time (UTC). For half of their forecasts, the NHC issues forecasts only three hours after that time, so some "early models — NHC90, BAM, and LBAR — are run using a 12-hour-old forecast for the current time. "Late models, such as the GFS and GFDL, finish after the advisory has already been issued. These models are interpolated to the current storm position for use in the following forecast cycle — for example, GFDI, the interpolated version of the GFDL model.

Consensus Methods

Top: WRF model simulation of Hurricane Rita tracks. *Bottom*: The spread of NHC multi-model ensemble forecast.

Using a consensus of forecast models reduces forecast error. Trackwise, the GUNA model is a consensus of the interpolated versions of the GFDL, UKMET with quality control applied to the cyclone tracker, United States Navy NOGAPS, and GFS models.

The version of the GUNA corrected for model biases is known as the CGUN. The TCON consensus is the GUNA consensus plus the Hurricane WRF model. The version of the TCON corrected for model biases is known as the TCCN. A lagged average of the last two runs of the members within the TCON plus the ECMWF model is known as the TVCN consensus. The version of the TVCN corrected for model biases is the TVCC consensus.

In early 2013, The NAVGEM replaced the NOGAPS as the Navy's primary operational global forecast model. For the 2013 season, and until model verification can occur, it is not being utilized in the development of any consensus forecasts.

For intensity, a combination of the LGEM, interpolated GFDL, interpolated HWRF, and DSHIPS models is known as the ICON consensus. The lagged average of the last two runs of models within the ICON consensus is called the IVCN consensus. Across the northwest Pacific and Southern Hemisphere, a ten-member STIPS consensus is formed from the output of the NOGAPS, GFS, the Japanese GSM, the Coupled Ocean/ Atmosphere Mesoscale Prediction System (COAMPS), the UKMET, the Japanese TYM, the GFDL with NOGAPS boundary conditions, the Air Force Weather Agency (AFWA) Model, the Australian Tropical Cyclone Local Area Prediction System, and the Weber Barotropic Model.

Ensemble Methods

No model is ever perfectly accurate because it is impossible to learn exactly everything about the atmosphere in a timely enough manner, and atmospheric measurements that are taken are not completely accurate. The use of the ensemble method of forecasting, whether it be a multi-model ensemble, or numerous ensemble members based on the global model, helps define the uncertainty and further limit errors.

The JMA has produced an 11-member ensemble forecast system for typhoons known as the Typhoon Ensemble Prediction System (TEPS) since February 2008, which is run out to 132 hours into the future. It uses a lower resolution version (with larger grid spacing) of its GSM, with ten perturbed members and one non-perturbed member. The system reduces errors by an average of 40 kilometres (25 mi) five days into the future when compared to its higher resolution GSM.

The Florida State Super Ensemble (FSSE) is produced from a suite of models which then uses statistical regression equations developed over a training phase to reduce their biases, which produces forecasts better than the member models or their mean solution. It uses 11 global models, including five developed at Florida State University, the Unified Model, the GFS, the NOGAPS, the United States Navy NOGAPS, the Australian Bureau of Meteorology Research Centre (BMRC) model, and Canadian Recherche en Prévision Numérique (RPN) model. It shows significant skill in track, intensity, and rainfall predictions of tropical cyclones.

The Systematic Approach Forecast Aid (SAFA) was developed by the Joint Typhoon Warning Center to create a selective consensus forecast which removed more erroneous forecasts at a 72hour time frame from consideration using the United States Navy NOGAPS model, the GFDL, the Japan Meteorological Agency›s global and typhoon models, as well as the UKMET. All the models improved during SAFA›s five-year history and removing erroneous forecasts proved difficult to do in operations.

Sunspot Theory

A 2010 report correlates low sunspot activity with high hurricane activity. Analyzing historical data, there was a 25% chance of at least one hurricane striking the continental United States during a peak sunspot year; a 64% chance during a low sunspot year. In June 2010, the hurricanes predictors in the US were not using this information.

Hurricane Forecast Model Accuracy

The accuracy of hurricane forecast models can vary significantly from storm to storm. For some storms, the factors affecting the hurricane track are relatively straightforward, and the models are not only accurate but they produce similar forecasts, while for other storms, the factors affecting the hurricane track and more complex, and different models produce very different forecasts.

REFERENCES

1. The First Climate Model". NOAA 200th Celebration. 2007. Retrieved 20 April 2010.
2. Phillips, Norman A. (April 1956). "The general circulation of the atmosphere: a numerical experiment. Quarterly Journal of the Royal Meteorological Society. **82** *(352): 123–154. Bibcode*:1956QJRMS..82..123P. doi:10.1002/qj.49708235202.
3. Cox, John D. (2002). Storm Watchers. John Wiley & Sons, Inc. p. 210. ISBN 0-471-38108-X.
4. Lynch, Peter (2006). "The ENIAC Integrations. The Emergence of Numerical Weather Prediction. Cambridge University Press. pp. 206–208. ISBN 978-0-521-85729-1.
5. Collins, William D.; et al. (June 2004). "Description of the NCAR Community Atmosphere Model (CAM 3.0) (PDF). University Corporation for Atmospheric Research. Retrieved 3 January 2011.
6. Xue, Yongkang & Michael J. Fennessey (20 March 1996). "Impact of vegetation properties on U.S. summer weather prediction (PDF). Journal of Geophysical Research. American Geophysical Union. **101** *(D3): 7419. Bibcode*:1996JGR...101.7419X. doi:10.1029/95JD02169. Retrieved 6 January 2011.
7. McGuffie, K. & A. Henderson-Sellers (2005). A climate modelling primer. John Wiley and Sons. p. 188. ISBN 978-0-470-85751-9.
8. "Pubs.GISS: Sun and Hansen 2003: Climate simulations for 1951-2050 with a coupled atmosphere-ocean model. pubs.giss.nasa.gov. 2003. Retrieved 2015-08-25.

9. Allen, Jeannie (February 2004). "Tango in the Atmosphere: Ozone and Climate Change. NASA Earth Observatory. Retrieved 20 April 2010.

10. Ken, Richard A (13 April 2001). "Global Warming: Rising Global Temperature, Rising Uncertainty. Science. **292** *(5515): 192–194. doi:10.1126/science.292.5515.192. PMID 11305301. Retrieved 20 April 2010.*

11. Atmospheric Model Intercomparison Project. The Program for Climate Model Diagnosis and Intercomparison, Lawrence Livermore National Laboratory. Retrieved 21 April 2010.

12. C. Jablonowski , M. Herzog , J. E. Penner , R. C. Oehmke , Q. F. Stout , B. van Leer, January 1960.5091 "Adaptive Grids for Weather and Climate Models Archived 13 April 2016 at the Wayback Machine. (2004). See also Christiane Jablonowski, Adaptive Mesh Refinement (AMR) for Weather and Climate Models Archived 28 August 2016 at the Wayback Machine. page (Retrieved 24 July 2010)

13. NCAR Command Language documentation: Non-uniform grids that NCL can contourArchived 3 March 2016 at the Wayback Machine. (Retrieved 24 July 2010)

14. High Resolution Global Environmental Modelling (HiGEM) home page. Natural Environment Research Council and Met Office. 18 May 2004. Retrieved 5 October 2010.

15. Mesoscale modelling. Retrieved 5 October 2010.

16. Climate Model Will Be First To Use A Geodesic Grid. Daly University Science News. 24 September 2001. Retrieved 3 May 2011.

17. Gridding the sphere. MIT GCM. Retrieved 9 September 2010.

18. IPCC Third Assessment Report - Climate Change 2001 - Complete online versions. IPCC. Retrieved 12 January 2014.

19. Arakawa›s Computation Device. Aip.org. Retrieved 2012-02-18.

20. COLA Report 27. Grads.iges.org. 1996-07-01. Retrieved 2012-02-18.

21. Table 2-10. Pcmdi.llnl.gov. Retrieved 2012-02-18.

22. Table of Rudimentary CMIP Model Features. Rainbow.llnl.gov. 2004-12-02. Retrieved 2012-02-18.

23. General Circulation Models of the Atmosphere. Aip.org. Retrieved 2012-02-18.

24. NOAA Geophysical Fluid Dynamics Laboratory (GFDL) (9 October 2012), NOAA GFDL Climate Research Highlights Image Gallery: Patterns of Greenhouse Warming, NOAA GFDL

25. Climate Change 2001: The Scientific Basis. Grida.no. Retrieved 2012-02-18.

26. "Chapter 3: Projected climate change and its impacts. IPCC Fourth Assessment Report: Climate Change 2007: Synthesis Report: Synthesis Report Summary for Policymakers., in IPCC AR4 SYR 2007

27. Pope, V. (2008). "Met Office: The scientific evidence for early action on climate change. Met Office website. Archived from the originalon 29 December 2010. Retrieved 7 March 2009.

28. Sokolov, A.P.; et al. (2009). "Probabilistic Forecast for 21st century Climate Based on Uncertainties in Emissions (without Policy) and Climate Parameters. Journal of Climate. **22** *(19): 5175–5204. Bibcode*:2009JCli...22.5175S. doi:10.1175/2009JCLI2863.1. Retrieved 12 January 2009.

29. IPCC, Summary for Policy MakersArchived 7 March 2016 at the Wayback Machine., Figure 4 Archived 21 October 2016 at the Wayback Machine., in IPCC TAR WG1 (2001), Houghton, J.T.; Ding, Y.; Griggs, D.J.; Noguer, M.; van der Linden, P.J.; Dai, X.; Maskell, K.; Johnson, C.A., eds., Climate Change 2001: The Scientific Basis, Contribution of Working Group I to the Third Assessment Report of the Intergovernmental Panel on Climate Change, Cambridge University Press, ISBN 0-521-80767-0 (pb: 0-521-01495-6).

30. Simulated global warming 1860–2000.

31. The National Academies Press website press release, Jan. 12, 2000: Reconciling Observations of Global Temperature Change.

32. Nasa Liftoff to Space Exploration Website: Greenhouse Effect. Archive.com. Recovered 1 Oct 2012.

33. Soden, Brian J.; Held, Isaac M. (2006). "An Assessment of Climate Feedbacks in Coupled Ocean–Atmosphere Models. J. Climate (19): 3354–3360. Bibcode:2006JCli...19.3354S. doi:10.1175/JCLI3799.1.

34. McAvaney et al., Chapter 8: Model Evaluation Archived 16 April 2016 at the Wayback Machine., in IPCC TAR WG1 (2001), Houghton, J.T.; Ding, Y.; Griggs, D.J.; Noguer, M.; van der Linden, P.J.; Dai, X.; Maskell, K.; Johnson, C.A., eds., Climate Change 2001: The Scientific Basis, Contribution of Working Group I to the Third Assessment Report of the Intergovernmental Panel on Climate Change, Cambridge University Press, ISBN 0-521-80767-0 (pb: 0-521-01495-6).

35. Cubasch et al., Chapter 9: Projections of Future Climate Change Archived 16 April 2016 at the Wayback Machine., Executive Summary, in IPCC TAR WG1 (2001), Houghton, J.T.; Ding, Y.; Griggs, D.J.; Noguer, M.; van der Linden, P.J.; Dai, X.; Maskell, K.; Johnson, C.A., eds., Climate Change 2001: The Scientific Basis, Contribution of Working Group I to the Third Assessment Report of the Intergovernmental Panel on Climate Change, Cambridge University Press, ISBN 0-521-80767-0 (pb: 0-521-01495-6).

36. Flato, Gregory (2013). "Evaluation of Climate Models (PDF). IPCC. pp. 768–769. Retrieved 22 February 2014.

37. Archived 3 May 2008 at the Wayback Machine.

38. What are general circulation models (GCM)?. Das.uwyo.edu. Retrieved 2012-02-18.

39. Meehl et al., Climate Change 2007 Chapter 10: Global Climate Projections Archived 15 April 2016 at the Wayback Machine.,in IPCC AR4 WG1 (2007), Solomon, S.; Qin, D.; Manning, M.; Chen, Z.; Marquis, M.; Averyt, K.B.; Tignor, M.; Miller, H.L., eds., Climate Change 2007: The Physical Science Basis, Contribution of Working Group I to the Fourth Assessment Report of the Intergovernmental Panel on Climate Change, Cambridge University Press,

ISBN 978-0-521-88009-1 (pb: 978-0-521-70596-7)

40. ARPEGE-Climat homepage, Version 5.1Archived 4 January 2016 at the Wayback Machine., 3 Sep 2009. Retrieved 1 Oct 2012. ARPEGE-Climat homepage Archived 19 February 2014 at the Wayback Machine., 6 Aug 2009. Retrieved 1 Oct 2012.

41. emics1. www.pik-potsdam.de. Retrieved 2015-08-25.

42. Wang, W.C.; P.H. Stone (1980). "Effect of ice-albedo feedback on global sensitivity in a one-dimensional radiative-convective climate model. J. Atmos. Sci. **37**: 545–52. Bibcode:1980JAtS...37..545W. doi:10.1175/1520-0469(1980)037<0545:EOIAFO>2.0.CO;2. Retrieved 2010-04-22.

CHAPTER 6

Regional Climatology

Chapter Contents

Climatology, branch of the atmospheric sciences concerned with both the description of climate and the analysis of the causes of climatic differences and changes and their practical consequences. Climatology treats the same atmospheric processes as meteorology, but it seeks as well to identify the slower-acting influences and longer-term changes of import, including the circulation of the oceans and the small yet measurable variations in the intensity of solar radiation.

TROPICAL CLIMATE

Ascent (negative values) is concentrated close to the solar equator; descent (positive values) is more diffuse.

A **tropical climate** is a climate typically found within the tropics, while a few locations outside the Tropics are considered to have a tropical climate. In the Köppen climate classification it is a non-arid climate in which all twelve months have mean temperatures of at least 18 °C (64 °F). Unlike the subtropics, where there are significant variations in day length and temperature to various degrees, with season, tropical temperature remains relatively constant throughout the year and seasonal variations are dominated by precipitation.

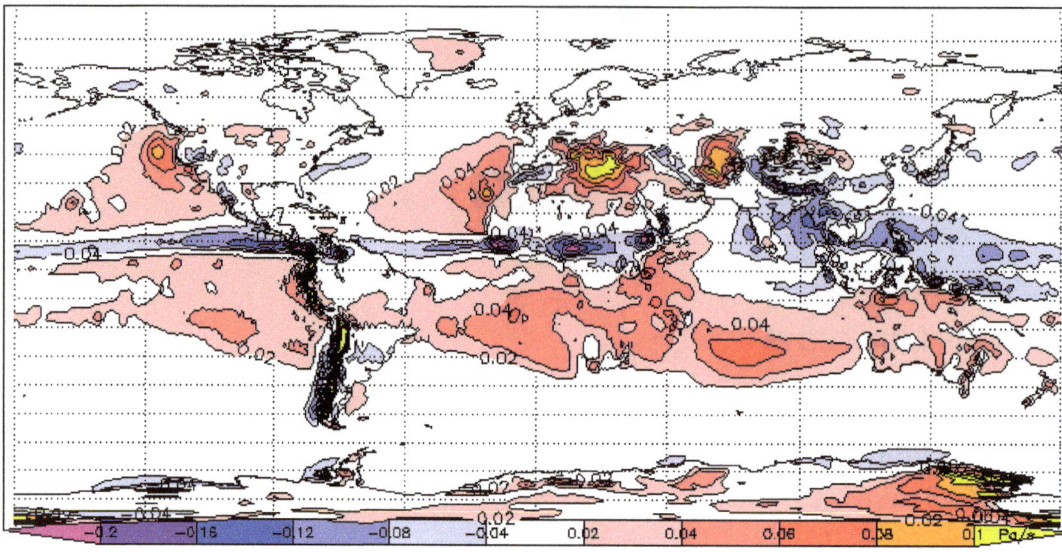

Intertropical Convergence Zone vertical velocity at 500 hPa, July average in units of pascals per second.

SUBTYPES

Within the tropical climate zone there are distinct varieties based on precipitation:

Tropical Rainforest Climate

In the Tropical rainforest climate (*Af*): all twelve months have an average precipitation of at least 60 mm (2.4 in). These climates usually occur within 5–10° latitude of the equator. In some eastern coast areas, they may extend to as much as over 26° away from the equator. This climate is dominated by the doldrums Low Pressure System all year round, and therefore has no natural seasons.

Examples:

- Apia, Samoa
- Mbandaka, DR Congo
- Bogor, Indonesia
- Davao, Philippines
- Klang, Malaysia
- Kuching, Malaysia
- Paramaribo, Suriname
- Quibdó, Colombia
- Ratnapura, Sri Lanka

- Singapore
- Fort Lauderdale, Florida, USA

Tropical Monsoon Climate

The Tropical monsoon climate (*Am*): is a type of climate that is common in South and Southeast Asia and in South and Central America, as well as part of Africa and Australia. It is dominated by the monsoon winds which change direction according to the seasons. This climate has a driest month (which nearly always occurs at or soon after the «winter» solstice for that side of the equator) with rainfall less than 60 millimetres (2.4 in), but more than (100 − [total annual precipitation {mm}/25]).

Examples:

- Abidjan, Ivory Coast
- Cairns, Queensland, Australia
- Chittagong, Bangladesh
- Guanare, Portuguesa, Venezuela
- Huế, Thừa Thiên–Huế, Vietnam
- Jakarta, Indonesia
- Macapá, Amapá, Brazil
- Miami, United States
- Port Harcourt, Rivers State, Nigeria
- Puerto Ayacucho, Amazonas, Venezuela
- Sinop, Mato Grosso, Brazil
- Yangon, Myanmar

Tropical Wet and Dry Climate

The Tropical wet and dry or savanna climate (*Aw*): These climates generally have a pronounced dry season, with the driest month having precipitation less than 60 millimetres (2.4 in) and less than (100 − [total annual precipitation {mm}/25]).

Beach in Naples, Florida lined with coconut trees is an example of a tropical climate. Although it lies in the subtropics over a hundred miles north of the tropic of cancer, the warm waters of the Gulf of Mexico give it a monthly mean temperature never under 18 °C (64 °F), classifying its climate as tropical.

Examples:

- Innisfail, Australia
- Barquisimeto, Venezuela
- Cartagena, Colombia
- Ciudad Guayana, Venezuela
- Dar es Salaam, Tanzania
- Darwin, Northern Territory, Australia
- Dili, East Timor
- Fortaleza, Ceará, Brazil
- Ho Chi Minh City, Vietnam
- Honolulu, Hawaii, United States
- Jijoca de Jericoacora, Ceará, Brazil
- Kaohsiung, Taiwan
- Kupang, Indonesia
- Lagos, Lagos State, Nigeria
- Medellín, Colombia

- Mumbai, Maharashtra, India
- West Palm Beach, United States
- Naples, Florida, United States
- Fort Myers, Florida, United States
- Port-au-Prince, Haiti
- Rio de Janeiro, Rio de Janeiro, Brazil
- São Luís, Maranhão, Brazil
- Valencia, Venezuela
- Veracruz, Veracruz, Mexico

Exceptions

Owing to different causes certain places within the tropics do not have a tropical climate, such as some desert areas and alpine regions.

Examples:

- The Sahara Desert.
- Alpine regions of the tropics. Mountaintops within the tropics, e.g. Mount Kenya, can be cold. However, like lowlands in the tropics (and unlike cold winter temperate zone regions), there is little seasonal variation of temperature.

Intertropical Convergence Zone

Because of the effect of sun angle on climate most areas within the tropics are hot year-round, with diurnal variations in temperature exceeding seasonal variations. Seasonal variations in tropical climate are dominated by changes in precipitation, which are in turn largely influenced by the tropical rain belt or Intertropical Convergence Zone (ITCZ), a portion of the Hadley cell. The ITCZ is shown, for July average, in the graphic. Areas of ascending air have heavy rainfall; areas of descending air are dry. The ITCZ somewhat follows the solar equator throughout the year, but with geographical variations, and in some areas (India) is heavily influenced by local large-scale monsoons.

TEMPERATE CLIMATE

In geography, **temperate** or **tepid**latitudes of Earth lie between the tropics and the polar regions.The temperatures in these regions are generally relatively moderate, rather than extremely hot or cold, and the changes between summer and winter are also usually moderate.

However, in certain areas, such as Asia and central North America, the variations between summer and winter can be extreme because these areas are far away from the sea, causing them to have a continental climate. In regions traditionally considered tropical, localities at high altitudes (e.g. parts of the Andes) may have a temperate climate.

Zones and Climate

The **north temperate zone** extends from the Tropic of Cancer (approximately 23.5° north latitude) to the Arctic Circle (approximately 66.5° north latitude). The **south temperate zone** extends from the Tropic of Capricorn (approximately 23.5° south latitude) to the Antarctic Circle (at approximately 66.5° south latitude).

In some climate classifications, the temperate zone is often broken down into several smaller zones based on latitude. These include Humid subtropical climate, Mediterranean climate, and oceanic.

Subtropical climates, located between 23.5 and 35.0 north or south latitude on the eastern or leeward sides of landmasses is the southernmost zone. This climate has long hot summers and short mild winters, with annual rainfall concentrated in the warmest part of the year. These climates occur in southern Asia, the southeastern United States, parts of eastern Australia, and in eastern coastal South America.

Mediterranean climate, occur between 30 and 42 north or south latitude on the western sides of landmasses. This climate has long hot summers and short mild winters, however seasonal rainfall is the opposite of that of the subtropical humid type, with a winter or cool season rainfall peak common. These climates occur near the rimlands of the Mediterranean Sea, in western Australia, in California, and in the southern most areas of South Africa.

The **maritime climate** occur in the higher middle latitudes between 45 and 60 north and south latitude. They are created by the onshore flow from the cool high latitude oceans to their west. This causes the climate to cool summers and cool, but not cold, winters (for their latitude). Annual rainfall is spread throughout the entire year. Regions with this climate include Western Europe, northwestern North America, and parts of New Zealand. They can also exist in the Eastern United States such as the east end of Long Island, New York and New England such as Block Island, Rhode Island and Chatham, Massachusetts and the high elevations along the Appalachian Mountains including Boone, North Carolina.

The vast majority of the world›s human population resides in temperate zones (if defined as comprising the subtropics as well), especially in the northern hemisphere because of its greater mass of land. The richest temperate flora in the world is found in southern Africa, where some 24,000 taxa (species and infraspecific taxa) have been described.

POLAR CLIMATE

The **polar climate** regions are characterized by a lack of warm summers. Every month in a polar climate has an average temperature of less than 10 °C (50 °F). Regions with polar climate cover more than 20% of the Earth. The sun shines for long hours in the summer, and for many fewer hours in the winter. A polar climate results in treeless tundra, glaciers, or a permanent or semi-permanent layer of ice. It has cool summers and very cold winters.

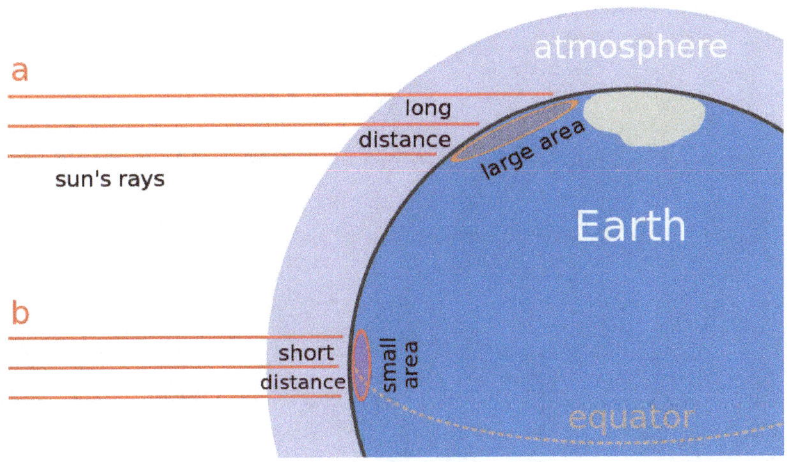

Solar radiation has a lower intensity in polar regions because the angle at which it hits the earth is not as direct as at the equator. Another effect is that sunlight has to go through more atmosphere to reach the ground.

Subtypes

There are two types of polar climate: **ET**, or tundra climate; and **EF**, or ice cap climate. A tundra climate is characterized by having at least one month whose average temperature is above 0 °C (32 °F), while an ice cap climate has no months above 0 °C (32 °F). In a tundra climate, trees cannot grow, but other specialized plants can grow. In an ice cap climate, no plants can grow, and ice gradually accumulates until it flows elsewhere. Many high altitude locations on Earth have a climate where no month has an average temperature of 10 °C (50 °F) or higher, but as this is due to elevation, this climate is referred to as Alpine climate. Alpine climate can mimic either tundra or ice cap climate.

Locations

On Earth, the only continent where the ice cap polar climate is predominant is Antarctica. All but a few isolated coastal areas on the island of Greenland also have the ice cap climate. Coastal regions of Greenland that do not have permanent ice sheets have the less extreme tundra climates. The northernmost part of the Eurasian land mass, from the extreme northeastern coast of Scandinavia and eastwards to the Bering Strait, large areas of northern Siberia and northern Iceland have tundra climate as well. Large areas in northern Canada and northern Alaska have tundra climate, changing to ice cap climate in the most northern parts of Canada. Southernmost South America (Tierra del Fuego where it abuts the Drake Passage) and such subantarctic islands such as the South Shetland

Islands and the Falkland Islands have tundra climates of slight thermal range in which no month is as warm as 10 °C (50 °F). These subantarctic lowlands are found closer to the equator than the coastal tundras of the Arctic basin.

A polar bear with cub.

Arctic

Some parts of the Arctic are covered by ice (sea ice, glacial ice, or snow) year-round, and nearly all parts of the Arctic experience long periods with some form of ice on the surface. Average January temperatures range from about −40 to 0 °C (−40 to 32 °F), and winter temperatures can drop below −50 °C (−58 °F) over large parts of the Arctic. Average July temperatures range from about −10 to 10 °C (14 to 50 °F), with some land areas occasionally exceeding 30 °C (86 °F) in summer.

The Arctic consists of ocean that is nearly surrounded by land. As such, the climate of much of the Arctic is moderated by the ocean water, which can never have a temperature below −2 °C (28 °F). In winter, this relatively warm water, even though covered by the polar ice pack, keeps the North Pole from being the coldest place in the Northern Hemisphere, and it is also part of the reason that Antarctica is so much colder than the Arctic. In summer, the presence of the nearby water keeps coastal areas from warming as much as they might otherwise, just as it does in temperate regions with maritime climates.

A map of the Arctic, the red line indicates the 10°C isotherm in July and the white area shows the average minimum extent of sea ice in summer as of 1975.

Antarctica

The climate of Antarctica is the coldest on the whole of Earth. Antarctica has the lowest temperature ever recorded: −89.2 °C (−128.6 °F) at Vostok Station. It is also extremely dry (technically a desert), averaging 166 millimetres (6.5 in) of precipitation per year. Even so, on most parts of the continent the snow rarely melts and is eventually compressed to become the glacial ice that makes up the ice sheet. Weather fronts rarely penetrate far into the continent.

Quantifying polar climate

There have been several attempts at quantifying what constitutes a polar climate.

Climatologist Wladimir Köppen demonstrated a relationship between the Arctic and Antarctic tree lines and the 10 °C (50 °F) summer isotherm; i.e., places where the average temperature in the warmest calendar month of the year is below 10 °C (50 °F) cannot support forests. See Köppen climate classification for more information.

Otto Nordenskjöld theorized that winter conditions also play a role: His formula is $W = 9 - 0.1\,C$, where W is the average temperature in the warmest month and C the average of the coldest month, both in degrees Celsius (this would mean, for example, that if a particular location had an average temperature of −20 °C (−4 °F) in its coldest month, the warmest month would need to average 11 °C (52 °F) or higher for trees to be able to survive there as 9 - 0.1(-20) = 11). Nordenskiöld›s line tends to run to the north of Köppen›s near the west coasts of the Northern Hemisphere continents, south of it in the interior sections, and at about the same latitude along the east coasts of both Asia and North America. In the Southern Hemisphere, all of Tierra del Fuego lies outside the polar region in Nordenskiöld›s system, but part of the island (including Ushuaia, Argentina) is reckoned as being within the Antarctic under Köppen›s.

In 1947, Holdridge improved on these schemes, by defining biotemperature: the mean annual temperature, where all temperatures below 0 °C (32 °F) *(and above 30 °C (86 °F))* are treated as 0 °C (32 °F) (because it makes no difference to plant life, being dormant). If the mean biotemperature is between 1.5 and 3 °C (34.7 and 37.4 °F), Holdridge quantifies the climate as subpolar (or alpine, if the low temperature is caused by altitude).

HIGHLAND

The term **highland** or **uplands** is used to denote any mountainous region or elevated mountainous plateau. Generally speaking, upland (or uplands) tends to refer to ranges of hills,typically up to 500-600 m. Highland (or highlands) is usually reserved for ranges of low mountains.

Highlands Worldwide

Probably the most known highlands in the anglophone world are the Scottish Highlands in the United Kingdom, the mountainous region north and west of the Highland Boundary Fault. The Highland council area is a local government area in the Scottish Highlands and Britain's largest local government area.

Many countries have areas that are officially or unofficially referred to as *highlands*. Other than Scotland, these include parts Ethiopia, Kenya, Eritrea, Yemen, Nigeria, Papua New Guinea, Syria, Cantabria and Nova Scotia (the latter being Latin for 'New Scotland' due to its resemblance to the country). Synonymous terms used in other countries include *high country*, used in New Zealand, New South Wales, Victoria, Tasmania and Southern Queensland in Australia, and parts of the United States (notably Western North Carolina), and highveld, used in South Africa.

The highlands in Australia are often above the elevation of 500 metres. These areas often receive snowfalls through winter. Most of the highlands lead up to large alpine or sub-alpine mountainous regions such as the Australian Alps, Snowy Mountains, Great Dividing Range, Northern Tablelands and Blue Mountains. The most mountainous region of Tasmania is the Central Highlands area, which covers most of the central western parts of the state. Many of these areas are highly elevated alpine regions.

A spine of mountains runs the length of the island of New Guinea, forming a populous highlands region.

The highlands of Iceland cover about 40% of the country and are mostly inhospitable to humans. They are generally referred to as land above 500 meters.

Additionally, the mountainous natural region of the Thai highlands is found in Northern Thailand.

The Cameron Highlands are a highland area and hill station in Northern Malaysia.

Other worlds

Highland continents – or terrae – are areas of topographically unstable terrain, with high peaks and valleys. They resemble highlands on Earth, but the term is applied to much larger areas on other planets. They can be found on Venus, Mercury, Mars and the Moon.

REFERENCES

1. Education Scotland. “Weather & climate Change Climates around the world”. educationscotland.gov.uk.
2. McColl, R. W. (2005). Encyclopedia of World Geography, Volume 1. (Facts on File Library of World Geography). New York: Facts on File. p. 919. ISBN 0-816-05786-9.
3. “Solar Illumination: Seasonal and Diurnal Patterns”. Encyclopedia.com. Retrieved October 4, 2012.
4. Cohen, Joel E.; Christopher Small (November 24, 1998). “Hypsographic demography: The distribution of human population by altitude”. Proceedings of the National Academy of Sciences of the United States of America, Vol. 95. Washington, D.C.: The Academy. pp. 14009–14014. Retrieved September 19, 2012.
5. Germishuizen, G.; Meyer, N. L. (2003). “Plants of southern Africa: An annotated checklist”. Strelitzia. 14: 1–1231. Retrieved 23 June 201

CHAPTER 7

Weather Forecasting

Chapter Contents

A daily weather forecast involves the work of thousands of observers and meteorologists all over the world, and the work of thousands of machines. Modern computers make forecasts more accurate than ever, and weather satellites orbiting the earth take photographs of clouds from space. Forecasters use the observations from ground and space, along with formulas and rules based on experience of what has happened in the past, and then make their forecast.

INTRODUCTION

Once an all-human endeavor based mainly upon changes in barometric pressure, current weather conditions, and sky condition, weather forecasting now relies on computer-based models that take many atmospheric factors into account. Human input is still required to pick the best possible forecast model to base the forecast upon, which involves pattern recognition skills, teleconnections, knowledge of model performance, and knowledge of model biases. The inaccuracy of forecasting is due to the chaotic nature of the atmosphere, the massive computational power required to solve the equations that describe the atmosphere, the error involved in measuring the initial conditions, and an incomplete understanding of atmospheric processes. Hence, forecasts become less accurate as the difference between current time and the time for which the forecast is being made (the *range* of the forecast) increases. The use of ensembles and model consensus help narrow the error and pick the most likely outcome.

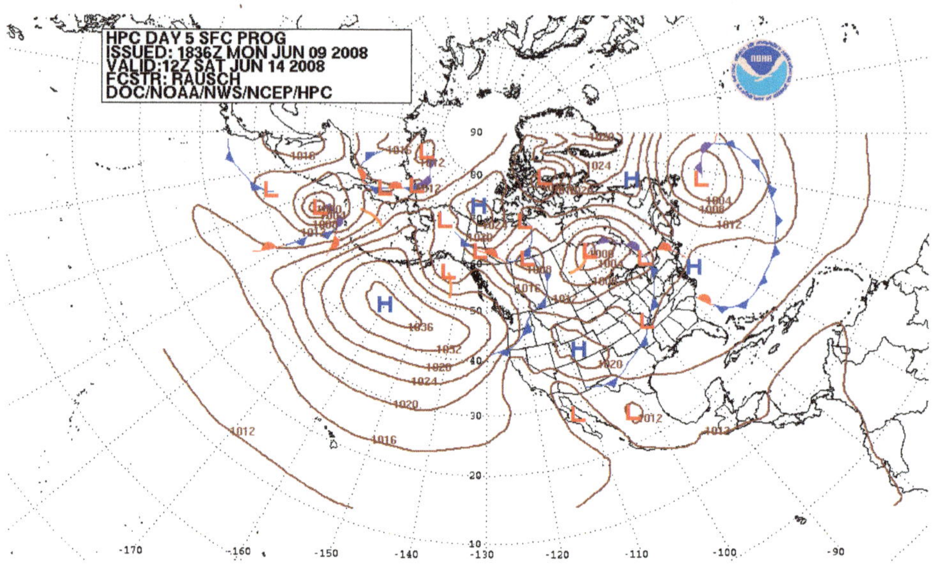

Forecast of surface pressures five days into the future for the north Pacific, North America, and north Atlantic Ocean

Weather forecasting is the application of science and technology to predict the state of the atmosphere for a given location. Human beings have attempted to predict the weather informally for millennia, and formally since the nineteenth century. Weather forecasts are made by collecting quantitative data about the current state of the atmosphere at a given place and using scientific understanding of atmospheric processes to project how the atmosphere will change.

There are a variety of end uses to weather forecasts. Weather warnings are important forecasts because they are used to protect life and property. Forecasts based on temperature and precipitation are important to agriculture, and therefore to traders within commodity markets. Temperature forecasts are used by utility companies to estimate demand over coming days. On an everyday basis, people use weather forecasts to determine what to wear on a given day. Since outdoor activities are severely curtailed by heavy rain, snow and wind chill, forecasts can be used to plan activities around these events, and to plan ahead and survive them. In 2014, the US spent $5.1 billion on weather forecasting.

HISTORY

Ancient forecasting

For millennia people have tried to forecast the weather. In 650 BC, the Babylonians predicted the weather from cloud patterns as well as astrology. In about 340 BC, Aristotle described weather patterns in *Meteorologica*. Later, Theophrastus compiled a book on weather forecasting, called the *Book of Signs*. Chinese weather prediction lore extends at least as far back as 300 BC, which was also around the same time ancient Indian astronomers developed weather-prediction methods. In New Testament times, Christ himself referred to deciphering and understanding local weather patterns, by saying, "When evening comes, you say, 'It will be fair weather, for the sky is red', and in the morning, 'Today it will be stormy, for the sky is red and overcast.' You know how to interpret the appearance of the sky, but you cannot interpret the signs of the times."

In 904 AD, Ibn Wahshiyya›s *Nabatean Agriculture* discussed the weather forecasting of atmospheric changes and signs from the planetary astral alterations; signs of rain based on observation of the lunar phases; and weather forecasts based on the movement of winds.

Ancient weather forecasting methods usually relied on observed patterns of events, also termed pattern recognition. For example, it might be observed that if the sunset was particularly red, the following day often brought fair weather. This experience accumulated over the generations to produce weather lore. However, not all of these predictions prove reliable, and many of them have since been found not to stand up to rigorous statistical testing.

Modern Methods

The *Royal Charter* sank in an 1859 storm, stimulating the establishment of modern weather forecasting.

It was not until the invention of the electric telegraph in 1835 that the modern age of weather forecasting began. Before that, the fastest that distant weather reports could travel was around 100 miles per day (160 km/d), but was more typically 40–75 miles per day (60–120 km/day) (whether by land or by sea). By the late 1840s, the telegraph allowed reports of weather conditions from a wide area to be received almost instantaneously, allowing forecasts to be made from knowledge of weather conditions further upwind.

The two men credited with the birth of forecasting as a science were officer of the Royal Navy Francis Beaufort and his protégé Robert FitzRoy. Both were influential men in British naval and governmental circles, and though ridiculed in the press at the time, their work gained scientific credence, was accepted by the Royal Navy, and formed the basis for all of today›s weather forecasting knowledge.

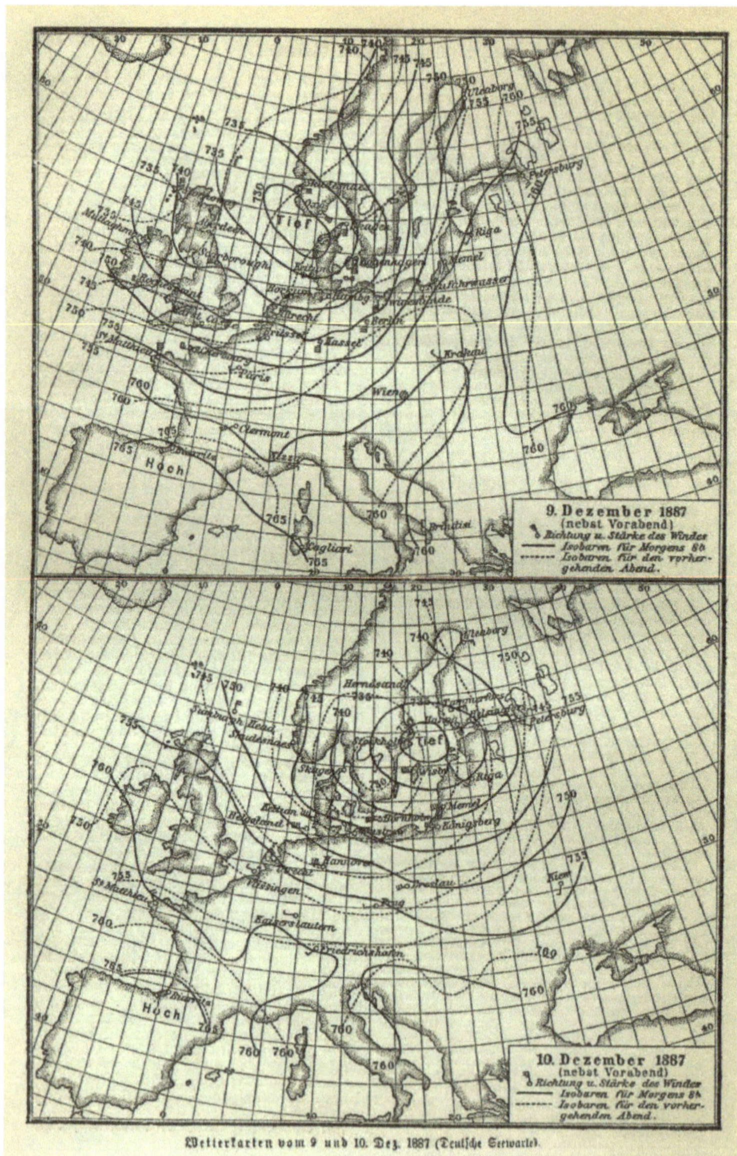

Weather map of Europe, 10 December 1887.

Beaufort developed the Wind Force Scale and Weather Notation coding, which he was to use in his journals for the remainder of his life. He also promoted the development of reliable tide tables around British shores, and with his friend William Whewell, expanded weather record-keeping at 200 British Coast guard stations. Robert FitzRoy was appointed in 1854 as chief of a new department within the Board of Trade to deal with the collection of weather data at sea as a service to mariners. This was the forerunner of the modern Meteorological Office. All ship captains were tasked with collating data on the weather and computing it, with the use of tested instruments that were loaned for this purpose. A storm in 1859 that caused the loss of the *Royal Charter* inspired FitzRoy to develop charts to allow predictions to be made, which he called *"forecasting the weather"*, thus coining the term "weather forecast". Fifteen land stations were established to use the new telegraph to transmit to him daily reports of weather at set times leading to the first gale warning service. His warning service for shipping was initiated in February 1861, with the use of telegraph communications. The first daily weather forecasts were published in *The Times* in 1861. In the following year a system was introduced of hoisting storm warning cones at the principal ports when a gale was expected. The *Weather Book* which FitzRoy published in 1863 was far in advance of the scientific opinion of the time.

As the electric telegraph network expanded, allowing for the more rapid dissemination of warnings, a national observational network was developed which could then be used to provide synoptic analyses. Instruments to continuously record variations in meteorological parameters using photography were supplied to the observing stations from Kew Observatory – these cameras had been invented by Francis Ronalds in 1845 and his barograph had earlier been used by FitzRoy.

To convey accurate information, it soon became necessary to have a standard vocabulary describing clouds; this was achieved by means of a series of classifications first achieved by Luke Howard in 1802, and standardized in the *International Cloud Atlas* of 1896.

Numerical prediction

It was not until the 20th century that advances in the understanding of atmospheric physics led to the foundation of modern numerical weather prediction. In 1922, English scientist Lewis Fry Richardson published "Weather Prediction By Numerical Process , after finding notes and derivations he worked on as an ambulance driver in World War I. He described therein how small terms in the prognostic fluid dynamics equations governing atmospheric flow could be neglected, and a finite differencing scheme in time and space could be devised, to allow numerical prediction solutions to be found.

Richardson envisioned a large auditorium of thousands of people performing the calculations and passing them to others. However, the sheer number of calculations required was too large to be completed without the use of computers, and the size of the grid and time steps led to unrealistic results in deepening systems. It was later found, through numerical analysis, that this was due to numerical instability. The first computerised weather forecast was performed by a team led by the mathematician John von Neumann; von Neumann publishing the paper *Numerical Integration of the Barotropic Vorticity Equation* in 1950. Practical use of numerical weather prediction began in 1955,spurred by the development of programmable electronic computers.

Broadcasts

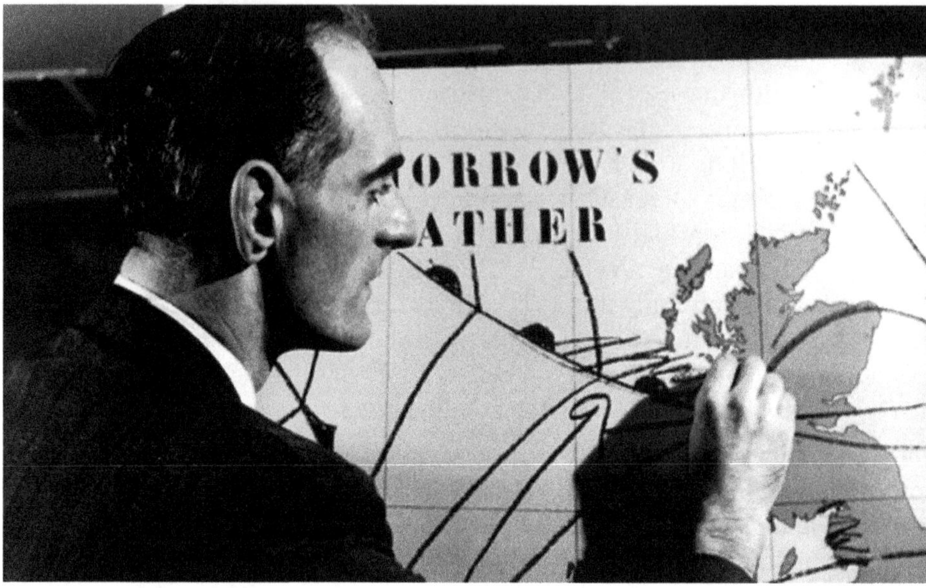

George Cowling (above) presented the first in-vision forecast on 11 January 1954 for the BBC.

The first ever daily weather forecasts were published in *The Times* on 1 August 1861, and the first weather maps were produced later in the same year. In 1911, the Met Office began issuing the first marine weather forecasts via radio transmission. These included gale and storm warnings for areas around Great Britain. In the United States, the first public radio forecasts were made in 1925 by Edward B. "E.B." Rideout, on WEEI, the Edison Electric Illuminating station in Boston.Rideout came from the U.S. Weather Bureau, as did WBZ weather forecaster G. Harold Noyes in 1931.

The world›s first televised weather forecasts, including the use of weather maps, were experimentally broadcast by the BBC in 1936. This was brought into practice in 1949 after World War II. George Cowling gave the first weather forecast while being televised in front of the map in 1954. In America, experimental television forecasts were made by James C Fidler in Cincinnati in either 1940 or 1947 on the DuMont Television Network. In the late 1970s and early 80s, John Coleman, the first weatherman on ABC-TV›s Good Morning America, pioneered the use of on-screen weather satellite information and computer graphics for television forecasts. Coleman was a co-founder of The Weather Channel (TWC) in 1982. TWC is now a 24-hour cable network. Some weather channels have started broadcasting on live broadcasting programs such as YouTube and Periscope to reach more viewers. One Meteorologist, Jeff Matthews, transformed the medium from forecasting live during television broadcasts to forecasting live on Periscope under the name of "Weathertainment , becoming the first meteorologist to deliver daily forecasts on a life-streaming application...and thus, making weather forecasting, interactive, for the public.

How models create forecasts

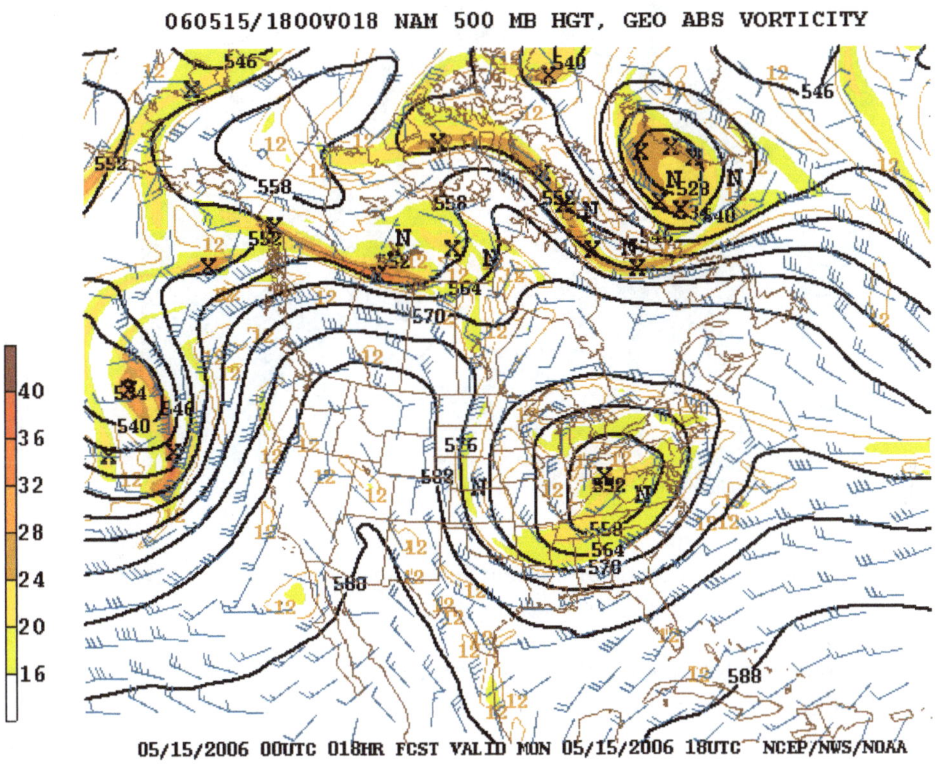

An example of 500 mbargeopotential height and absolute vorticity prediction from a numerical weather prediction model.

The basic idea of numerical weather prediction is to sample the state of the fluid at a given time and use the equations of fluid dynamics and thermodynamics to estimate the state of the fluid at some time in the future. The main inputs from country-based weather services are surface observations from automated weather stations at ground level over land and from weather buoys at sea. The World Meteorological Organization acts to standardize the instrumentation, observing practices and timing of these observations worldwide. Stations either report hourly in METAR reports, or every six hours in SYNOP reports.Sites launch radiosondes, which rise through the depth of the troposphere and well into the stratosphere. Data from weather satellites are used in areas where traditional data sources are not available. Compared with similar data from radiosondes, the satellite data has the advantage of global coverage, however at a lower accuracy and resolution. Meteorological radar provide information on precipitation location and intensity, which can be used to estimate precipitation accumulations over time. Additionally, if a pulse Doppler weather radar is used then wind speed and direction can be determined.

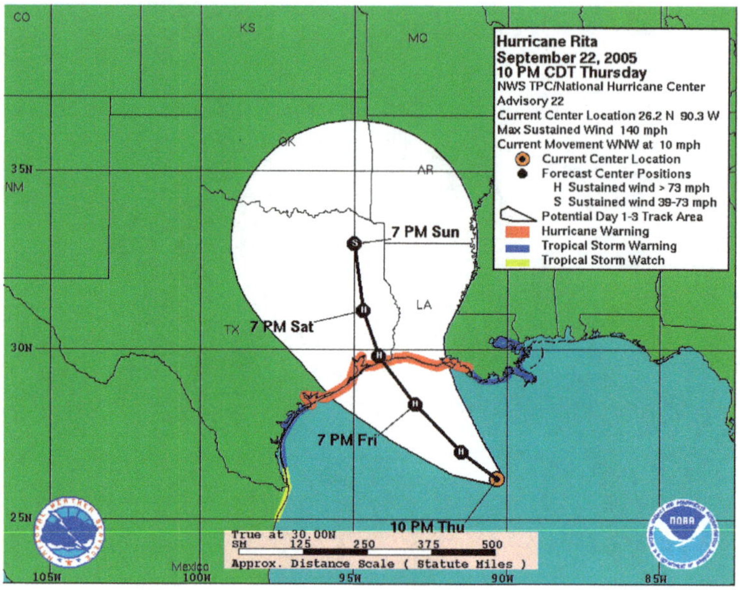

Modern weather predictions aid in timely evacuations and potentially save lives and prevent property damage.

Commerce provides pilot reports along aircraft routes, and ship reports along shipping routes. Research flights using reconnaissance aircraft fly in and around weather systems of interest such as tropical cyclones. Reconnaissance aircraft are also flown over the open oceans during the cold season into systems which cause significant uncertainty in forecast guidance, or are expected to be of high impact 3–7 days into the future over the downstream continent.

Models are *initialized* using this observed data. The irregularly spaced observations are processed by data assimilation and objective analysis methods, which perform quality control and obtain values at locations usable by the model's mathematical algorithms (usually an evenly spaced grid). The data are then used in the model as the starting point for a forecast. Commonly, the set of equations used to predict the known as the physics and dynamics of the atmosphere are called primitive equations. These equations are initialized from the analysis data and rates of change are determined. The rates of change predict the state of the atmosphere a short time into the future. The equations are then applied to this new atmospheric state to find new rates of change, and these new rates of change predict the atmosphere at a yet further time into the future. This *time stepping* procedure is continually repeated until the solution reaches the desired forecast time. The length of the time step is related to the distance between the points on the computational grid.

The length of the time step chosen within the model is related to the distance between the points on the computational grid, and is chosen to maintain numerical stability. Time steps for global models are on the order of tens of minutes, while time steps for regional models are between one and four minutes. The global models are run at varying times into the future. The UKMET Unified Model is run six days into the future, the European

Centre for Medium-Range Weather Forecasts model is run out to 10 days into the future, while the Global Forecast System model run by the Environmental Modeling Center is run 16 days into the future. The visual output produced by a model solution is known as a prognostic chart, or *prog*. The raw output is often modified before being presented as the forecast. This can be in the form of statistical techniques to remove known biases in the model, or of adjustment to take into account consensus among other numerical weather forecasts. MOS or model output statistics is a technique used to interpret numerical model output and produce site-specific guidance. This guidance is presented in coded numerical form, and can be obtained for nearly all National Weather Service reporting stations in the United States. As proposed by Edward Lorenz in 1963, long range forecasts, those made at a range of two weeks or more, are impossible to definitively predict the state of the atmosphere, owing to the chaotic nature of the fluid dynamics equations involved. In numerical models, extremely small errors in initial values double roughly every five days for variables such as temperature and wind velocity.

Essentially, a model is a computer program that produces meteorological information for future times at given locations and altitudes. Within any modern model is a set of equations, known as the primitive equations, used to predict the future state of the atmosphere. These equations—along with the ideal gas law—are used to evolve the density, pressure, and potential temperature scalar fields and the velocity vector field of the atmosphere through time. Additional transport equations for pollutants and other aerosols are included in some primitive-equation mesoscale models as well. The equations used are nonlinear partial differential equations which are impossible to solve exactly through analytical methods, with the exception of a few idealized cases. Therefore, numerical methods obtain approximate solutions. Different models use different solution methods: some global models use spectral methods for the horizontal dimensions and finite difference methods for the vertical dimension, while regional models and other global models usually use finite-difference methods in all three dimensions.

TECHNIQUES

Persistence

The simplest method of forecasting the weather, persistence, relies upon today's conditions to forecast the conditions tomorrow. This can be a valid way of forecasting the weather when it is in a steady state, such as during the summer season in the tropics. This method of forecasting strongly depends upon the presence of a stagnant weather pattern. It can be useful in both short range forecasts and long range forecasts.

Use of a barometer

Measurements of barometric pressure and the pressure tendency (the change of pressure over time) have been used in forecasting since the late 19th century. The larger the change in pressure, especially if more than 3.5 hPa (2.6 mmHg), the larger the change in weather can be expected. If the pressure drop is rapid, a low pressure system is approaching, and there is a greater chance of rain. Rapid pressure rises are associated with improving weather conditions, such as clearing skies.

Looking at the sky

Marestail shows moisture at high altitude, signalling the later arrival of wet weather.

Along with pressure tendency, the condition of the sky is one of the more important parameters used to forecast weather in mountainous areas. Thickening of cloud cover or the invasion of a higher cloud deck is indicative of rain in the near future. At night, high thin cirrostratus clouds can lead to halos around the moon, which indicates an approach of a warm front and its associated rain. Morning fog portends fair conditions, as rainy conditions are preceded by wind or clouds which prevent fog formation. The approach of a line of thunderstorms could indicate the approach of a cold front. Cloud-free skies are indicative of fair weather for the near future. A bar can indicate a coming tropical cyclone. The use of sky cover in weather prediction has led to various weather lore over the centuries.

Nowcasting

The forecasting of the weather within the next six hours is often referred to as **nowcasting**. In this time range it is possible to forecast smaller features such as individual showers and thunderstorms with reasonable accuracy, as well as other features too small to be resolved by a computer model. A human given the latest radar, satellite and observational data will be able to make a better analysis of the small scale features present and so will be able to make a more accurate forecast for the following few hours. However, there are now expert systems using those data and mesoscale numerical model to make better extrapolation, including evolution of those features in time.

Use of forecast models

In the past, the human forecaster was responsible for generating the entire weather forecast based upon available observations. Today, human input is generally confined to choosing a model based on various parameters, such as model biases and performance.Using a consensus of forecast models, as well as ensemble members of the various models, can help reduce forecast error. However, regardless how small the average error becomes with any individual system, large errors within any particular piece of guidance are still possible on any given model run. Humans are required to interpret the model data into weather forecasts that are understandable to the end user. Humans can use knowledge of local effects which may be too small in size to be resolved by the model to add information to the forecast. While increasing accuracy of forecast models implies that humans may no longer be needed in the forecast process at some point in the future, there is currently still a need for human intervention.

Analog technique

The analog technique is a complex way of making a forecast, requiring the forecaster to remember a previous weather event which is expected to be mimicked by an upcoming event. What makes it a difficult technique to use is that there is rarely a perfect analog for an event in the future. Some call this type of forecasting pattern recognition. It remains a useful method of observing rainfall over data voids such as oceans, as well as the forecasting of precipitation amounts and distribution in the future. A similar technique is used in medium range forecasting, which is known as teleconnections, when systems in other locations are used to help pin down the location of another system within the surrounding regime. An example of teleconnections are by using El Niño-Southern Oscillation (ENSO) related phenomena.

QUALITY ASSESSMENT

Mathematical models are often developed to map the forecast skill of a weather prediction methodology.

Communicating forecasts to the public.

An example of a two-day weather forecast in the visual style that an American newspaper might use. Temperatures are given in Fahrenheit.

Most end users of forecasts are members of the general public. Thunderstorms can create strong winds and dangerous lightning strikes that can lead to deaths, power outages, and widespread hail damage. Heavy snow or rain can bring transportation and commerce to a stand-still, as well as cause flooding in low-lying areas. Excessive heat or cold waves can sicken or kill those with inadequate utilities, and droughts can impact water usage and destroy vegetation.

Several countries employ government agencies to provide forecasts and watches/warnings/advisories to the public in order to protect life and property and maintain commercial interests. Knowledge of what the end user needs from a weather forecast must be taken into account to present the information in a useful and understandable way. Examples include the National Oceanic and Atmospheric Administration›s National Weather Service (NWS) and Environment Canada›s Meteorological Service (MSC).Traditionally, newspaper, television, and radio have been the primary outlets for presenting weather forecast information to the public. Increasingly, the internet is being used due to the vast amount of specific information that can be found. In all cases, these outlets update their forecasts on a regular basis.

Severe Weather Alerts and Advisories

A major part of modern weather forecasting is the severe weather alerts and advisories which the national weather services issue in the case that severe or hazardous weather is expected. This is done to protect life and property. Some of the most commonly known of severe weather advisories are the severe thunderstorm and tornado warning, as well as the severe thunderstorm and tornado watch. Other forms of these advisories include winter weather, high wind, flood, tropical cyclone, and fog. Severe weather advisories and alerts are broadcast through the media, including radio, using emergency systems as the Emergency Alert System which break into regular programming.

Low Temperature Forecast

The low temperature forecast for the current day is calculated using the lowest temperature found between 7pm that evening through 7am the following morning. So, in short, today's forecasted low is most likely tomorrow's low temperature.

SPECIALIST FORECASTING

There are a number of sectors with their own specific needs for weather forecasts and specialist services are provided to these users.

Air Traffic

Because the aviation industry is especially sensitive to the weather, accurate weather forecasting is essential. Fog or exceptionally low ceilings can prevent many aircraft from landing and taking off. Turbulence and icing are also significant in-flight hazards. Thunderstorms are a problem for all aircraft because of severe turbulence due to their updrafts and outflow boundaries, icing due to the heavy precipitation, as well as large hail, strong winds, and lightning, all of which can cause severe damage to an aircraft in flight. Volcanic ash is also a significant problem for aviation, as aircraft can lose engine power within ash clouds. On a day-to-day basis airliners are routed to take advantage of the jet stream tailwind to improve fuel efficiency. Aircrews are briefed prior to takeoff on the conditions to expect en route and at their destination. Additionally, airports often change which runway is being used to take advantage of a headwind. This reduces the distance required for takeoff, and eliminates potential crosswinds.

Ash cloud from the 2008 eruption of Chaitén volcano stretching across Patagonia from the Pacific to the Atlantic Ocean.

Marine

Commercial and recreational use of waterways can be limited significantly by wind direction and speed, wave periodicity and heights, tides, and precipitation. These factors can each influence the safety of marine transit. Consequently, a variety of codes have been established to efficiently transmit detailed marine weather forecasts to vessel pilots via radio, for example the MAFOR (marine forecast). Typical weather forecasts can be received at sea through the use of RTTY, Navtex and Radiofax.

Agriculture

Farmers rely on weather forecasts to decide what work to do on any particular day. For example, drying hay is only feasible in dry weather. Prolonged periods of dryness can ruin cotton, wheat, and corn crops. While corn crops can be ruined by drought, their dried remains can be used as a cattle feed substitute in the form of silage. Frosts and freezes play havoc with crops both during the spring and fall. For example, peach trees in full bloom can have their potential peach crop decimated by a spring freeze. Orange groves can suffer significant damage during frosts and freezes, regardless of their timing.

Forestry

Weather forecasting of wind, precipitations and humidity is essential for preventing and controlling wildfires. Different indices, like the *Forest fire weather index* and the *Haines Index*, have been developed to predict the areas more at risk to experience fire from natural or human causes. Conditions for the development of harmful insects can be predicted by forecasting the evolution of weather, too.

Utility Companies

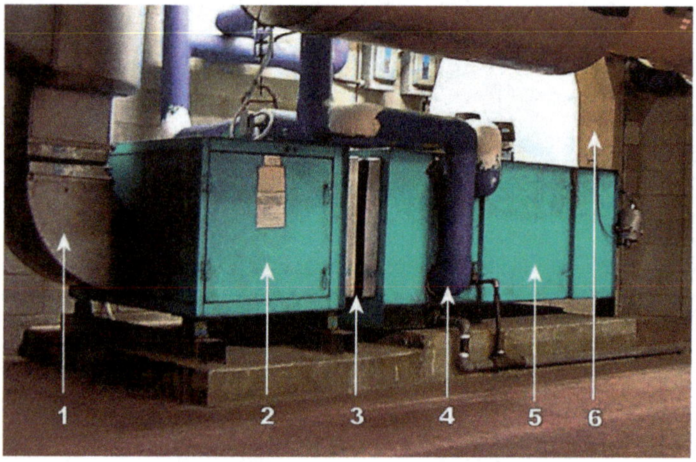

An air handling unit is used for the heating and cooling of air in a central location (click on image for legend).

Electricity and gas companies rely on weather forecasts to anticipate demand which can be strongly affected by the weather. They use the quantity termed the degree day to determine how strong of a use there will be for heating (heating degree day) or cooling (cooling degree day). These quantities are based on a daily average temperature of 65 °F (18 °C). Cooler temperatures force heating degree days (one per degree Fahrenheit), while warmer temperatures force cooling degree days. In winter, severe cold weather can cause a surge in demand as people turn up their heating. Similarly, in summer a surge in demand can be linked with the increased use of air conditioning systems in hot weather. By anticipating a surge in demand, utility companies can purchase additional supplies of power or natural gas before the price increases, or in some circumstances, supplies are restricted through the use of brownouts and blackouts.

Other Commercial Companies

Increasingly, private companies pay for weather forecasts tailored to their needs so that they can increase their profits or avoid large losses. For example, supermarket chains may change the stocks on their shelves in anticipation of different consumer spending habits in different weather conditions. Weather forecasts can be used to invest in the commodity market, such as futures in oranges, corn, soybeans, and oil.

Military Applications

United Kingdom Armed Forces

Royal Navy

The UK Royal Navy, working with the UK Met Office, has its own specialist branch of weather observers and forecasters, as part of the Hydrographic and Meteorological (HM) specialisation, who monitor and forecast operational conditions across the globe, to provide accurate and timely weather and oceanographic information to submarines, ships and Fleet Air Arm aircraft.

Royal Air Force

A mobile unit in the RAF, working with the UK Met Office, forecasts the weather for regions in which British, allied servicemen and women are deployed. A group based at Camp Bastion provides forecasts for the British armed forces in Afghanistan.

United States Armed Forces

US Navy

Emblem of JTWC Joint Typhoon Warning Center.

Similar to the private sector, military weather forecasters present weather conditions to the war fighter community. Military weather forecasters provide pre-flight and in-flight weather briefs to pilots and provide real time resource protection services for military installations. Naval forecasters cover the waters and ship weather forecasts. The United

States Navy provides a special service to both themselves and the rest of the federal government by issuing forecasts for tropical cyclones across the Pacific and Indian Oceans through their Joint Typhoon Warning Center.

US Air Force

Within the United States, Air Force Weather provides weather forecasting for the Air Force and the Army. Air Force forecasters cover air operations in both wartime and peacetime operations and provide Army support; United States Coast Guard marine science technicians provide ship forecasts for ice breakers and other various operations within their realm; and Marine forecasters provide support for ground- and air-based United States Marine Corps operations. All four military branches take their initial enlisted meteorology technical training at Keesler Air Force Base. Military and civilian forecasters actively cooperate in analyzing, creating and critiquing weather forecast products.

REFERENCES

1. Fostering Innovation, Creating Jobs, Driving Better Decisions: The Value of Government Data. Economics and Statistics Administration Office of the Chief Economist. July 2014. p. 15.
2. Meteorology by Lisa Alter
3. Weather: Forecasting from the Beginning
4. University of California Museum of Paleontology. Aristotle (384-322 B.C.E.). Retrieved on 2008-01-12.
5. David Pingree. THE INDIAN AND PSEUDO-INDIAN PASSAGES IN GREEK AND LATIN ASTRONOMICAL AND ASTROLOGICAL TEXTS (PDF). pp. 141–195 [143–4]. Retrieved 2010-03-01.
6. Bible Gateway passage: Matthew 16:2-3 - English Standard Version . Bible Gateway. Retrieved 2016-12-01.
7. Fahd, Toufic. : 842. Missing or empty |title= (help), in Rashed, Roshdi; Morelon, Régis (1996). Encyclopedia of the History of Arabic Science. **3**. Routledge. pp. 813–852. ISBN 0-415-12410-7.
8. Jerry Wilson. Skywatch Signs of the Weather. Retrieved on 2007-04-15.
9. Joseph Henry: Inventor of the Telegraph? Smithsonian Institution . Archived from the original on June 26, 2006. Retrieved 2006-06-29.
10. Ausman, Megaera. USPS Historian . About the United States Postal Service. USPS. Retrieved 28 April 2013.
11. Mail, Royal. (UK) . British Postal Museum. Postal Heritage Trust. Retrieved 28 April 2013.
12. Encyclopædia Britannica. Telegraph. Retrieved on 2007-05-05.
13. Eric D. Craft. An Economic History of Weather Forecasting . Archived from the original on 3 May 2007. Retrieved 15 April 2007.

14. The birth of the weather forecast . BBC. 30 April 2015. Retrieved 30 April 2015.
15. Mellersh, H. E. L. (1968). FitzRoy of the Beagle. Hart-Davis. ISBN 0-246-97452-4
16. Kington, John (1997). Mike Hulme and Elaine Barrow, ed. Climates of the British Isles: Present, Past and Future. Routledge. p. 147.
17. Ronalds, B.F. (2016). Sir Francis Ronalds: Father of the Electric Telegraph. London: Imperial College Press. ISBN 978-1-78326-917-4.
18. Ronalds, B.F. (June 2016). Sir Francis Ronalds and the Early Years of the Kew Observatory . Weather. doi:10.1002/wea.2739.
19. Richardson, Lewis Fry, Weather Prediction by Numerical Process (Cambridge, England: Cambridge University Press, 1922). Available on-line at: Internet Archive.org.
20. Lynch, Peter (2006). The Emergence of Numerical Weather Prediction. Cambridge U.P.
21. Charney, Fjörtoft and von Neumann, 1950, Numerical Integration of the Barotropic Vorticity Equation Tellus, 2, 237-254
22. Paul N. Edwards. Atmospheric General Circulation Modeling. Retrieved on 2007-02-16.
23. Helen Czerski. Orbit: Earth›s Extraordinary Journey: 150 years since the first UK weather forecast . BBC. Retrieved November 5, 2013.
24. Met Office (2012). National Meteorological Library and Fact Sheet 8 -- The Shipping Forecast (PDF). 1. pp. 3–5. Retrieved 2013-04-10.
25. Meteorology Facts, information, pictures | Encyclopedia.com articles about meteorology . Encyclopedia.com. Retrieved 2014-02-21.
26. BBC - Weather - A history of TV weather forecasts BBC Weather
27. Hunt, Roger, "The end of weather forecasting at Met Office London", Weather, June 2007, v.62, no.6, pp.143-146
28. Answers: Understanding weather forecasts . USA Today. 2006-02-08.
29. CJR Rewind: Hot Air, Columbia Journalism Review, reprint, first published in the January/February 2010 issue.
30. National Climatic Data Center. Key to METAR Surface Weather Observations. Retrieved on 2008-03-09.
31. UNISYS. SYNOP Data Format (FM-12): Surface Synoptic Observations. Retrieved on 2008-05-25.
32. Gaffen, Dian J. (2007-06-07). Radiosonde Observations and Their Use in SPARC-Related Investigations. Retrieved on 2008-05-25.
33. NASA. Interactive Global Composite Weather Satellite Images. Retrieved on 2008-05-25.
34. NOAA. Goes Eastern US Sector Infrared Image. Retrieved on 2008-05-25.
35. Met Office. Satellite applications. Retrieved on 2008-05-25.

36. Tony Reale. ATOVS Sounding Products (ITSVC-12). Retrieved on 2008-05-25.
37. Andrew Treloar and Peter Brookhouse. The use of accumulated rainfall maps from weather radar systems to assist wildfire detection reconnaissance. Retrieved on 2008-05-25.
38. University of Washington. An improving forecast. Retrieved on 2007-04-15
39. Ballish, Bradley A. and V. Krishna Kumar (2008-05-23). Investigation of Systematic Differences in Aircraft and Radiosonde Temperatures with Implications for NWP and Climate Studies. Retrieved on 2008-05-25.
40. 403rd Wing (2011). The Hurricane Hunters . 53rd Weather Reconnaissance Squadron. Retrieved 2006-03-30.
41. Lee, Christopher. Drone, Sensors May Open Path Into Eye of Storm . The Washington Post. Retrieved 2008-02-22.
42. National Oceanic and Atmospheric Administration (2010-11-12). NOAA Dispatches High-Tech Research Plane to Improve Winter Storm Forecasts . Retrieved 2010-12-22.
43. University Corporation for Atmospheric Research (2007-08-14). The WRF Variational Data Assimilation System (WRF-Var). Retrieved on 2008-05-25.
44. Pielke, Roger A. (2002). Mesoscale Meteorological Modeling. Academic Press. pp. 285–287. ISBN 0-12-554766-8.
45. Sunderam, V. S.; van Albada, G. Dick; Peter, M. A.; Sloot, J. J. Dongarra (2005). Computational Science – ICCS 2005: 5th International Conference, Atlanta, GA, USA, May 22–25, 2005, Proceedings, Part 1. Springer. p. 132. ISBN 978-3-540-26032-5. Retrieved 2011-01-02.
46. Zwieflhofer, Walter; Kreitz, Norbert; European Centre for Medium Range Weather Forecasts (2001). Developments in teracomputing: proceedings of the ninth ECMWF Workshop on the Use of High Performance Computing in Meteorology. World Scientific. p. 276. ISBN 978-981-02-4761-4. Retrieved 2011-01-02.
47. Chan, Johnny C. L. & Jeffrey D. Kepert (2010). Global Perspectives on Tropical Cyclones: From Science to Mitigation. World Scientific. pp. 295–296. ISBN 978-981-4293-47-1. Retrieved 2011-02-24.
48. Holton, James R. (2004). An introduction to dynamic meteorology, Volume 1. Academic Press. p. 480. ISBN 978-0-12-354015-7. Retrieved 2011-02-24.
49. Brown, Molly E. (2008). Famine early warning systems and remote sensing data. Springer. p. 121. ISBN 978-3-540-75367-4. Retrieved 2011-02-24.
50. Ahrens, C. Donald (2008). Essentials of meteorology: an invitation to the atmosphere. Cengage Learning. p. 244. ISBN 978-0-495-11558-8. Retrieved 2011-02-11.
51. Daniel Andersson. Improved accuracy of surrogate models using output postprocessing. Retrieved on 2008-05-25.
52. Cox, John D. (2002). Storm Watchers. John Wiley & Sons, Inc. pp. 222–224. ISBN 0-471-38108-X.

53. Pielke, Roger A. (2002). Mesoscale Meteorological Modeling. Academic Press. pp. 48–49. ISBN 0-12-554766-8.

54. Pielke, Roger A. (2002). Mesoscale Meteorological Modeling. Academic Press. pp. 18–19. ISBN 0-12-554766-8.

55. Strikwerda, John C. (2004). Finite difference schemes and partial differential equations. SIAM. pp. 165–170. ISBN 978-0-89871-567-5. Retrieved 2010-12-31.

56. Pielke, Roger A. (2002). Mesoscale Meteorological Modeling. Academic Press. p. 65. ISBN 0-12-554766-8.

57. University of Illinois at Urbana-Champaign. Persistence Forecasting: Today equals Tomorrow. Retrieved on 2007-02-16.

58. USA Today. Understanding air pressure. Retrieved on 2008-05-25.

59. Weather Doctor. Applying The Barometer To Weather Watching. Retrieved on 2008-05-25.

60. Dennis Eskow (March 1983). Make Your Own Weather Forecasts . Popular Mechanics. **159** *(3): 148. Retrieved 2011-04-02.*

61. Mark Moore (2009-03-25). Field Forecasting - A Short Summary. Retrieved on 2012-02-15.

62. Jerry Wilson. Skywatch: Signs of the Weather. Retrieved on 2008-05-25.

63. Glossary of Meteorology. Retrieved on 2015-05-26.

64. E-notes.com. Weather and Climate | What Is Nowcasting? Retrieved on 2011-09-08.

65. NASA. Weather Forecasting Through the Ages. Retrieved on 2008-05-25.

66. Klaus Weickmann, Jeff Whitaker, Andres Roubicek and Catherine Smith (2001-12-01). The Use of Ensemble Forecasts to Produce Improved Medium Range (3-15 days) Weather Forecasts. Climate Diagnostics Center. Retrieved on 2007-02-16.

67. Todd Kimberlain. Tropical cyclone motion and intensity talk (June 2007). Retrieved on 2007-07-21.

68. Richard J. Pasch, Mike Fiorino, and Chris Landsea. TPC/NHC'S Review of the NCEP Production Suite for 2006. Retrieved on 2008-05-05.

69. Roebber P. J. and Bosart L. F. The complex relationship between forecast skill and forecast value : A real-world analysis. Retrieved on 2008-05-25.

70. Other Forecasting Methods: climatology, analogue and numerical weather prediction. Retrieved on 2006-02-16.

71. Kenneth C. Allen. Pattern Recognition Techniques Applied to the NASA-ACTS Order-Wire Problem.Retrieved on 2007-02-16.

72. Weather Associates, Inc. The Role of Teleconnections & Ensemble Forecasting in Extended- to Medium-Range Forecasting. Retrieved on 2007-02-16.

73. Thinkquest.org. Teleconnections: Linking El Niño with Other Places. Retrieved on 2007-02-16.

74. University of Illinois at Urbana-Champaign. Lightning. Retrieved on 2007-02-16.

75. Associated Press. Upstate N.Y. residents dig out from heavy snow. Retrieved on 2008-05-25.
76. National Flood Insurance Program. Flood Risk Scenarios: Flash Flood. Retrieved on 2008-05-25.
77. National Weather Service. About NOAA's National Weather Service. Retrieved on 2007-02-16.
78. Environment Canada. Main website. Retrieved on 2007-02-16.
79. Canadian Heritage. Primary Sources of Local Information. Retrieved on 2008-05-26.
80. National Weather Service. National Weather Service Mission Statement. Retrieved on 2008-05-25.
81. Environment Canada. Canadian Weather Alerts. Retrieved on 2008-05-26.
82. Federal Communications Commission. Emergency Alert System. Retrieved on 2008-05-26.
83. Weather Channel - Calculation of Low Temperature Forecast
84. Government Printing Office. Title 14: Aeronautics and Space. Retrieved on 2008-05-26.
85. Aircraft Owners and Pilots Association. Aircraft Icing. Retrieved on 2008-05-26.
86. National Weather Service Forecast Office Dodge City, Kansas. Aviation Hazards They Didn't Tell You About. Retrieved on 2008-05-26.
87. Bureau of Meteorology. Aviation Hazards: Thunderstorms and Deep Convection. Retrieved on 2008-05-26.
88. Volcanic Ash Aviation Hazard. Retrieved on 2008-05-26.
89. Ned Rozell. Amazing flying machines allow time travel. Retrieved on 2008-05-08.
90. National Weather Service. A Pilot's Guide to Aviation Weather Services. Retrieved on 2008-05-26.
91. Eric C. King. Takeoff Tools Crosswind Calculator Instructions. Retrieved on 2008-05-26.
92. Great Lakes and Seaway Shipping. MAFOR Weather Code. Retrieved on 2008-05-27.
93. Blair Fannin. Dry weather conditions continue for Texas. Retrieved on 2008-05-26.
94. Dr. Terry Mader. Drought Corn Silage. Retrieved on 2008-05-26.
95. Kathryn C. Taylor. Peach Orchard Establishment and Young Tree Care. Retrieved on 2008-05-26.
96. Associated Press. After Freeze, Counting Losses to Orange Crop. Retrieved on 2008-05-26.
97. Climate Prediction Center. Degree Day Explanation. Retrieved on 2008-05-25.
98. The New York Times. Futures/Options; Cold Weather Brings Surge in Prices of Heating Fuels. Retrieved on 2008-05-25.

99. BBC. Heatwave causes electricity surge. Retrieved on 2008-05-25.
100. Toronto Catholic Schools. The Seven Key Messages of the Energy Drill Program. Retrieved on 2008-05-25.
101. CSIRO. Providing specialized weather forecasts. Retrieved on 2008-05-25.
102. Stephen Jewson and Rodrigo Caballero. The Use of Weather Forecasts in the Pricing of Weather Derivatives. Retrieved on 2008-05-25.
103. Met Office. Weather forecasting for military operations. Retrieved on 2012-10-23.
104. Joint Typhoon Warning Center. Joint Typhoon Warning Center Mission Statement. Retrieved on 2008-05-27.
105. United States Air Force.Air Force Weather Agency. Retrieved on 2008-05-26.
106. United States Military. US Coast Guard Jobs - Enlisted Occupations. Retrieved on 2008-05-26.
107. Rod Powers. United States Marine Corps Enlisted Job Descriptions and Qualification Factors: Field 68 - Meteorology and Oceanography (METOC). Retrieved on 2008-05-26.
108. Keesler Air Force Base. Military officers usually received their education from a civilian institution. Keesler News: March 9, 2006. United States Air Force Retrieved on 2008-05-26.

CHAPTER 8

Paleoclimatology

Chapter Contents

- **Introduction**
- **Reconstructing Ancient Climates**
- **Climate Forcings**

Climatology is the scientific study of climates, which is defined as the mean weather conditions over a period of time. A branch of study within atmospheric sciences, it also takes into account the variables and averages of short-term and long-term weather conditions. Climatology is different than meteorology and can be divided into different areas of study.

Various approaches to this field can be taken, including paleoclimatology, which focuses on studying the climate over the course of the Earth's existence by examining records of tree rings, rocks and sediment, and ice cores. Historical climatology focuses primarily on climate changes throughout history and the effects of the climate on people and events over time.

INTRODUCTION

Paleoclimatology (in British spelling, palaeoclimatology) is the study of changes in climate taken on the scale of the entire history of Earth. It uses a variety of proxy methods from the Earth and life sciences to obtain data previously preserved within things such as rocks, sediments, ice sheets, tree rings, corals, shells and microfossils. It then uses the records to determine the past states of the Earth›s various climate regions and its atmospheric system. Studies of past changes in the environment and biodiversity often reflect on the current situation, specifically the impact of climate on mass extinctions and biotic recovery. The scientific study field of paleoclimate began to form in the early 19th century, when discoveries about glaciations and natural changes in Earth›s past climate helped to understand the greenhouse effect.

RECONSTRUCTING ANCIENT CLIMATES

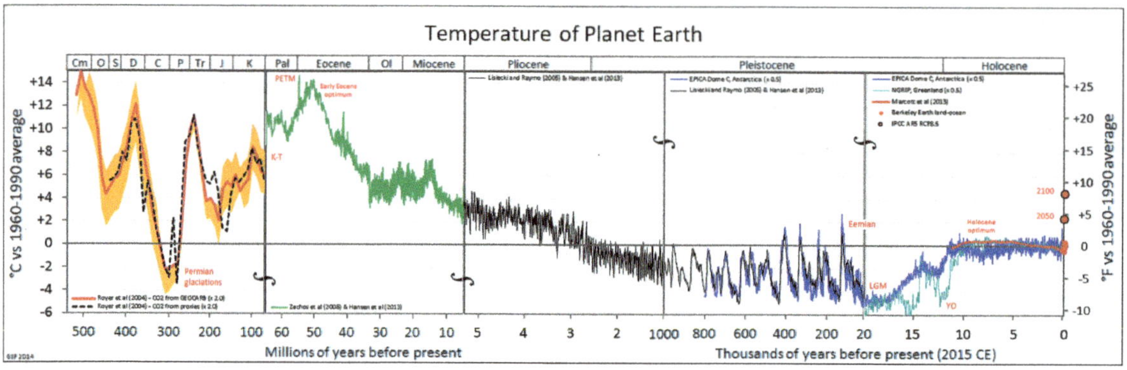

Palaeotemperature graphs compressed together.

Paleoclimatologists employ a wide variety of techniques to deduce ancient climates.

Ice

Mountain glaciers and the polar ice caps/ice sheets provide much data in paleoclimatology. Ice-coring projects in the ice caps of Greenland and Antarctica have yielded data going back several hundred thousand years, over 800,000 years in the case of the EPICA project.

- Air trapped within fallen snow becomes encased in tiny bubbles as the snow is compressed into ice in the glacier under the weight of later years' snow. The trapped air has proven a tremendously valuable source for direct measurement of the composition of air from the time the ice was formed.
- Layering can be observed because of seasonal pauses in ice accumulation and can be used to establish chronology, associating specific depths of the core with ranges of time.
- Changes in the layering thickness can be used to determine changes in precipitation or temperature.
- Oxygen-18 quantity changes ($\delta^{18}O$) in ice layers represent changes in average ocean surface temperature. Water molecules containing the heavier O-18 evaporate at a higher temperature than water molecules containing the normal Oxygen-16 isotope. The ratio of O-18 to O-16 will be higher as temperature increases. It also depends on other factors such as the water's salinity and the volume of water locked up in ice sheets. Various cycles in those isotope ratios have been detected.
- Pollen has been observed in the ice cores and can be used to understand which plants were present as the layer formed. Pollen is produced in abundance and

its distribution is typically well understood. A pollen count for a specific layer can be produced by observing the total amount of pollen categorized by type (shape) in a controlled sample of that layer. Changes in plant frequency over time can be plotted through statistical analysis of pollen counts in the core. Knowing which plants were present leads to an understanding of precipitation and temperature, and types of fauna present. Palynology includes the study of pollen for these purposes.

- Volcanic ash is contained in some layers, and can be used to establish the time of the layer's formation. Each volcanic event distributed ash with a unique set of properties (shape and color of particles, chemical signature). Establishing the ash's source will establish a range of time to associate with layer of ice.

Dendroclimatology

Climatic information can be obtained through an understanding of changes in tree growth. Generally, trees respond to changes in climatic variables by speeding up or slowing down growth, which in turn is generally reflected a greater or lesser thickness in growth rings. Different species, however, respond to changes in climatic variables in different ways. A tree-ring record is established by compiling information from many living trees in a specific area.

Older intact wood that has escaped decay can extend the time covered by the record by matching the ring depth changes to contemporary specimens. By using that method, some areas have tree-ring records dating back a few thousand years. Older wood not connected to a contemporary record can be dated generally with radiocarbon techniques. A tree-ring record can be used to produce information regarding precipitation, temperature, hydrology, and fire corresponding to a particular area.

Sedimentary Content

On a longer time scale, geologists must refer to the sedimentary record for data.

- Sediments, sometimes lithified to form rock, may contain remnants of preserved vegetation, animals, plankton or pollen, which may be characteristic of certain climatic zones.
- Biomarker molecules such as the alkenones may yield information about their temperature of formation.
- Chemical signatures, particularly Mg/Ca ratio of calcite in Foraminifera tests, can be used to reconstruct past temperature.
- Isotopic ratios can provide further information. Specifically, the $\delta^{18}O$ record responds to changes in temperature and ice volume, and the $\delta^{13}C$ record reflects a range of factors, which are often difficult to disentangle.

Sea floor core sample labelled to identify the exact spot on the sea floor where the sample was taken. Slight variations in location can make a significant difference in the chemical and biological composition of the sediment sample.

Sedimentary facies

On a longer time scale, the rock record may show signs of sea level rise and fall, and features such as “fossilised sand dunes can be identified. Scientists can get a grasp of long term climate by studying sedimentary rock going back billions of years. The division of earth history into separate periods is largely based on visible changes in sedimentary rock layers that demarcate major changes in conditions. Often, they include major shifts in climate.

Sclerochronology

Corals (see also sclerochronology)

Coral “rings” are similar to tree rings except that they respond to different things, such as the water temperature, freshwater influx, pH changes, and wave action. From there, certain equipment can be used to derive the sea surface temperature and water salinity from the past few centuries. The $\delta^{18}O$ of coralline red algae provides a useful proxy of the combined sea surface temperature and sea surface salinity at high latitudes and the tropics, where many traditional techniques are limited.

Limitations

A multinational consortium, the European Project for Ice Coring in Antarctica (EPICA), has drilled an ice core in Dome C on the East Antarctic ice sheet and retrieved ice from roughly 800,000 years ago. The international ice core community has, under the auspices of International Partnerships in Ice Core Sciences (IPICS), defined a priority project to obtain the oldest possible ice core record from Antarctica, an ice core record reaching back to or towards 1.5 million years ago. The deep marine record, the source of most isotopic data, exists only on oceanic plates, which are eventually subducted: the oldest remaining material is 200 million years old. Older sediments are also more prone to corruption by diagenesis. Resolution and confidence in the data decrease over time.

Notable climate events in Earth history

Knowledge of precise climatic events decreases as the record goes back in time, but some notable climate events are known

- Faint young Sun paradox (start)
- Huronian glaciation (~2400 Mya Earth completely covered in ice probably due to Great Oxygenation Event)
- Later Neoproterozoic Snowball Earth (~600 Mya, precursor to the Cambrian Explosion)
- Andean-Saharan glaciation (~450 Mya)
- Carboniferous Rainforest Collapse (~300 Mya)
- Permian–Triassic extinction event (251.4 Mya)
- Oceanic Anoxic Events (~120 Mya, 93 Mya, and others)
- Cretaceous–Paleogene extinction event (66 Mya)
- Paleocene–Eocene Thermal Maximum (Paleocene–Eocene, 55Mya)
- Younger Dryas/The Big Freeze (~11,000 BC)
- Holocene climatic optimum (~7000–3000 BC)
- Climate changes of 535-536 (535–536 AD)
- Medieval warm period (900–1300)
- Little Ice Age (1300–1800)
- Year Without a Summer (1816)

History of the Atmosphere

Earliest Atmosphere

The first atmosphere would have consisted of gases in the solar nebula, primarily hydrogen. In addition, there would probably have been simple hydrides such as those now found in gas giants like Jupiter and Saturn, notably water vapor, methane and ammonia. As the solar nebula dissipated, the gases would have escaped, partly driven off by the solar wind.

Second Atmosphere

The next atmosphere, consisting largely of nitrogen and carbon dioxide and inert gases, was produced by outgassing from volcanism, supplemented by gases produced during the late heavy bombardment of Earth by huge asteroids. A major part of carbon dioxide emissions were soon dissolved in water and built up carbonate sediments.

Water-related sediments have been found dating from as early as 3.8 billion years ago. About 3.4 billion years ago, nitrogen was the major part of the then stable second atmosphere . An influence of life has to be taken into account rather soon in the history of the atmosphere because hints of early life forms have been dated to as early as 3.5 billion years ago. The fact that it is not perfectly in line with the 30% lower solar radiance (compared to today) of the early Sun has been described as the faint young Sun paradox .

The geological record, however, shows a continually relatively warm surface during the complete early temperature record of Earth with the exception of one cold glacial phase about 2.4 billion years ago. In the late Archaean eon, an oxygen-containing atmosphere began to develop, apparently from photosynthesizing cyanobacteria (see Great Oxygenation Event) which have been found as stromatolite fossils from 2.7 billion years ago. The early basic carbon isotopy (isotope ratio proportions) was very much in line with what is found today, suggesting that the fundamental features of the carbon cycle were established as early as 4 billion years ago.

Third Atmosphere

The constant rearrangement of continents by plate tectonics influences the long-term evolution of the atmosphere by transferring carbon dioxide to and from large continental carbonate stores. Free oxygen did not exist in the atmosphere until about 2.4 billion years ago, during the Great Oxygenation Event, and its appearance is indicated by the end of the banded iron formations. Until then, any oxygen produced by photosynthesis was consumed by oxidation of reduced materials, notably iron. Molecules of free oxygen did not start to accumulate in the atmosphere until the rate of production of oxygen began to exceed the availability of reducing materials. That point was a shift from a reducing atmosphere to an oxidizing atmosphere. O_2 showed major variations until reaching a steady state of more than 15% by the end of the Precambrian. The following time span was the Phanerozoic eon, during which oxygen-breathing metazoan life forms began to appear. The amount of oxygen in the atmosphere has fluctuated over the last 600 million years, reaching a peak of 35% during the Carboniferous period, significantly higher than today's 21%. Two main processes govern changes in the atmosphere: plants use carbon dioxide from the atmosphere, releasing oxygen and the breakdown of pyrite and volcanic eruptions release sulfur into the atmosphere, which oxidizes and hence reduces the amount of oxygen in the atmosphere. However, volcanic eruptions also release carbon dioxide, which plants can convert to oxygen. The exact cause of the variation of the amount of oxygen in the atmosphere is not known. Periods with much oxygen in the atmosphere are associated with rapid development of animals. Today's atmosphere contains 21% oxygen, which is high enough for rapid development of animals. Currently, anthropogenic greenhouse gases are accumulating in the atmosphere, which is the main cause of global warming.

Climate During Geological Ages

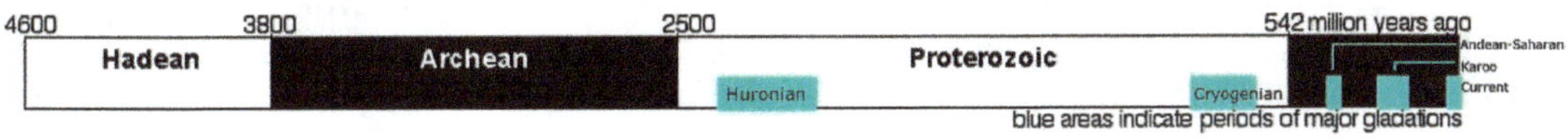

Timeline of glaciations, shown in blue

- The Huronian glaciation, is the first known glaciation in Earth's history, and lasted from 2400-2100 million years ago.
- The Cryogenian glaciation lasted from 850-635 million years ago.
- The Andean-Saharan glaciation lasted from 450–420 million years ago.
- The Karoo glaciation lasted from 360–260 million years ago.
- The Quaternary glaciation is the current glaciation period and begun 2.58 million years ago.

Precambrian Climate

The climate of the late Precambrian showed some major glaciation events spreading over much of the earth. At this time the continents were bunched up in the Rodinia supercontinent. Massive deposits of tillites are found and anomalous isotopic signatures are found, which gave rise to the Snowball Earth hypothesis. As the Proterozoic Eon drew to a close, the Earth started to warm up. By the dawn of the Cambrian and the Phanerozoic, life forms were abundant in the Cambrian explosion with average global temperatures of about 22 °C.

Phanerozoic Climate

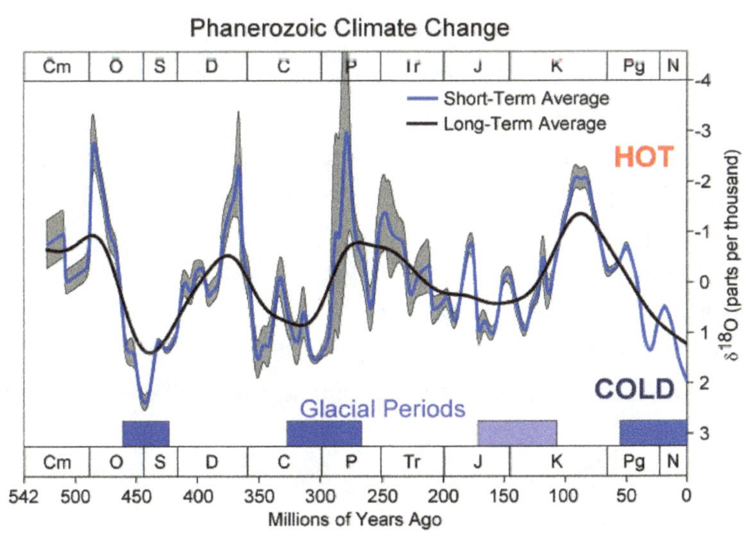

500 million years of climate change.

Major drivers for the preindustrial ages have been variations of the sun, volcanic ashes and exhalations, relative movements of the earth towards the sun and tectonically induced effects as for major sea currents, watersheds and ocean oscillations. In the early Phanerozoic, increased atmospheric carbon dioxide concentrations have been linked to driving or amplifying increased global temperatures.Royer et al. 2004 found a climate sensitivity for the rest of the Phanerozoic which was calculated to be similar to today's modern range of values.

The difference in global mean temperatures between a fully glacial Earth and an ice free Earth is estimated at approximately 10 °C, though far larger changes would be observed at high latitudes and smaller ones at low latitudes . One requirement for the development of large scale ice sheets seems to be the arrangement of continental land masses at or near the poles. The constant rearrangement of continents by plate tectonics can also shape long-term climate evolution. However, the presence or absence of land masses at the poles is not sufficient to guarantee glaciations or exclude polar ice caps. Evidence exists of past warm periods in Earth's climate when polar land masses similar to Antarctica were home to deciduous forests rather than ice sheets.

The relatively warm local minimum between Jurassic and Cretaceous goes along with an increase of subduction and mid-ocean ridge volcanism due to the breakup of the Pangea supercontinent.

Superimposed on the long-term evolution between hot and cold climates have been many short-term fluctuations in climate similar to, and sometimes more severe than, the varying glacial and interglacial states of the present ice age. Some of the most severe fluctuations, such as the Paleocene-Eocene Thermal Maximum, may be related to rapid climate changes due to sudden collapses of natural methane clathrate reservoirs in the oceans.

A similar, single event of induced severe climate change after a meteorite impact has been proposed as reason for the Cretaceous–Paleogene extinction event. Other major thresholds are the Permian-Triassic, and Ordovician-Silurian extinction events with various reasons suggested.

Quaternary Climate

The Quaternary sub-era includes the current climate. There has been a cycle of ice ages for the past 2.2–2.1 million years (starting before the Quaternary in the late Neogene Period).

Note in the graphic on the right the strong 120,000-year periodicity of the cycles, and the striking asymmetry of the curves. This asymmetry is believed to result from complex interactions of feedback mechanisms. It has been observed that ice ages deepen by progressive steps, but the recovery to interglacial conditions occurs in one big step.

The graph below shows the temperature change over the past 12 000 years, from various sources. The thick black curve is an average.

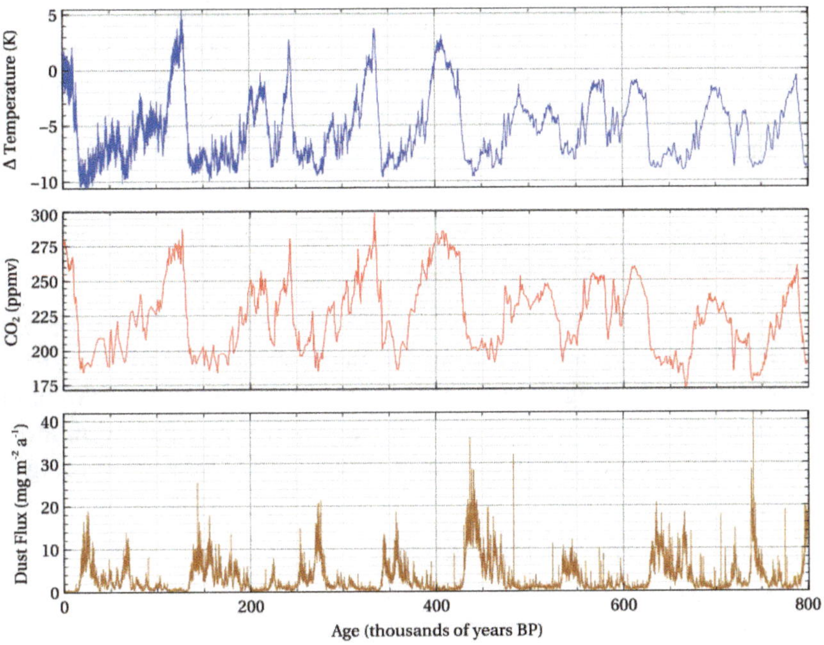

Ice core data for the past 800,000 years . Note length of glacial-interglacial cycles averages ~100,000 years. Blue curve is temperature,red curve is atmospheric CO_2concentrations, and brown curve is dust fluxes. Today's date is on the left side of the graph because the x-axis values represent "age before 1950".

CLIMATE FORCINGS

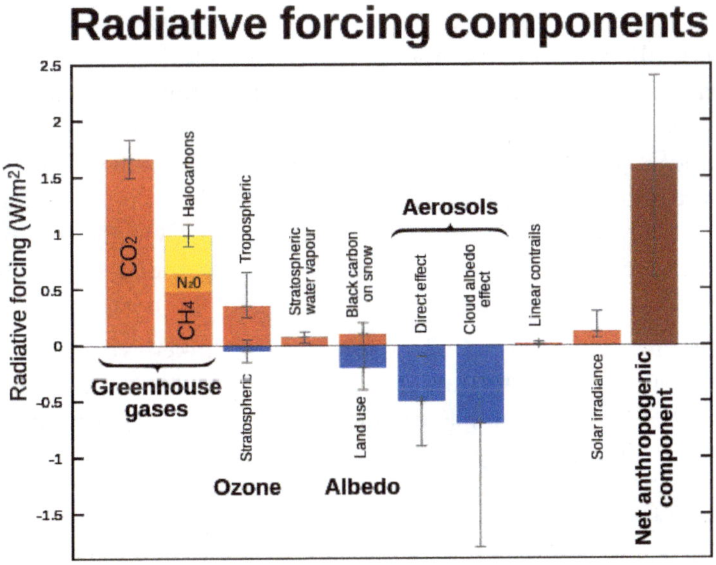

Radiative forcings, IPCC (2007).

The climate forcing is the difference of radiant energy (sunlight) received by the Earth and the outgoing longwave radiation back to space. The radiative forcing is quantified based on the CO_2 amount in the tropopause, in units of watts per square meter to the Earth›s surface. Dependent on the radiative balance of incoming and outgoing energy, the Earth either warms up or cools down. Earth radiative balance originates from changes in solar insolation and the concentrations of greenhouse gases and aerosols. Climate change may be due to internal processes in Earth sphere›s and/or following external forcings.

Internal Processes and Forcings

The Earth's climate system involves the study of the atmosphere, biosphere, cryosphere, hydrosphere, and lithosphere, and the sum of these processes from Earth sphere's is considered the processes affecting the climate. Greenhouse gases act as the internal forcing of the climate system. Particular interests in climate science and paleoclimatology focuses on the study of Earth climate sensitivity, in response to the sum of forcings.

Examples:

- Thermohaline circulation (Hydrosphere)
- Life (Biosphere)

External forcings

- The Milankovitch cycles determine Earth distance and position to the Sun. The solar insolation, is the total amount of solar radiation received by Earth.
- Volcanic eruptions, are considered an external forcing.
- Human changes of the composition of the atmosphere or land use.

Mechanisms

On timescales of millions of years, the uplift of mountain ranges and subsequently weathering processes of rocks and soils and the subduction of tectonic plates, are an important part of the carbon cycle. The weathering sequesters CO_2, by the reaction of minerals with chemicals (especially silicate weathering with CO_2) and thereby removing CO_2 from the atmosphere and reducing the radiative forcing. The opposite effect is volcanism, responsible for the natural greenhouse effect, by emitting CO_2into the atmosphere, thus affecting glaciation (Ice Age) cycles. James Hansen suggested that humans emit CO_2 10,000 times faster than natural processes have done in the past.

Ice sheet dynamics and continental positions (and linked vegetation changes) have been important factors in the long term evolution of the earth's climate. There is also a close correlation between $CO_2$and temperature, where CO_2 has a strong control over global temperatures in Earth history.

REFERENCES

1. Sahney, S. & Benton, M.J. (2008). "Recovery from the most profound mass extinction of all time" (PDF). Proceedings of the Royal Society: Biological. 275 *(1636): 759–65. doi:10.1098/rspb.2007.1370. PMC 2596898. PMID 18198148.*
2. Halfar, J.; Steneck, R.S.; Joachimski, M.; Kronz, A.; Wanamaker, A.D. (2008). Coralline red algae as high-resolution climate recorders . Geology. 36 *(6): 463. doi:10.1130/G24635A.1.*
3. Cobb, K.; Charles, C. D.; Cheng, H; Edwards, R. L. (2003). El Nino/Southern Oscillation and tropical Pacific climate during the past millennium . Nature. 424 *(6946): 271–6.* doi:10.1038/nature01779. PMID 12867972.
4. Jouzel, Jean; Masson-Delmotte, V.; Cattani, O.; Dreyfus, G.; Falourd, S.; Hoffmann, G.; Minster, B.; Nouet, J.; et al. (10 August 2007). Orbital and Millennial Antarctic Climate Variability over the Past 800,000 Years . Science. 317 *(5839): 793–796. Bibcode*:2007Sci...317..793J. doi:10.1126/science.1141038. PMID 17615306.
5. Page 1 1 International Partnerships in Ice Core Sciences (IPICS) The oldest ice core: A 1.5 million year record of climate and greenhouse gases from Antarctica (PDF). Retrieved 22 September 2011.
6. Zahnle, K.; Schaefer, L.; Fegley, B. (2010). Earth›s Earliest Atmospheres . Cold Spring Harbor Perspectives in Biology. 2 *(10): a004895. doi:10.1101/cshperspect.a004895. PMC 2944365. PMID 20573713.*
7. B. Windley: The Evolving Continents. Wiley Press, New York 1984
8. J. Schopf: Earth›s Earliest Biosphere: Its Origin and Evolution. Princeton University Press, Princeton, N.J., 1983
9. Christopher R. Scotese, Back to Earth History : Summary Chart for the Precambrian, Paleomar Project
10. Beerling, David (2007). The emerald planet: how plants changed Earth›s history. Oxford University press. p. 47. ISBN 9780192806024.
11. Peter Ward:[1] Out of Thin Air: Dinosaurs, Birds, and Earth's Ancient Atmosphere
12. Summary for Policymakers (PDF). Climate Change 2007: The Physical Science Basis. Contribution of Working Group I to the Fourth Assessment Report of the Intergovernmental Panel on Climate Change. Intergovernmental Panel on Climate Change. 5 February 2007.
13. Rosemarie E. Came, John M. Eiler, Jan Veizer, Karem Azmy, Uwe Brand & Christopher R. Weidman; Eiler; Veizer; Azmy; Brand; Weidman (September 2007). Coupling of surface temperatures and atmospheric CO_2concentrations during the Palaeozoic era". Nature. 449 *(7159): 198–201. Bibcode*:2007Natur.449..198C. doi:10.1038/nature06085. PMID 17851520.
14. Royer, Dana L. and Robert A. Berner, Isabel P. Montañez, Neil J. Tabor, David J. Beerling *(July 2004). "CO_2 as a primary driver of Phanerozoic climate". GSA Today.* 14 *(3): 4–10.* doi:10.1130/1052-5173(2004)014<4:CAAPDO>2.0.CO;2.

15. Douwe G. Van Der Meer, Richard E. Zeebe, Douwe J. J. van Hinsbergen, Appy Sluijs, Wim Spakman, and Trond H. Torsvik (February 2014). Plate tectonic controls on atmospheric CO2 levels since the Triassic . PNAS. 111 *(12): 4380–4385. doi:10.1073/pnas.1315657111.*

16. Jouzel, J.; Masson-Delmotte, V.; Cattani, O.; Dreyfus, G.; Falourd, S.; Hoffmann, G.; Minster, B.; Nouet, J.; Barnola, J. M. (2007-08-10). Orbital and Millennial Antarctic Climate Variability over the Past 800,000 Years . Science. 317 *(5839): 793–796. doi:10.1126/science.1141038. ISSN 0036-8075. PMID 17615306.*

17. Lüthi, Dieter; Le Floch, Martine; Bereiter, Bernhard; Blunier, Thomas; Barnola, Jean-Marc; Siegenthaler, Urs; Raynaud, Dominique; Jouzel, Jean; Fischer, Hubertus (2008-05-15). High-resolution carbon dioxide concentration record 650,000–800,000 years before present . Nature. 453 *(7193): 379–382.* doi:10.1038/nature06949. ISSN 0028-0836. PMID 18480821.

18. Lambert, F.; Delmonte, B.; Petit, J. R.; Bigler, M.; Kaufmann, P. R.; Hutterli, M. A.; Stocker, T. F.; Ruth, U.; Steffensen, J. P. (2008-04-03). Dust-climate couplings over the past 800,000 years from the EPICA Dome C ice core . Nature. 452 *(7187): 616–619.* doi:10.1038/nature06763. ISSN 0028-0836.

19. Lambert, F.; Bigler, M.; Steffensen, J. P.; Hutterli, M.; Fischer, H. Centennial mineral dust variability in high-resolution ice core data from Dome C, Antarctica . Climate of the Past. 8 *(2): 609–623.* doi:10.5194/cp-8-609-2012.

20. Annex III: Glossary (PDF). IPCC AR5. Climate change may be due to natural internal processes or external forcings such as modulations of the solar cycles, volcanic eruptions and persistent anthropogenic changes in the composition of the atmosphere or in land use.

21. Caldeira, Ken (June 18, 1992). Enhanced Cenozoic chemical weathering and the subduction of pelagic carbonate . Nature. 357 *(6379): 578–581.* doi:10.1038/357578a0.

22. Cin-Ty Aeolus Lee, Douglas M. Morton, Mark G. Little, Ronald Kistler, Ulyana N. Horodyskyj, William P. Leeman, and Arnaud Agranier (January 28, 2008). Regulating continent growth and composition by chemical weathering . PNAS. 105 *(13): 4981–4986. doi:10.1073/pnas.0711143105.*

23. van der Meer, Douwe (March 25, 2014). Plate tectonic controls on Atmospheric CO2 since the Triassic . PNAS. 111 *(12): 4380–4385. doi:10.1073/pnas.1315657111.*

24. James Hansen (2009). The 8 Minute Epoch 65 million Years with James Hansen . University of Oregon.

25. ROYER, D. L.; PAGANI, M.; BEERLING, D. J. (1 July 2012). Geobiological constraints on Earth system sensitivity to CO2 during the Cretaceous and Cenozoic . Geobiology. 10 *(4): 298–310. doi:10.1111/j.1472-4669.2012.00320.x. PMID 22353368.*

26. Royer, Dana L. (1 December 2006). CO2-forced climate thresholds during the Phanerozoic . Geochimica et Cosmochimica Acta. 70 *(23): 5665–5675. Bibcode*:2006GeCoA..70.5665R. doi:10.1016/j.gca.2005.11.031.

Index